INTERMEDIATE FLUID
MECHANICS

INTERMEDIATE FLUID MECHANICS

Robert H. Nunn

Naval Postgraduate School
Monterey, California

● **HEMISPHERE PUBLISHING CORPORATION**
A member of the Taylor & Francis Group

New York   Washington   Philadelphia   London

INTERMEDIATE FLUID MECHANICS

1 2 3 4 5 6 7 8 9 0   B R B R   8 9 8 7 6 5 4 3 2 1 0 9

This book was set in Times Roman by Edwards Brothers, Inc.
The editors were Sandra Tamburrino and Lynne Lackenbach.
Cover design by Sharon DePass.
Braun-Brumfield, Inc. was the printer and binder.

Library of Congress Cataloging-in-Publication Data

Nunn, R. H. (Robert H.)
   Intermediate fluid mechanics.

   Includes index.
   1. Fluid mechanics. I. Title.
TA357.N88   1989   620.1'06   88-34746
ISBN 0-89116-647-5

This book is dedicated to my wife,
Caroline,
and to my three sons.

CONTENTS

Part 2 One-Dimensional Compressible Flow

Part 3   Simple Viscous Flows

13   Turbulent Flows

Appendixes

Epilogue: Computer Applications and Computational Fluid Dynamics

Index

PREFACE

Programs in engineering education have expanded considerably in the last few decades, and a master's degree is now considered by many to represent the nominal level for practicing engineers. Whereas a single introductory course in fluid mechanics was once considered to be adequate for first-degree engineers, many institutions now include a second course in the upper-division undergraduate curriculum. This is particularly true in universities that expect a significant portion of their students to continue their formal education into the graduate level.

At the Naval Postgraduate School this midrange course has been viewed as a "terminal" course for some and, for others, as a "transition" course. For those who do not plan to pursue a greater depth of study in fluid mechanics, I have sought to provide an *overview*. For students whose interests are likely to require further study in the area, the goal has been to provide a *preview* of coming attractions. These course characteristics are not inherently compatible, and the corresponding course objectives present a number of challenges to the teacher as well as to the students (the distinction between a "challenge" and a "pain in the neck" being somewhat obscure).

One of these challenges has been the designation of a suitable textbook. Introductory texts, by their nature, provide careful development of fundamental concepts, such as fluid properties, fluid statics, the notion of conserved entities, the control volume approach, and energy and momentum methods. Because of the necessity of providing practical working tools, these texts often emphasize integral formulations, based upon steady, incompressible, one-dimensional hypotheses and backed by empiricism. Beyond fluid statics, the focus is usually upon internal flows, with viscous effects and turbulence treated in an overall sense. From a practical point of view, in an introductory text it is not possible (and perhaps undesirable) to go to any great lengths concerning such subjects as potential flow and the effects of compressibility.

At the other end of the spectrum are a number of excellent works that are most useful as reference texts for in-depth but separate studies of viscous flows, compressible flows, or ideal flows. The result, I feel (and it is my experience), is that these important subareas of fluid mechanics appear to the student as independent and distinct, and somehow out of balance in their contribution to an overall view of fluid mechanics. This situation has made it difficult to accomplish the objective of providing a second course that serves as an overview. From the preview point of view, I have found it necessary to extend the treatments of these subareas, especially potential flows and compressible flows.

In short, texts that have been evaluated are either too basic, too advanced, or too something or other. For this reason the growth of the course has been accompanied by the generation of rather copious amounts of handout material, relying less and less upon existing texts and more and more upon the instructor's notes.

Turning to a discussion of the scope and organization of this text, I begin with an acknowledgment of its limitations. By its very nature, this book begins at the end of an introductory level of fluid mechanics and ends at a juncture with advanced treatment of the subject. Some may feel that the coverage starts too far into the subject (or not far enough) and/or that there is not enough (or too much) of an extension to first-course material.

From the point of view of limiting the material up front, readers will not find the usual preliminaries involving such important matters as units and fluid properties. By the same token, the appendixes do not include tabular material on such items as properties and unit conversions. This is because it is *assumed* that readers will have at hand a good introductory text with which they are familiar.

Other than shear exhaustion, the stopping point has again been determined by the nature of the text. This material is more than enough to provide the overview and preview aspects of the objectives previously discussed. To go further would be to infringe too deeply upon the territory of the many advanced texts now in print. I must also emphasize that this book is not meant to stand by itself in the absence of an instructor. There are many points, not just at the beginning and end of the coverage, that invite the insertion of extensions to the theory and applications, as well as presentations of other points of view.

The presentation is organized into three parts: ideal fluid flow (potential flow), compressible flow, and viscous flow. The categorization is more or less standard, but the best order of presentation is by no means obvious. In deciding which to teach first, one should keep in mind the makeup of the curriculum, as well as the goals of the course. For instance, compressible flow should probably not be left until last unless there is additional course work on the subject available in the curriculum—material planned for presentation near the end of a course often receives short shrift.

On the other hand, the principles of ideal fluid motion form mathematical and conceptual foundations that are quite useful for development of the relationships governing viscous flow. (It seems better to add shear stresses to the Euler equation than delete them from the Navier-Stokes equation—the commonality of the expressions for fluid acceleration provides a useful link between ideal flow and viscous flow.) For these and other reasons, there does seem to be some logic to support the presentation of this material in the order given above: ideal, compressible, and viscous. In any case either viscous flow or compressible flow can follow ideal flow, according to the preference of the teacher.

In my use of this material, giving about equal time to each of the three parts, I have found that in an 11-week quarter there is little difficulty in covering Chapters 1–4, 6–8, and 10–12. Chapters 5, 9, and 13 provide extensions to material in each of the three parts and are often treated in a descriptive way. The treatment of turbulent flows (Chapter 13) is given priority for whatever "extra" time is available.

Following some introductory remarks, Part One of the text develops the Euler equation as the governing differential equation of motion for ideal flows. Both kinetic and kinematic aspects of ideal flow are emphasized, and the Bernoulli equation is derived in extended form to show the various special cases to which it is applicable. The notions of vorticity, circulation, and irrotational motion are described, and formulations of the velocity potential and stream function lead to the Laplace equation. In Chapter 4 the usual hydrodynamic singularities are defined, and some examples of the use of superposition are explored. (Only planar two-dimensional flows are treated in any detail—axisymmetric geometries are mentioned as simple alternatives employing the same methods) At the end of Part One, Chapter 5 provides an introduction to applications of the theory of complex variables, and methods of distributed singularities are demonstrated. Some recent results in these areas are presented to give the reader an appreciation of the power of these methods.

The procedure adopted in this text is to replace one complication with another so that the level of sophistication remains somewhat even. Accordingly, in Part Two the treatment of compressible flows is restricted to one-dimensional cases. The introductory material establishes the ground rules for Part Two and provides a link to the fluid kinematics of Part One by showing the connection between vorticity and entropy change in a flow. Following some definitions (stagnation state, speed of sound, and so on), the working relationships for isentropic flow are developed and applied. Application of the theory to converging-diverging nozzles leads to the implementation of pressure boundary conditions and the concomitant necessity for discontinuities (shock waves) under certain circumstances. This leads into Chapter 8, where normal and oblique shock-wave theories are presented along with the formulas for expansion processes (Prandtl-Meyer flows).

By way of extension to the theory of compressible flows, Chapter 9 presents methods for treating flows with other sources of entropy change—irreversibilities due to friction (Fanno flows), flows with heat transfer (Rayleigh flows), and isothermal flows. The assumption of constant area is common to these developments, and the combination of these effects with those due to area change is left as a problem that is best solved with the computer.

Finally, in Part Three viscous effects are treated. Chapter 10 begins with a presentation on fluid viscosity that is meant to instill an appreciation for the meaning and impact of momentum transport. Following a general discussion of the state of stress of a fluid element (including the use of index notation), the Navier-Stokes equations are developed in the most primitive way—these relationships are formidable enough without the generalities that accompany an approach via tensor calculus. Such an approach may be the choice of the instructor, but in this text only the fluid-solid analogy method will be found. A few exact solutions to the Navier-Stokes equations are presented next; they are exercises in interpreting the various terms, specifying boundary conditions, and showing that the Navier-Stokes equations contain within them the solutions to the elementary problems treated in an introductory course. The presentation of Stokes's first problem then leads directly to the concept of the boundary layer.

The boundary layer simplification begins Chapter 12 and leads to the boundary layer equation. Following a discussion of boundary layer separation, both the similarity and integral methods of solution are developed for laminar flat plate flows. Boundary layer characteristics and frictional resistance are calculated, and, finally, it becomes necessary to broach the subject of turbulence.

I have already spoken of challenges, and foremost among these has been the presentation of the subject of turbulent flows in a single chapter. It was something of a temptation to "solve" this problem by leaving it to the authors of more-advanced texts. To do so, however, would have led to a premature departure from the treatment of viscous flows. In Chapter 13, therefore, I have tried to provide enough information on this subject to serve the goals of overview and preview. Neither fluids nor authors can get from laminar to turbulent without passing through transition, so I have attempted a qualitative presentation of some of the more important aspects of this extremely complex process. This is followed by an elementary explanation of the rules of averaging and their application to the equations of motion for turbulent flows. Some of the more common constitutive laws for turbulent flows are also discussed, and these are applied in calculations for pipe flows and the flat plate boundary layer. Combined laminar-turbulent boundary layers are also discussed, and, near the end of the chapter, there is included some preliminary insights into the analysis of separated flows and form drag.

In closing, I wish to acknowledge my great debt to all those who have contributed to the literature in fluid mechanics in the form of introductory texts and advanced reference works. What I have done is to try to work the middle ground of this vast field of information. In doing so, I have borrowed heavily from "above" and from "below." I can make no particular claim to creativity in this effort, but I do hope to have performed a useful service in broadening the perspective from which students of fluid mechanics will view their subject.

Robert H. Nunn

BACKGROUND AND INTRODUCTORY MATERIAL

This text is meant to serve as an extension to the many introductory books on fluid mechanics. For this reason the level and nature of the material presented presuppose some understanding on the reader's part of the fundamental aspects of fluid mechanics. It is the author's intention to build on this understanding; for those who may feel a bit "rusty" in this regard, a review of one or more of the many excellent texts on the subject is highly recommended. The bibliography at the end of this chapter lists several texts suggested for purposes of review.

Without dwelling on the definition of a fluid, which is assumed to be one of those concepts retained from a first course in fluid mechanics, it may be useful to review what is meant by "mechanics." Figure 1.1 illustrates this and points out the breadth implied by the general use of the term "mechanics." In this text we shall not treat the subject of *fluid statics* except as it comes under discussion as a special case of fluid dynamics, as shown in Fig. 1.1.

To further explain the nature and scope of what follows, Table 1.1 shows some of the major categories into which fluid mechanics has been subdivided over the years.

Whether a particular problem in fluid mechanics falls into one or the other of these categories depends on the relative influence of the two main physical properties defining a fluid: density and viscosity. If the effect of viscosity is negligible (we shall see in some detail what is meant by effect), then the problem may be one in "hydrodynamics." If the density variations in the flow are domi-

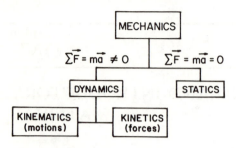

Figure 1.1 Branches of mechanics.

nant, or at least significant, then the methods of compressible flow analysis must be brought to bear. Few problems involving *real* fluids are totally devoid of the effects of viscosity and variations in density, so the matter is often one requiring a subjective judgment tempered (we always hope) by some knowledge and experience as to what are the important factors in a given flow situation. In fact, many of the most challenging and frustrating problems in fluid mechanics involve flow situations that reside in the "gray areas" that span the categories shown in Table 1.1.

The organization of this text follows that of Table 1.1. The order of presentation of the material is based on convenience and experience. There are, of course, other categorizations possible: unsteady vs. steady, internal vs. external, one-dimensional vs. multidimensional, etc. Descriptions such as these are often used to further define the groupings of Table 1.1. The reader should not infer from these remarks that the solution of problems in fluid mechanics is solely one of choosing the right "category." The likelihood that a particular problem can be solved is greatly increased if it can be confined to such straightforward definitions. The usual case, however, is one in which the engineer is forced to neglect the less important facets of a flow for the simple reason that the problem cannot be solved with all possible effects taken into account. Necessity is the mother of simplification!

Table 1.1 Main divisions of fluid mechanics and some of their features

Ideal fluid flow	Compressible flow	Viscous flow
Hydrodynamics	Gas dynamics	Emphasis on viscous effects
Negligible viscous effects	Density varies from point to point in flow	Density often held constant
Considerations mainly kinematic	Viscous effects often neglected or greatly simplified	Boundary layer theory applies
Mathematically complex, multidimensional	Simple geometries	Mixing and diffusion important
Density often held constant	Often treated as one-dimensional	Turbulence

1.1 REVIEW OF SOME IMPORTANT MATHEMATICAL TOOLS

In engineering analysis, mathematics is an essential language for expressing physical relationships in compact form. For instance, $F = ma$ is a simple mathematical form that imposes many far-reaching influences on all of nature. Engineers, in their studies as well as in practical applications, must keep in mind both sides of such relationships as well as the symbology by which they are linked (in this case, the equals sign). In subsequent chapters we try to maintain an emphasis on the essential physical implications of the governing mathematical relationships. This is not always easy, however, because the language of mathematics can become so encumbered with sophisticated nuances that the physics is lost in a forest of symbology.

Probably the only way to avoid mathematical obfuscation is to develop a familiarity with the language that allows one to "see through" the mathematics to the underlying principles. To assist in gaining this familiarity, this section provides a brief review of some of the mathematical tools that are frequently employed in investigation of fluid flows. Other mathematical operations will be described where they are most directly applicable.

Derivatives of Vectors

Quantities of interest in fluid mechanics often depend on many variables. One common dependency is on location in space, and this location may be specified by denoting three characteristic distances, or coordinates, relative to some point of reference. This reference point is the origin of our coordinate system. Two popular coordinate systems—Cartesian and polar-cylindrical—are illustrated in Fig. 1.2.

To completely specify the location of a point it is necessary to provide both the magnitude and direction of the coordinates. The resulting quantity is termed a vector. (Vectors, requiring two specifications, may be viewed as a subset of a more general set called tensors. Part Three of this text, which deals with viscous flows, includes more information on tensors.) In the Cartesian coordinate system, for instance, the location vector of the point $P(x, y, z)$ is given by

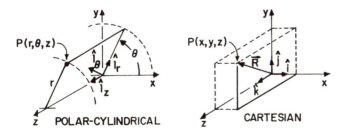

Figure 1.2 Cartesian and polar-cylindrical coordinate systems.

$$\mathbf{R} = x\hat{\imath} + y\hat{\jmath} + z\hat{k}$$

If we wish to express the rate of change of the position of a point P relative to the origin, we write

$$\frac{d\mathbf{R}}{dt} = \frac{d}{dt}(x\hat{\imath} + y\hat{\jmath} + z\hat{k})$$

$$= \frac{dx}{dt}\hat{\imath} + \frac{dy}{dt}\hat{\jmath} + \frac{dz}{dt}\hat{k} + \left(x\frac{d\hat{\imath}}{dt} + y\frac{d\hat{\jmath}}{dt} + z\frac{d\hat{k}}{dt}\right)$$

But since the unit vectors are of constant (unit) magnitude *and* constant direction (they are aligned with the fixed coordinate axes), the last term above involving the rates of change of the unit vectors is zero and

$$\frac{d\mathbf{R}}{dt} = \frac{dx}{dt}\hat{\imath} + \frac{dy}{dt}\hat{\jmath} + \frac{dz}{dt}\hat{k}$$

When one or more of the coordinates are curvilinear, as in the polar-cylindrical system of Fig. 1.2, we must be careful to account for changes in direction of the unit vectors. Thus, in the polar-cylindrical system,

$$\mathbf{R} = r\hat{1}_r + z\hat{1}_z$$

and

$$\frac{d\mathbf{R}}{dt} = r\frac{d}{dt}(\hat{1}_r) + \frac{dr}{dt}\hat{1}_r + \frac{dz}{dt}\hat{1}_z$$

Note that the unit vectors $\hat{1}_r$ and $\hat{1}_\theta$ are constrained to lie in the (x, y) plane, but they rotate as the angle θ changes due to the motion of point P. Thus though the magnitude of $\hat{1}_r$ is constant, its orientation changes at the rate $d\theta/dt$, and the direction of this change is that of $\hat{1}_\theta$. We find, therefore, that

$$\frac{d}{dt}(\hat{1}_r) = \frac{d\theta}{dt}\hat{1}_\theta$$

and

$$\frac{d\mathbf{R}}{dt} = r\frac{d\theta}{dt}\hat{1}_\theta + \frac{dr}{dt}\hat{1}_r + \frac{dz}{dt}\hat{1}_z$$

As a further example, consider the velocity vector $\mathbf{q}$ with components v_r, v_θ, and v_z in polar-cylindrical notation:

$$\mathbf{q} = v_r\hat{1}_r + v_\theta\hat{1}_\theta + v_z\hat{1}_z$$

The acceleration $d\mathbf{q}/dt$ may be expressed as

$$\frac{d\mathbf{q}}{dt} = \dot{\mathbf{q}} = \dot{v}_r\hat{1}_r + \dot{v}_\theta\hat{1}_\theta + \dot{v}_z\hat{1}_z + v_r\dot{\theta}\hat{1}_\theta - v_\theta\dot{\theta}\hat{1}_r$$

Both the r and θ components of acceleration include additional terms due to the rotation of the unit vectors $\hat{1}_\theta$ and $\hat{1}_r$, respectively. Since $\dot{\theta} = v_\theta/r$, these may be expressed in the more familiar forms as $v_\theta v_r/r$ and $-v_\theta^2/r$.

Exercise Note that the unit vectors in polar-cylindrical coordinates may be expressed as follows:

$$\hat{1}_r = \hat{i}\cos\theta + \hat{j}\sin\theta$$

$$\hat{1}_\theta = -\hat{i}\sin\theta + \hat{j}\cos\theta$$

$$\hat{1}_z = \hat{k}$$

From the first of these,

$$\frac{d}{dt}(\hat{1}_r) = -\hat{i}\dot{\theta}\sin\theta + \hat{j}\dot{\theta}\cos\theta = \dot{\theta}\hat{1}_\theta \tag{1.1}$$

and from the second,

$$\frac{d}{dt}(\hat{1}_\theta) = -\hat{i}\dot{\theta}\cos\theta - \hat{j}\dot{\theta}\sin\theta = -\dot{\theta}\hat{1}_r \tag{1.2}$$

Vector Differential Operations

The vector differential operator ∇, (called del or nabla) is defined in rectangular coordinates as

$$\nabla = \hat{i}\frac{\partial}{\partial x} + \hat{j}\frac{\partial}{\partial y} + \hat{k}\frac{\partial}{\partial z} \tag{1.3}$$

The operator ∇ must operate *on* something (scalar, vector, or tensor function), and it is a *vector operator* because it has two components in each of the directions. When the vector differential operator is applied to a scalar function S, we have the *gradient*

$$\text{grad } S = \nabla S = \hat{i}\frac{\partial S}{\partial x} + \hat{j}\frac{\partial S}{\partial y} + \hat{k}\frac{\partial S}{\partial z} \tag{1.4}$$

Note that the components of ∇S express the rates of change of S along the three coordinate directions and grad S is the vector sum of these rates.

Though the del operator may be applied directly to vectors and tensors (e.g., $\nabla \mathbf{q}$), it is more frequently found in applications involving products. The *scalar product* (or dot product) produces the *divergence* of a vector. Thus with $\mathbf{q} = u\hat{i} + v\hat{j} + w\hat{k}$,

$$\text{div } \mathbf{q} = \nabla \cdot \mathbf{q} = \frac{\partial u}{\partial x} + \frac{\partial v}{\partial y} + \frac{\partial w}{\partial z} \tag{1.5}$$

The divergence of a vector is a scalar. The *vector product* (or cross product) produces the *curl*:

$$\text{curl } \mathbf{q} = \nabla \times \mathbf{q} = \hat{\imath}\left(\frac{\partial w}{\partial y} - \frac{\partial v}{\partial z}\right) + \hat{\jmath}\left(\frac{\partial u}{\partial z} - \frac{\partial w}{\partial x}\right) + \hat{k}\left(\frac{\partial v}{\partial x} - \frac{\partial u}{\partial y}\right)$$

$$= \begin{vmatrix} \hat{\imath} & \hat{\jmath} & \hat{k} \\ \dfrac{\partial}{\partial x} & \dfrac{\partial}{\partial y} & \dfrac{\partial}{\partial z} \\ u & v & w \end{vmatrix} \tag{1.6}$$

It is extremely important to note that these forms involving the operator ∇, including its definition, are *valid in general only for Cartesian coordinate systems*. In curvilinear coordinate systems the derivatives of the unit vectors become involved and, as we have seen, significant differences from the Cartesian results are to be expected.

All of the previous expressions—gradient, divergence, and curl—have important physical interpretations when applied to fluid flow problems. These will be brought out in due course. None of these operations is commutative (e.g., $\nabla S \neq S\nabla$) or associative (e.g., $\nabla \cdot S\mathbf{q} \neq \nabla S \cdot \mathbf{q}$). They are, however, distributive (e.g., $\text{curl}(\mathbf{q} + \mathbf{r}) = \text{curl } \mathbf{q} + \text{curl } \mathbf{r}$). Further discussion of vector differential operations may be found in Bird et al. (1960, pp. 715–742) and Schey (1973).

Integration of Functions of More than One Variable

Often in the analysis of fluid flows it is necessary to deduce the nature of a function from its partial derivatives. Thus, given

$$\frac{\partial F}{\partial x} = P(x, y) \quad \text{and} \quad \frac{\partial F}{\partial y} = Q(x, y) \tag{1.7}$$

it is necessary to find $F(x, y)$.

The first of the expressions above will be satisfied by

$$F(x, y) = \int P(x, y)\, dx + f(y) \tag{1.8}$$

where the y appearing under the integral sign is treated as a parameter that does not vary in the integration with respect to x. The notation $f(y)$ means that f depends *only* on y (and *not* on x or anything else). If we now differentiate with respect to y,

$$\frac{\partial F}{\partial y} = \frac{\partial}{\partial y}\int P(x, y)\, dx + f'(y) = Q(x, y)$$

where $f'(y) \equiv df/dy$. From this,

$$f'(y) = Q(x, y) - \frac{\partial}{\partial y}\int P(x, y)\, dx$$

and

$$f(y) = \int \left[Q(x, y) - \frac{\partial}{\partial y} \int P(x, y)\, dx \right] dy + C$$

where C is a constant of integration. Substituting this result into Eq. (1.8), the required function is

$$F(x, y) = \int P(x, y)\, dx + \int \left[Q(x, y) - \frac{\partial}{\partial y} \int P(x, y)\, dx \right] dy + C \qquad (1.9)$$

Example Consider the functions

$$P(x, y) = 2xy + 1 \qquad \text{and} \qquad Q(x, y) = x^2 + 4y$$

from which

$$\int P\, dx = x^2 y + x$$

and

$$\frac{\partial}{\partial y} \int P\, dx = x^2$$

Applying Eq. (1.9),

$$F(x, y) = x^2 y + x + \int (x^2 + 4y - x^2)\, dy + C$$
$$= x^2 y + x + 2y^2 + C$$

Note that in this example,

$$\frac{\partial P}{\partial y} = \frac{\partial Q}{\partial x} = 2x$$

and from Eq. (1.7),

$$\frac{\partial^2 F}{\partial y\, \partial x} = \frac{\partial^2 F}{\partial x\, \partial y}$$

This result is an expression of the necessary continuous nature of $F(x, y)$ and its first derivatives. It allows the expression of dF as an exact differential

$$dF = P\, dx + Q\, dy$$

Functions of this type are especially common in the analysis of ideal fluid flows (Part One of this text).

1.2 THE NOTION OF A CONTINUUM

Throughout this text we refer to conditions at a point in a field of flow. This implies a location in space that can be defined to an unlimited degree of precision.

Physically, however, there is always a limit to which the "smallness" of dimensions may be taken without the occurrence of unwanted discontinuities. For instance, the distance between molecules of a gas at standard conditions is of the order of 3×10^{-9} m. If the size of our point is of a similar magnitude, we cannot even be sure that it will contain a molecule. The conditions at such a point are likely to be totally misleading in terms of the average properties of a larger volume surrounding it. The imposition of the continuum hypothesis requires that molecularly averaged values at a point are meaningful.

In most cases this is not a serious limitation. In the standard gas, for instance, there are about 10 million billion (10^{16}) molecules in a cubic millimeter. Average properties of the volume would truly represent the characteristics and activity of the molecules therein—such a volume is quite large, from a microscopic point of view. On the other hand, a cubic millimeter might be quite small from the perspective of the overall field of analysis so that the volume could be considered to be a point for purposes of such analysis. In addition, conditions could be expected to vary continuously from one millimeter to another in such a gas, unless some very sudden events occur. The continuum hypothesis can also break down in cases where the characteristic distance between molecules becomes relatively large. At the upper reaches of the stratosphere, which extends to about 20 miles above the earth's surface, the density of air decreases to about 1/100th of its sea-level value. In regions such as this, precise calculations must begin to consider the microscopic or statistical approaches to analysis.

The concept of a continuum also bears upon the mathematical notion of the derivative. By definition, the derivative of a function $f(x)$ is

$$f'(x) = \frac{df}{dx} = \lim_{\Delta x \to 0} \frac{f(x + \Delta x) - f(x)}{\Delta x}$$

In our application of this definition to physical problems, we must think of the limiting process as not really proceeding to $\Delta x = 0$ because in the end we would expect severe discontinuities in $f'(x)$ as the molecular level is approached. Our definition of zero in this limit must be that volume of space over which the derivative approaches a value that does not change as Δx is decreased. That this volume is small enough to represent a point, for purposes of analysis, is the crux of the continuum hypothesis.

1.3 BIBLIOGRAPHY

As has been noted, students are encouraged to review material related to their previous courses in fluid mechanics. This area of the published literature abounds with excellent texts, and we suggest that students review those with which they are most familiar. The following list contains both introductory and intermediate texts, with much overlap in the scope and level of coverage. In addition to valuable review material, these texts, and others like them, contain presentations that provide alternative points of view relating to much of the subject matter of this book.

Allen, T. A., Jr., and R. L. Ditsworth: *Fluid Mechanics,* McGraw-Hill, New York, 1972.

Bertin, J. J.: *Engineering Fluid Mechanics,* Prentice-Hall, Englewood Cliffs, N.J., 1984.

Daugherty, R. L., and J. B. Franzini: *Fluid Mechanics with Engineering Applications,* 7th ed., McGraw-Hill, New York, 1977.

Fox, J. A.: *An Introduction to Engineering Fluid Mechanics,* McGraw-Hill, New York, 1974.

Fox, R. W., and A. T. McDonald: *Introduction to Fluid Mechanics,* 3rd ed., John Wiley & Sons, New York, 1985.

Gerhart, P. M., and R. J. Gross: *Fundamentals of Fluid Mechanics,* Addison-Wesley, Reading, Mass., 1985.

John, J. E. A., and W. L. Haberman: *Introduction to Fluid Mechanics,* 2nd ed., Prentice-Hall, Englewood Cliffs, N.J., 1980.

Kreider, J. F.: *Principles of Fluid Mechanics,* Allyn & Bacon, Boston, 1985.

Mironer, A.: *Engineering Fluid Mechanics,* McGraw-Hill, New York, 1979.

Olson, R. M.: *Essentials of Engineering Fluid Mechanics,* 4th ed., Harper & Row, New York, 1980.

Pao, R. H. F.: *Fluid Mechanics,* John Wiley & Sons, New York, 1961.

Prasuhn, A. L.: *Fundamentals of Fluid Mechanics,* Prentice-Hall, Englewood Cliffs, N.J., 1980.

Roberson, J. A., and C. T. Crowe: *Engineering Fluid Mechanics,* Houghton Mifflin, Boston, 1975.

Rouse, H., and J. W. Howe: *Basic Mechanics of Fluids,* John Wiley & Sons, New York, 1953.

Shames, I. H.: *Mechanics of Fluids,* 2nd ed., McGraw-Hill, New York, 1982.

White, F. M.: *Fluid Mechanics,* McGraw-Hill, New York, 1979.

REFERENCES

Bird, R. B., W. E. Stewart, and E. N. Lightfoot: *Transport Phenomena,* John Wiley & Sons, New York, 1960.

Schey, H. M.: *Div, Grad, Curl, and All That,* W. W. Norton, New York, 1973.

PART
ONE

IDEAL FLUID FLOWS

FORCES AND MOTIONS IN IDEAL FLOWS

Shear forces, however small they may be, cause fluids to deform. This, or some version of it, defines substances called fluids—they may be either liquids or gases, or mixtures of phases. When the motion of a *solid* is studied, it is often sufficient to consider the solid as a rigid particle on which all attention is focused. Fluid motions may also be analyzed by the particle approach (the method is sometimes called the Lagrangian approach), but most fluid motions consist of an enormous number of particles, and this treatment of each of them is often impractical. In such cases the more usual control volume or Eulerian point of view is adopted in which the changes within a region of fluid flow, or a flow field, are considered.

Whatever approach we take to the problem of fluid flow analysis, we must take into account the fact that changes in the properties of the fluid, or in the flow itself, can occur for two fundamental reasons: *the passage of time and the motion from point to point in the flow*. The mathematical expression of the *total* change of any quantity in the flow must therefore include these dependencies. For example, if Q is any such quantity, then $Q = Q(\mathbf{s}, t)$, where $\mathbf{s}$ is a vector specifying location. If the components of $\mathbf{s}$ are x, y, and z, then we may write $Q = Q(x, y, z, t)$. A change in Q due to differences $\delta \mathbf{s}$ and δt may be written

$$\delta Q = Q(x + \delta x, y + \delta y, z + \delta z, t + \delta t) - Q(x, y, z, t)$$

(Except where otherwise noted, the symbol δ is used to identify a quantity that will eventually be limited to a differential size.) Denoting the change in Q due to differentially small changes in time and distance as

$$dQ = \lim_{\substack{\delta \mathbf{s} \to d\mathbf{s} \\ \delta t \to dt}} \delta Q$$

then

$$dQ = \frac{\partial Q}{\partial x} dx + \frac{\partial Q}{\partial y} dy + \frac{\partial Q}{\partial z} dz + \frac{\partial Q}{\partial t} dt$$

Division by the time increment dt gives

$$\frac{dQ}{dt} = \frac{\partial Q}{\partial x}\frac{dx}{dt} + \frac{\partial Q}{\partial y}\frac{dy}{dt} + \frac{\partial Q}{\partial z}\frac{dz}{dt} + \frac{\partial Q}{\partial t} \tag{2.1}$$

This total rate-of-change expression is a general result for anything that depends on x, y, z, and t. In solid mechanics it is often sufficient to consider only the time variation because the spacial dependency is negligible. In such cases the distinction between dQ/dt and $\partial Q/\partial t$ in Eq. (2.1) is not necessary. In problems in fluid mechanics, on the other hand, the spacial terms in Eq. (2.1) are often the most important ones. The problem sometimes can be treated as one in which Q is not dependent on time, if the flow is viewed from a broad perspective (that is, when time-dependent motions are on a scale that is small enough to make them difficult to differentiate). In a general approach to fluid mechanics, however, it is important always to bear in mind the influences of both time and space.

2.1 DERIVATIVE MOVING WITH THE FLOW

A particular form of Eq. (2.1) is obtained if the velocities dx/dt, dy/dt, and dz/dt are given special meaning—that is, these terms are taken to represent the *velocity components of the flow*. In such a case the observed rate, dQ/dt, will be that particular rate due to *moving with the fluid*. Specifying the components of the velocity of the flow as u, v, and w in the x, y, and z directions, respectively, the result is

$$\frac{DQ}{Dt} = u\frac{\partial Q}{\partial x} + v\frac{\partial Q}{\partial y} + w\frac{\partial Q}{\partial z} + \frac{\partial Q}{\partial t} \tag{2.2}$$

The rate of change given in Eq. (2.2) is called the *derivative moving with the flow* (also known as the substantive derivative, the particle derivative, or the material derivative). It is simply a shorthand way of saying that when we move from point to point in a field of flow, we move *with* the flow and not independently of it. The capital D notation is used to identify this special case of the total rate of change. The first three terms on the right-hand side of Eq. (2.2) constitute the "convective" rate of change of the quantity Q (in the earth sciences the term "diffusive" is often used) and is due solely to the fluid motion. If there is no motion, then u, v, and w are all zero and there is no convective change. There may still be a change in a property due to the passage of time at a fixed location. A common example is the change of internal energy in a motionless body of water undergoing heating. This change would be called a local change (because it occurs at a particular point in the flow) and is due to time dependency as distinct from spacial dependency. If *all* properties of a flow are invariant with time ($\partial Q/\partial t = 0$), then the flow is said to be *steady*. Most of the flows treated in this text are of the steady kind, but it is nevertheless vital to remember that both local and convective changes can be important in a flow of a fluid. From a notational point of view, this fact can be kept in view by using the D notation. For instance, the acceleration of a fluid particle may be written

$$\frac{D\mathbf{q}}{Dt} = u\frac{\partial \mathbf{q}}{\partial x} + v\frac{\partial \mathbf{q}}{\partial y} + w\frac{\partial \mathbf{q}}{\partial z} + \frac{\partial \mathbf{q}}{\partial t} \tag{2.3}$$

where $\mathbf{q}$ is the velocity vector at each point in the flow.

2.2 VECTOR EXPRESSIONS FOR FORCES ON A FLUID PARTICLE

Suppose that a force may be taken to act at the center of mass of a particle. (The most frequently encountered force of this type is weight.) Such a force is referred to as a *body force* and can be described in two dimensions by

$$\mathbf{B} = \hat{\imath}B_x + \hat{\jmath}B_y = \hat{\imath}B \sin \alpha + \hat{\jmath}B \cos \alpha$$

The angle α is defined in Fig. 2.1, and B_x and B_y are the Cartesian components of the vector of magnitude B.

Let us now define a distance h that is measured *opposite* to the direction of the force $\mathbf{B}$. The effect on h of traversing in the x and y directions is illustrated in Fig. 2.2. If we move away from the point of application of the force in the x direction (that is, with constant y), then the *change* in h is $\delta h = -\delta x \sin \alpha$ or, since y is held constant, $\partial h/\partial x = -\sin \alpha$. If we move in the y direction, we find that the corresponding rate of change of h is $\partial h/\partial y = -\cos \alpha$. The body force

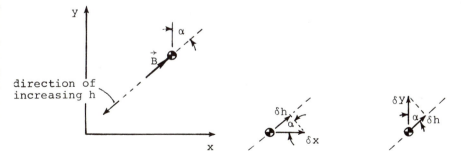

Figure 2.1 Body force acting in the *x-y* plane.   **Figure 2.2**

can now be expressed in terms of the direction h instead of the angle α:

$$\mathbf{B} = -B\left(\hat{\imath}\,\frac{\partial h}{\partial x} + \hat{\jmath}\,\frac{\partial h}{\partial y}\right)$$

and the logical extension to three dimensions is

$$\mathbf{B} = -B\left(\hat{\imath}\,\frac{\partial h}{\partial x} + \hat{\jmath}\,\frac{\partial h}{\partial y} + \hat{k}\,\frac{\partial h}{\partial z}\right) = -B\,\boldsymbol{\nabla}h \tag{2.4}$$

The most commonly felt body force is the force of gravity acting on a fluid element. This force is the "weight" of the element and acts in a direction that we like to think of as "down." Since gravity acts downward, the direction h is what we call "up" because it is by definition opposite to the direction of the body force (weight, in this case). The magnitude of the weight of a mass m, in a location where the acceleration of gravity is g, is $B = mg$, and the vector expression, Eq. (2.4), is

$$\mathbf{B} = -mg\,\boldsymbol{\nabla}h \tag{2.5}$$

Let us now consider the other main category of forces that can act on a fluid element. These are called *surface forces* (as distinguished from body forces) and on a per-unit-area basis they are termed stresses. In fluid mechanics the component of the surface force per unit area that acts *normal* to a face of a fluid element (the "normal" stress) is called the *pressure*. Pressure is considered to be positive if it acts in a compressive sense (opposite to the sign convention for normal stresses in solid mechanics). Pressure forces act throughout a field of fluid and, if the pressure is uniform, there will be no net pressure force on a fluid element within the field. Unbalanced pressure forces, and therefore fluid motions, can exist if there are pressure differences in a substance. Figure 2.3 illustrates such a situation in the x direction. Surface forces that act parallel to the surface of a fluid element form *shear* stresses. These stresses, not considered in Part One of this text, are evidence of viscous action. They receive much attention in Part Three.

The x component of the pressure force is seen from Fig. 2.3 to be

$$(F_p)_x = \left[P - \left(P + \frac{\partial P}{\partial x}\,\delta x\right)\right]\delta y\,\delta z = -\frac{\partial P}{\partial x}\,\delta x\,\delta y\,\delta z$$

and, if the contributions in the other coordinate directions are taken into account,

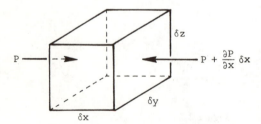

Figure 2.3 Unbalanced pressures acting on the surface of a fluid element.

the pressure-force vector may be written

$$\mathbf{F}_p = \hat{\imath}\left(-\frac{\partial p}{\partial x}\,\delta x\,\delta y\,\delta z\right) + \hat{\jmath}\left(-\frac{\partial p}{\partial y}\,\delta y\,\delta x\,\delta z\right) + \hat{k}\left(-\frac{\partial p}{\partial z}\,\delta z\,\delta x\,\delta y\right)$$

But the quantity $\delta x\,\delta y\,\delta z$ is simply the volume of the fluid element, δV, so that if the fluid density is denoted by ρ,

$$\mathbf{F}_p = -\delta V\,\nabla P = -\frac{m}{\rho}\,\nabla P \tag{2.6}$$

(The reader should be careful to note that the operator

$$\nabla = \hat{\imath}\,\frac{\partial}{\partial x} + \hat{\jmath}\,\frac{\partial}{\partial y} + \hat{k}\,\frac{\partial}{\partial z}$$

is a *vector* operator, and the gradient of the pressure, ∇P, is a vector quantity in the direction of the resultant pressure force.)

2.3 EULER EQUATION

If the two forces described above are the only important forces in a fluid flow problem, then the application of Newton's second law to the fluid element gives

$$\mathbf{B} + \mathbf{F}_p = m\mathbf{a}$$

With the applications of Eqs. (2.3), (2.5), and (2.6),

$$\frac{D\mathbf{q}}{Dt} = -\frac{1}{\rho}\,(\nabla P + \gamma\,\nabla h) \tag{2.7}$$

where γ is the specific weight of the fluid ($\gamma = \rho g$). Again, convective as well as unsteady effects are taken into account through the use of the D notation.

Equation (2.7) is a very famous relationship in fluid mechanics and is known as the Euler equation in honor of the Swiss mathematician Leonhard Euler (1707–1783). This relationship was first presented in the mid-eighteenth century and, as the previous development has indicated, is valid for fluid motion caused by pressure differences and body forces. Other forces, if influential, would invalidate the Euler equation.

Exercise If Eq. (2.7) is evaluated for motionless fluids, we obtain

$$\nabla P = -\gamma\,\nabla h$$

This is the law of hydrostatic variation of pressure expressed for three dimensions. Observe that the Euler equation contains within it the governing relation for fluid statics.

In many instances, coordinate systems are oriented so that the y axis points "up" opposite to the direction of gravity. In such cases we may write (see Fig. 2.4)

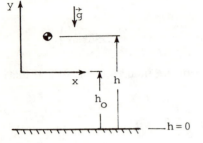

—$h = 0$ **Figure 2.4**

$$h = h_0 + y$$

so that

$$\nabla h = \hat{j}\,\frac{\partial h}{\partial y} = \hat{j}(1)$$

For the hydrostatic case, then,

$$\nabla P = -\gamma j \qquad \text{or} \qquad \frac{dP}{dy} = -\gamma$$

and, after integration,

$$P = -\gamma y + C = -\gamma(h - h_0) + C$$

The constant of integration, C, must be obtained by defining a reference point for evaluating the pressure variation in the fluid. For instance, if we measure pressure relative to the origin of h ($P = 0$ at $h = 0$), then

$$C = -\gamma h_0 \qquad \text{and} \qquad P = -\gamma h$$

It is inherent in the nature of fluids that pressures are important only insomuch as they differ from point to point. Thus, it is important to note that the pressure P in the expression above is really a *difference* in pressure relative to a reference level ($h = 0$ in this case).

To ensure that the physical meaning of Eq. (2.7) is not obscured by the elegance of the vector notation, it is recommended that the reader extract the three components of this relationship from the general form. For instance, in the y direction:

$$\underbrace{\frac{\partial v}{\partial t}}_{\substack{\text{local} \\ \text{acceleration}}} + \underbrace{u\,\frac{\partial v}{\partial x} + v\,\frac{\partial v}{\partial y} + w\,\frac{\partial v}{\partial z}}_{\substack{\text{convective} \\ \text{acceleration}}} = \underbrace{-\frac{1}{\rho}\left(\frac{\partial P}{\partial y} + \gamma\,\frac{\partial h}{\partial y}\right) + 0}_{\substack{\text{forces per} \\ \text{unit mass}}}$$

These terms may be considered to be the effects (accelerations on the left-hand side) of the causes (forces on the right-hand side). All, of course, are in the y

direction. The zero with emphasis on the right-hand side of this expression serves as a reminder that there are other forces that can act on a fluid element which, if influential, must appear in appropriate form with the terms on the right-hand side. Chief among these other causes of acceleration (or, more likely, deceleration) are those due to the effects of fluid viscosity. When these effects are considered, the equation of motion is no longer termed the Euler equation—it is then in a form called the Navier-Stokes equation.

For the time being we consider ideal fluid motion as that described by Eq. (2.7). This gives us a *kinetic* definition (concerned with forces) of what is meant by ideal fluid flow: *Ideal fluid flow is one that is influenced only by pressure forces and/or body forces*. It might also be mentioned here that these forces give no tendency for the fluid element to rotate as it traverses a flow field. Such absence of rotation leads to a *kinematic* definition of ideal fluid flow, as we shall see later.

Even though we neglect the *effects* of fluid viscosity in Part One of this text, this should not give the impression that the results are valid only for fluids that have no viscosity. Such "inviscid" fluids do not exist in fact (although zero viscosity can be nearly achieved in some rather esoteric circumstances), and if such an assumption were necessary this part of the text would have little more than academic interest to engineers. In actual practice there are many fluid flow situations in which major *portions* of the flow field are beyond the *influences* of viscosity. Thus, *the analysis of ideal fluid flows requires that the flow behave ideally, not the fluid*.

Description in Streamline Coordinates

Because flows do not generally occur in straight lines, the Cartesian frame of reference does not provide a particularly useful framework for interpreting the messages contained in the equations of motion. It is more reasonable to consider the components of motion along and normal to fluid pathlines, and coordinate systems thus erected are called streamline, natural, or intrinsic. Figure 2.5 illustrates such a system in which the dimension *s* is measured in the direction tangent to a streamline, in the flow direction, at point *P*. The coordinate *n* is measured in the direction of the local center of curvature of the streamline at point *P*, and *m* completes the mutually orthogonal system. In unsteady flow the diagram in

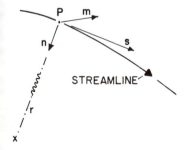

Figure 2.5 Streamline coordinate system.

Fig. 2.5 must be thought of as a "snapshot" of the flow because, as time proceeds, the flow direction at point P, and hence the direction of the streamline, may change.

Let us consider the components of the acceleration of a fluid element at point P:

$$a_s = \frac{\partial v_s}{\partial t} + v_s \frac{\partial v_s}{\partial s} + v_n \frac{\partial v_s}{\partial n} + v_m \frac{\partial v_s}{\partial m} = \frac{\partial v_s}{\partial t} + \frac{\partial}{\partial s}\left(\frac{v_s^2}{2}\right)$$

$$a_n = \frac{\partial v_n}{\partial t} + v_s \frac{\partial v_n}{\partial s} + v_n \frac{\partial v_n}{\partial n} + v_m \frac{\partial v_n}{\partial m} = \frac{\partial v_n}{\partial t} + v_s \frac{\partial v_n}{\partial s}$$

The simplified forms on the right-hand sides of these expressions result from the fact that at point P the only nonzero velocity component is v_s in the direction of the streamline. (In the m direction, the only nonzero component of acceleration is the local acceleration, $\partial v_m / \partial t$, since the velocity vector lies in the s-n plane at that point.) The last term in the acceleration expression for the n direction may be further reduced by noting the geometry sketched in Fig. 2.6. The symbol δ indicates changes from conditions at point P to conditions at a point slightly displaced from P along the streamline. If r is used to denote the radius of curvature of the streamline at P, then the geometry of Fig. 2.6 indicates that

$$\tan \alpha = \frac{\delta v_n}{v_s + \delta v_s} \quad \text{and} \quad \alpha = \frac{\delta s}{r}$$

For small angles, therefore ($\tan \alpha \approx \alpha$),

$$\frac{\delta v_n}{\delta s} = \frac{v_s + \delta v_s}{r}$$

and in the limit as δs and δv_s approach zero,

$$\frac{\partial v_n}{\partial s} = \frac{v_s}{r}$$

Excluding the unsteady terms from the previous relationships, the equations of motion become:

In the streamline direction:

$$\frac{\partial}{\partial s}\left(\frac{v_s^2}{2}\right) = -\frac{1}{\rho}\left(\frac{\partial p}{\partial s} + \gamma \frac{\partial h}{\partial s}\right) \tag{2.7s}$$

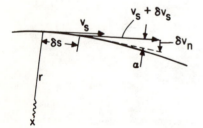

Figure 2.6

Normal to the streamlines:

$$\frac{v_s^2}{r} = -\frac{1}{\rho}\left(\frac{\partial P}{\partial n} + \gamma\frac{\partial h}{\partial n}\right) \tag{2.7n}$$

These two forms of the Euler equation are the reason we have gone to all this trouble. They give us the following physical insights for *steady ideal fluid flow* with no change in *h*:

Pressure changes are opposite in sign to velocity changes along a streamline. Pressure decreases in the direction of the center curvature of a streamline. Straight streamlines have no pressure variation normal to themselves.

In addition to these far-reaching statements, we may easily integrate Eq. (2.7s) if we restrict the path of integration to lie along a streamline of the flow. Such restriction is the integral sense of the constraints implied by the partial derivative notation used in Eq. (2.7s): "If *s* and only *s* is varied, that is, if you don't stray off of the streamline on which you are located, then, with constant density, the rate of change in $v_s^2/2$ is equal in magnitude but opposite in sign to the change in $P/\rho + gh$."

2.4 BERNOULLI EQUATION

Before we formally integrate Eq. (2.7s) along a streamline, we must do something about the density. For now, we shall simply assume that the density is constant so that Eq. (2.7s) may be written

$$\frac{\partial}{\partial s}\left(\frac{v_s^2}{2} + \frac{P}{\rho} + gh\right) = 0$$

It is now easy to see that if we integrate along a streamline in *ideal, steady, incompressible flow*,

$$\frac{v_s^2}{2} + \frac{P}{\rho} + gh = \text{constant along a streamline} \tag{2.8}$$

This is the most simple form of the famous expression attributed to Daniel Bernoulli and about which we shall have much more to say. The Bernoulli equation allows the calculation of pressures in a flow field if the velocities are known. The pressures, in turn, allow the determination of forces acting on bodies in or around the flow—one of the main goals of fluid mechanical analysis. For now it suffices to emphasize that this expression is valid only under the conditions stated: along a streamline in ideal, steady, incompressible flows.

Exercise In fully developed flow in a pipe, the fluid moves parallel to the pipe walls at a velocity that does not change in the direction of the flow (the streamline direction). In such circumstances v_s does not vary along the stream-lines and, if the pipe is horizontal, the Bernoulli equation says that the pressure will not vary along the pipe. Clearly this contradicts experience and fact. What is the dilemma? There is none, really—application of Eq. (2.8) to this situation is in violation of the stated conditions. Such pipe flow is influenced by viscous effects. If this were not so, there would in fact be no pressure drop along the pipe (with severe economic impact on the pump industry).

PROBLEMS

2.1 An engineer works at a job located 2 km from his residence. He decides to measure the air temperature one morning as he walks to work and, on the morning in question, the temperature is increasing everywhere along his path at the rate of 8°C per hour. In addition to this change, because of local climactic conditions, the temperature increases 4°C between his home and work. What rate of change of temperature will the engineer observe if he walks at a rate of 4 km/h? What would be the observed rate if he were to walk from work to home under the same conditions?

2.2 The concentration of pollutants in a certain river is increasing at the rate of 150 ppm (parts per million) per hour. In addition, the pollution concentration increases with distance in the direction of the river flow due to an influx from sewer lines. This latter rate of increase can be expressed as 50 ppm/mi at a certain point in the river. The river flows at 5 mi/h.

(a) A boat is used to survey the pollution concentration. Find the apparent rates of pollution concentration (in parts per million per hour) if the boat is headed upstream or downstream and cruises at 8 mi/h relative to the water. Repeat for the case where the boat is just drifting with the current.

(b) Which, if any, of the above results corresponds to the derivative moving with the flow, given by $D(\text{pollution})/Dt$?

2.3 For the following flow velocities, find the derivative moving with the flow (Du/Dt):

(a) $u = \dfrac{x^2 y^2}{zt}$

(b) $u = yz + t$

2.4 A velocity field is given by

$$u = yz + t \qquad v = xz + t \qquad w = xy$$

Find the acceleration components at $(1, 1, 1)$ in terms of time t.

2.5 A velocity field is given by $\mathbf{q} = 2xy\hat{\imath} + x^2\hat{\jmath}$. Find the acceleration $D\mathbf{q}/Dt$, and use the Euler equation, Eq. (2.7), to find the pressure gradient vector. Repeat using the Bernoulli equation to find the components of the pressure gradient, $\partial P/\partial x$ and $\partial P/\partial y$.

2.6 Repeat problem 2.5 but with $\mathbf{q} = 2xy\hat{\imath} - x^2\hat{\jmath}$. Does the Bernoulli equation give the same result as the Euler equation? If not, why not? Note that in problem 2.5, $\partial u/\partial y = \partial v/\partial x$, whereas in this problem this is not true.

2.7 Derive the equation of continuity for two-dimensional incompressible flow in polar-cylindrical coordinates. Do this by equating the rate of flow out of a small cylindrical element of space to the rate of flow in less the rate of accumulation of mass within the element. (See Fig. 2.7.)

2.8 The result of problem 2.7 is

$$\frac{\partial}{\partial r}(rv_r) + \frac{\partial v_\theta}{\partial \theta} = 0$$

Show that for a continuous circular path, the velocity can vary only with r. Find the pressure variation in such a flow in the directions tangent and normal to the streamlines, with and without hydrostatic effects.

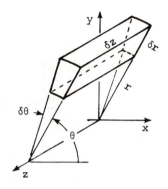

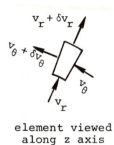

element viewed
along z axis

Figure 2.7 Geometry for problem 2.7.

2.9 A fluid (specific weight $= 60 \text{ lb/ft}^3$) flows upward at an inclination of 30° above the horizontal. If the flow is at uniform and constant speed along a straight path, find the pressure variation in the directions tangent and normal to the streamlines. Does the pressure variation depend on the flow speed?

2.10 The Bernoulli expression has the dimensions of energy per unit mass and is sometimes derived from energy considerations. Identify the various forms of energy in Eq. (2.8) and discuss the conditions under which the equation is valid from an energy point of view.

THREE

TOOLS FOR USE IN IDEAL FLUID FLOWS

In the previous chapter we began looking at the nature of ideal motions of fluids and, in particular, the kinetic concepts arising from the fact that in such motions pressures are the dominant surface forces acting on a fluid element. There are also *kinematic* concepts that apply to ideal flows, and the mathematical relationships that result from these are particularly useful in the analysis of such flows. In this chapter we examine these kinematic aspects, which, in combination with the stream function, provide the fundamental tools and methods commonly used in the analysis of ideal fluid flows.

3.1 FLOWS WITH AND WITHOUT ROTATIONAL MOTION

In developing the kinetic notions of ideal fluid flow, we were obliged to neglect the effects of fluid friction, which, were they important, would have led to shear stresses acting on the surface of a fluid element. These shear stresses would have invalidated the expressions that were used for the total forces acting in ideal fluid flow. In addition, they could have led to rotation of the fluid element, and this rotation, or the absence thereof, is related directly to the kinematic definition of ideal fluid flow. We now develop the mathematical devices by which we may precisely define what is meant by rotation.

Let us consider the velocities that a fluid element possesses when it is part of a field of flow. For purposes of illustration we shall consider those components of the motion that would be viewed in the *x*-*y* plane. Similar results are easily obtained from analyses in the other two planes. In Fig. 3.1 the notation $(\delta v)_x$ is to be interpreted as "the change in v when moving in the x direction," for ex-

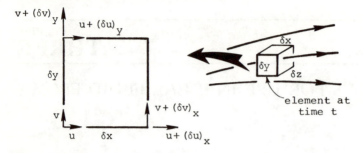

Figure 3.1 Element in a field of flow.

ample. In Fig. 3.1 the corners of the element at δx and δy are shown to have positive velocities $(\delta v)_x$ and $(\delta u)_y$, respectively, relative to the lower left-hand corner of the element. Because of these relative velocities, the sides of the element will rotate as time passes and, after a small increment of time δt, the element will be distorted into a different shape similar to that shown in Fig. 3.2.

For clarity, only the effects of $(\delta v)_x$ and $(\delta u)_y$ are shown here. The average rate of angular motion of the two sides is

$$\frac{\dot{\gamma}_1 + \dot{\gamma}_2}{2} = \frac{1}{2}\left[\frac{(\delta v)_x}{\delta x} - \frac{(\delta u)_y}{\delta y}\right]$$

Here we have used the right-hand rule, which assigns a positive sense to counterclockwise rotations. The *rate of rotation,* ω, is defined as this average, in the limit as the element becomes differentially small. In this limit the velocity increments δv and δu may be expressed as $(\partial v/\partial x)\,\delta x$ and $(\partial u/\partial y)\,\delta y$, respectively, so that

$$\omega_z = \frac{1}{2}\left(\frac{\partial v}{\partial x} - \frac{\partial u}{\partial y}\right) \tag{3.1z}$$

The subscript $(\ )_z$ denotes the direction of the normal to the plane of the rotation, which in this case is the x-y plane. Similar analyses in the two orthogonal planes lead to the following results:

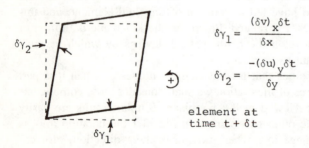

$$\delta\gamma_1 = \frac{(\delta v)_x\,\delta t}{\delta x}$$

$$\delta\gamma_2 = \frac{-(\delta u)_y\,\delta t}{\delta y}$$

element at
time $t + \delta t$

Figure 3.2 Element subjected to velocity differences.

$$\omega_x = \frac{1}{2}\left(\frac{\partial w}{\partial y} - \frac{\partial v}{\partial z}\right) \tag{3.1x}$$

$$\omega_y = \frac{1}{2}\left(\frac{\partial u}{\partial z} - \frac{\partial w}{\partial x}\right) \tag{3.1y}$$

These three components may be combined, using the definition of the curl in Cartesian coordinates, as follows:

$$\boldsymbol{\omega} = \frac{1}{2}(\nabla \times \mathbf{q}) \tag{3.2}$$

Exercise Construct a sketch showing positive velocity increments $(\delta u)_x$ and $(\delta v)_y$ relative to a point in the x-y plane where the velocity components are u and v. The sides of the element move away from each other with relative velocities $(\partial u/\partial x)\,\delta x$ and $(\partial v/\partial y)\,\delta y$ in the x and y directions, respectively. The rate of increase in the volume of the element is

$$\frac{\partial u}{\partial x}\,\delta x(\delta y\,\delta z) + \frac{\partial v}{\partial y}\,\delta y(\delta x\,\delta z) = \left(\frac{\partial u}{\partial x} + \frac{\partial v}{\partial y}\right)(\delta x\,\delta y\,\delta z)$$

If the fluid is incompressible, the density of the element cannot change. Since the element has a fixed mass, its volume must be constant. Thus, we see that $\partial u/\partial x + \partial v/\partial y = 0$ in such a case, and the expression for mass continuity in two-dimensional steady incompressible flow is obtained. Physically, if the element grows in one direction, it must shrink in the other by an equal and opposite amount. In three dimensions the net rate of growth in the three directions must add up to zero.

Vorticity

The *vorticity*, ζ, is defined as twice the rate of rotation of a fluid particle

$$\zeta = \nabla \times \mathbf{q} \tag{3.3}$$

with components expressed as twice the values given in Eqs. (3.1x, y, z). Note that from Eq. (3.3) the vorticity vector satisfies the relationship $\nabla \cdot \zeta = 0$. This divergence relationship is similar to that for the velocity when conservation of mass is expressed for steady incompressible flow. Thus, vorticity is conserved, and there are a number of interesting and important theorems regarding its behavior in a fluid flow. The reader may wish to consult the literature in this regard. See, for instance, Robertson (1965, chap. 4) and Li and Lam (1976, chap. 8).

Circulation

The circulation Γ is defined as

$$\Gamma = \int_C \mathbf{q} \cdot d\mathbf{s} \tag{3.4}$$

In words the circulation is the closed line integral over any path C in a flow field of the component of the velocity tangent to that path (Fig. 3.3).

If A is a surface area bounded by C, then the integral above may be converted to an integral over the surface A by means of Stokes's theorem (Schey, 1973, p. 92):

$$\Gamma = \int_A (\nabla \times \mathbf{q}) \cdot \hat{n} \, dA$$

where $\hat{n}$ is the unit outward normal to dA at each point on the surface. Note, however, that $\nabla \times \mathbf{q}$ is the vorticity vector at each of the area elements and $(\nabla \times \mathbf{q}) \cdot \hat{n}$ is the component of vorticity, ζ_n, normal to dA at each point on the surface. Thus we may write

$$\Gamma = \int_A \zeta_n \, dA$$

This expression is known as Kelvin's relation and indicates that the circulation associated with a velocity field bounded by a closed curve is equal to the surface integral of the normal component of the vorticity over any surface bounded by that curve. Because the surface bounded by the curve may be warped at will without changing its bounding curve, particles that have a certain amount of rotation (vorticity) when passing through one surface will have rotation when passing through another, provided that both surfaces are bounded by the same curve in space. For surfaces having the same bounding curve, the circulation remains constant, requiring a change in vorticity inversely with normal cross-sectional area. This result is another view of the conservation of vorticity. It is analogous to the observation that incompressible volume flow through a surface encompassed by streamlines will be at the same rate through any other surface bounded by the same streamlines, with velocity increasing through decreasing areas and vice versa. Many important relationships have been developed regarding circulation and vorticity; the interested reader may wish to consult the references suggested previously.

Kelvin's relation may be expressed in differential form as

$$\frac{d\Gamma}{dA} = \zeta_n$$

dA

C

Figure 3.3

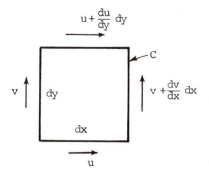

Figure 3.4

This result provides another method of determining the vorticity or rotation in a flow field. For example, if we orient a differentially small surface dA normal to the z axis and indicate the velocity components on the path C bounding that area, we have (Fig. 3.4)

$$d\Gamma = u\,dx + \left(v + \frac{\partial v}{\partial x}dx\right)dy - \left(u + \frac{\partial u}{\partial y}dy\right)dx - v\,dy$$

(In evaluating $d\Gamma$, only the components of velocity tangent to the curve C are of importance.) Finally, with $dA = dx\,dy$,

$$\zeta_z = \frac{d\Gamma}{dA} = \frac{\partial v}{\partial x} - \frac{\partial u}{\partial y}$$

in accordance with previous results, Eq. (3.1z).

Irrotational Motion and Velocity Potential

If the rate of rotation (and therefore the vorticity) is zero in a fluid flow, it is said to be irrotational. Each fluid particle moves about without rotating—like the gondolas on a Ferris wheel. The particles may be distorted, but the average rotation must be zero in irrotational flow. Motion such that $\partial v/\partial x - \partial u/\partial y$ is zero is irrotational in the x-y plane, and is sometimes also called *pure angular strain*—this is similar to the deformation shown in the Fig. 3.2. On the other hand, if there is no angular strain, then $\partial v/\partial x = -\partial u/\partial y$ and the motion is called pure rotation or *rigid body rotation* in the x-y plane. Similar special cases exist, of course, in the other coordinate directions.

From Eq. (3.2) or (3.3) we see that the mathematical expression of the *kinematic* notion of irrotational flow (i.e., to do with motions) is

$$\nabla \times \mathbf{q} = 0 \tag{3.5}$$

If this is the case, as it is in ideal fluid flow, then the operator ∇ must be parallel or antiparallel to the velocity vector $\mathbf{q}$. The gradient of any scalar, ∇ (anything), must be a vector aligned with ∇ and hence aligned with $\mathbf{q}$ according to Eq. (3.5).

If we choose a particular scalar ϕ, such that the magnitude of its gradient is equal to that of the velocity, then $\nabla\phi$ will be a vector with the same direction and magnitude as the velocity vector. Here we shall adopt a sign convention giving a value of ϕ that decreases in the direction of the flow. That is, ϕ is *defined* according to

$$-\nabla\phi = \mathbf{q} \tag{3.6}$$

The scalar quantity ϕ is called the *velocity potential*. It is related to flow in irrotational motion in the same way that temperature is related to heat flux in the conduction of heat. A rapid decrease in ϕ indicates a high velocity in the direction of that decrease. Just as the existence of irrotational motion requires that the velocity potential exist, the existence of a velocity potential means that the flow is irrotational. This latter statement may be easily proved by inserting Eq. (3.6) into Eq. (3.2)—the result is Eq. (3.5), which defines irrotational flow.

> **Exercise** Another approach to the existence of the velocity potential stems from $\partial u/\partial y - \partial v/\partial x = 0$. Let $F(x, y)$ be given such that $u = \partial F/\partial x$. Then
>
> $$\frac{\partial}{\partial y}\left(\frac{\partial F}{\partial x}\right) = \frac{\partial u}{\partial y} \quad \text{and} \quad \frac{\partial v}{\partial x} = \frac{\partial u}{\partial y} = \frac{\partial}{\partial y}\left(\frac{\partial F}{\partial x}\right) = \frac{\partial}{\partial x}\left(\frac{\partial F}{\partial y}\right)$$
>
> from which it follows that $v = \partial F/\partial y$ and, therefore,
>
> $$dF = \frac{\partial F}{\partial x}\,dx + \frac{\partial F}{\partial y}\,dy = u\,dx + v\,dy \ldots$$
>
> an exact differential expression. Setting $F = -\phi$, we have $u = -\partial\phi/\partial x$, $v = -\partial\phi/\partial y$, and the extension is easily developed; that is, $\mathbf{q} = -\nabla\phi$.
>
> *One way or another, if the flow is irrotational, then ϕ exists; and if ϕ exists, then the flow must be irrotational!*

3.2 BERNOULLI EQUATION REVISITED

In Chapter 2 we found that if the path of integration is confined to lie along a streamline, it is possible to integrate the Euler equation to obtain the Bernoulli equation, Eq. (2.8), for ideal, steady, incompressible flow. With the concept of vorticity we may now express the integral of the Euler equation in a much more general form. Moreover, we shall see that the existence of the velocity potential in ideal flow obviates the necessity for adherence to a streamline in the integration. We begin by expressing the acceleration, given by Eq. (2.3), in a divergence form (vector manipulations used here are left as exercises for the reader):

$$\frac{D\mathbf{q}}{Dt} = \frac{\partial\mathbf{q}}{\partial t} + (\mathbf{q}\cdot\nabla)\mathbf{q}$$

From the definition of the vorticity, $\zeta = \nabla \times \mathbf{q}$, the convective part of the acceleration becomes

$$(\mathbf{q} \cdot \nabla)\mathbf{q} = \frac{1}{2} \nabla(\mathbf{q} \cdot \mathbf{q}) - \mathbf{q} \times \boldsymbol{\zeta}$$

and, when these two expressions are combined, the Euler equation may be written as

$$\frac{\partial \mathbf{q}}{\partial t} + \frac{1}{2} \nabla(\mathbf{q} \cdot \mathbf{q}) - \mathbf{q} \times \boldsymbol{\zeta} = -\frac{1}{\rho} \nabla P - g \nabla h$$

or

$$\frac{\partial \mathbf{q}}{\partial t} + \frac{1}{2} \nabla q^2 + g \nabla h + \frac{1}{\rho} \nabla P = \mathbf{q} \times \boldsymbol{\zeta}$$

This is the "vorticity form" of the Euler equation. The appearance of the vorticity must not be construed to mean that it is valid for nonideal flows. The Euler equation only allows for pressure and body forces. The quantity $\mathbf{q} \times \boldsymbol{\zeta}$ in this expression is often called the Bernoulli constant. In order to proceed further, we shall replace this term by defining the function π such that $\nabla \pi = \mathbf{q} \times \boldsymbol{\zeta}$. In addition, we shall define a path $\mathbf{s}$ such that the differential change of any scalar quantity $Q(x, y, z)$ is given by

$$dQ = \frac{\partial Q}{\partial x} dx + \frac{\partial Q}{\partial y} dy + \frac{\partial Q}{\partial z} dz = \nabla Q \cdot d\mathbf{s}$$

If, now, the vorticity form of the Euler equation is dotted into $d\mathbf{s}$, we have

$$\frac{\partial \mathbf{q}}{\partial t} \cdot d\mathbf{s} + d\left(\frac{q^2}{2} + gh\right) + \frac{dP}{\rho} = d\pi$$

This expression is now in a form that can be integrated along the path s with the result

$$\int_s \frac{\partial \mathbf{q}}{\partial t} \cdot d\mathbf{s} + \frac{q^2}{2} + gh + \int_s \frac{dP}{\rho} = \pi + f(t) \tag{3.7}$$

The $f(t)$ term is necessary to account for time variations that may be overlayed on the spacial integration. Equation (3.7) is the general form of the Bernoulli equation and is indeed much more comprehensive than the result arrived at in Eq. (2.8). In particular, there is no constraint to streamlines, and the flow may be unsteady and compressible. The application of this general form is not restricted to flows in which there is zero rotation. The generalized Bernoulli theorem is seldom evaluated in its entirety. Instead, any one of several cases contained within it may be considered separately, as is done in the following section.

Special Cases

1. If π is a constant, then the vorticity and the velocity vectors must be parallel ($\mathbf{q} \times \boldsymbol{\zeta} = 0$) or the vorticity is itself zero. The former case is called a Beltrami

flow and is, for instance, a rough model of portions of the vortical flow in the wake of a propeller. The latter case is, of course, one of irrotational flow.

2. If the flow is steady, then the partial derivative with respect to time is zero and $f(t)$ is at most a constant.

3. If the flow is incompressible, then the integral of the pressure-density term becomes simply P/ρ.

4. If the flow is irrotational, then not only is π a constant, but the velocity potential may be incorporated. Thus,

$$\mathbf{q} \cdot d\mathbf{s} = -\nabla\phi \cdot d\mathbf{s} = -d\phi$$

and

$$\int_s \frac{\partial \mathbf{q}}{\partial t} \cdot d\mathbf{s} = \frac{\partial}{\partial t} \int_s \mathbf{q} \cdot d\mathbf{s} = \frac{\partial}{\partial t} \int_s -d\phi = -\frac{\partial \phi}{\partial t}$$

5. For incompressible irrotational flow (combine cases 3 and 4):

$$-\frac{\partial \phi}{\partial t} + \frac{q^2}{2} + \frac{P}{\rho} + gh = \text{constant} + f(t)$$

6. If the steady flow constraint is added to case 5, we obtain

$$\frac{q^2}{2} + \frac{P}{\rho} + gh = \text{constant} \qquad (3.8)$$

Even though each of these cases is valid only for flows in which frictional effects (i.e., shear stresses) are negligible, the generality of Eq. (3.7) is remarkable. In case 5 above we have come back to the *form* of the Bernoulli equation derived in Chapter 2, but we have now been relieved of the necessity of adhering to the streamlines in its application. This important distinction is made possible by introducing the constraint of irrotational flow and the concomitant existence of the velocity potential.

What are the practical ramifications of irrotational flow? To appreciate this fully we must introduce another concept of fluid mechanics, that of the stream function.

3.3 STREAM FUNCTION

Consider a three-dimensional region of a flow field denoted by V and bounded by the surface S, as in Fig. 3.5. At any point on the surface, the rate of outflow of mass may be expressed as

$$\delta \dot{m} = \rho \mathbf{q} \cdot \hat{n} \, \delta S$$

where n is the unit vector in the direction of the *outward* normal to δS. This definition of $\hat{n}$ means that at points on the surface where the velocity vector $\mathbf{q}$ has an inward normal component, the quantity $\delta \dot{m}$ will have a negative value and,

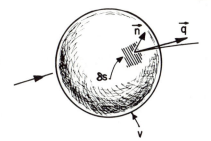

Figure 3.5

if there is an outward component of velocity, the value of $\delta\dot{m}$ there will be positive. Summing over all of the incremental areas on the surface S will give the *net* rate of *outflow* (outflow less inflow) so that

$$\dot{m} = \int_S \rho\mathbf{q}\cdot\hat{n}\,dS$$

If this quantity is positive, then more mass is leaving the volume than is entering, so the mass of the volume must be decreasing:

$$\int_S \rho\mathbf{q}\cdot\hat{n}\,dS = -\frac{\partial}{\partial t}\int_V \rho\,dV$$

The surface integral may be converted to a volume integral with the aid of Gauss's theorem, which states that, for any vector $\mathbf{A}$,

$$\int_V \boldsymbol{\nabla}\cdot\mathbf{A}\,dV = \int_S \mathbf{A}\cdot\hat{n}\,dS$$

This theorem, also called the divergence theorem, is a general mathematical expression. With $A = \rho\mathbf{q}$,

$$-\frac{\partial}{\partial t}\int_V \rho\,dV = \int_V \boldsymbol{\nabla}\cdot(\rho\mathbf{q})\,dV$$

[The divergence theorem is a general mathematical expression that is developed, for instance, in Schey (1973).] On the left-hand side of this expression, differentiation with respect to time may be written so that it precedes the integration over space, and we obtain, after rearrangement:

$$\boldsymbol{\nabla}\cdot(\rho\mathbf{q}) + \frac{\partial\rho}{\partial t} = 0 \tag{3.9}$$

It is left as an exercise to show that this may also be written as $D\rho/Dt + \rho\boldsymbol{\nabla}\cdot\mathbf{q} = 0$.

Equation (3.9) is the well-known continuity equation for continuous flow without mass sources. For incompressible flow this becomes $\boldsymbol{\nabla}\cdot\mathbf{q} = 0$, and for two-dimensional incompressible flow we obtain

$$\frac{\partial u}{\partial x} + \frac{\partial v}{\partial y} = 0 \tag{3.10}$$

Now, if we consider a continuous function $\psi(x, y)$ with continuous partial derivatives so that

$$\frac{\partial}{\partial x}\left(\frac{\partial \psi}{\partial y}\right) = \frac{\partial}{\partial y}\left(\frac{\partial \psi}{\partial x}\right)$$

then

$$\frac{\partial}{\partial x}\left(\frac{\partial \psi}{\partial y}\right) - \frac{\partial}{\partial y}\left(\frac{\partial \psi}{\partial x}\right) = 0$$

and comparison with Eq. (3.10) gives

$$u = \frac{\partial \psi}{\partial y} \quad \text{and} \quad v = -\frac{\partial \psi}{\partial x} \tag{3.11}$$

The function ψ, whose existence depends only on the continuity relation, is called the *stream function*.

The stream function is *not* a kinematic idea, as is velocity potential, even though it is denoted by a Greek symbol that looks very much like ϕ (phi, sounds like "fee"). For the stream function ψ (sounds like "sigh" or "psee") we look to the mathematical expressions for the continuity of mass in a fluid flow. Flows that satisfy continuity have a stream function and vice versa. The restriction of these notions to incompressible and/or steady flows is not severe, and other stream function definitions may easily be proposed for such cases. The restriction to *two-dimensional* flows is indeed a practical necessity because other definitions, valid in three dimensions, are so complicated that their use is seldom justified. More on three-dimensional stream functions may be found, for instance, in Rouse (1959, pp. 42–45).

Note again that even though both ϕ and ψ are scalar functions whose derivatives in coordinate directions give velocity components, the two notions come from entirely different physical rationales. The mathematics governing their use is fundamentally different. In fact, we shall see that from a geometric point of view they are about as different as they can be. Let us look further at the stream function we have defined and consider its variation as we move from one point to another in a field of flow. Since $\psi = \psi(x, y)$,

$$d\psi = \frac{\partial \psi}{\partial x} dx + \frac{\partial \psi}{\partial y} dy = -v\, dx + u\, dy \tag{3.12}$$

On a line in space that connects points having a constant value of ψ, therefore, $d\psi = 0$ and

$$v\, dx = u\, dy \quad \text{or} \quad \left(\frac{dy}{dx}\right)_{d\psi=0} = \frac{v}{u}$$

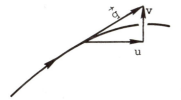

Figure 3.6 Geometry of a streamline.

The slope of a constant ψ line is v/u and, therefore, such a line is tangent to the velocity vector (Fig. 3.6). This is what is meant by a *streamline,* and it may be concluded that lines of constant stream function are the same as streamlines. In three-dimensional and axisymmetric flows, there are *surfaces* of constant stream function, called *streamtubes.*

Another very useful property of the stream function is obtained if we examine the change in ψ when we cross from one streamline to another in a flow. In Fig. 3.7 the increment of flow passing between two streamlines is seen to be

$$dQ' = u \, dy - v \, dx$$

Here the ($'$) notation is used to denote a quantity that is expressed per unit depth normal to the plane of the two-dimensional flow. With Eq. (3.12) we see that dQ' is equal to $d\psi$, the difference in stream function between the two streamlines. (The sign convention on ψ implies that flow is from left to right when looking in the direction of increasing ψ.) A very important result stemming from this observation is that if we draw streamlines having equal differences in stream function, the rate of flow will be the same between adjacent lines. Accordingly, such streamlines will come closer together as the velocity increases, and they will diverge in regions of decreasing flow velocity. A field of streamlines together with their stream function values will completely describe the velocity vector—its magnitude and direction.

Physical Construction of Stream Function

The definition of the stream function can be constructed for various coordinate systems if the following ground rules are observed:

1. Stand at a point in the flow.
2. Consider flow from left to right as positive.

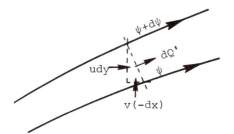

Figure 3.7 Flow and stream function increments between two streamlines.

3. Set the volume flow per unit depth equal to the difference in stream function, $\delta\psi$.
4. Repeat in each of the coordinate directions.

In plane Cartesian coordinates (Fig. 3.8):
Looking in the x direction,

$$\delta Q' = -v\,\delta x = \delta\psi$$

$$v = -\frac{\partial\psi}{\partial x}$$

Looking in the y direction,

$$\delta Q' = u\,\delta y = \delta\psi$$

$$u = \frac{\partial\psi}{\partial y}$$

In plane-cylindrical coordinates (Fig. 3.9):
Looking in the r direction,

$$\delta Q' = -v_\theta\,\delta r = \delta\psi$$

$$v_\theta = -\frac{\partial\psi}{\partial r}$$

Looking in the θ direction,

$$\delta Q' = v_r(r\,\delta\theta) = \delta\psi$$

$$v_r = \frac{1}{r}\frac{\partial\psi}{\partial\theta}$$

Here we have considered only the Lagrange stream function, which is defined for two-dimensional plane flows (i.e., there is no motion normal to the plane, and flow patterns described in the x-y or r-θ planes are repeated in all parallel planes). The Stokes stream function is developed in a similar way for axisymmetric two-dimensional flows.

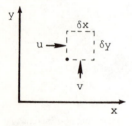

Figure 3.8

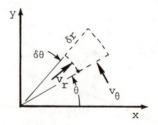

Figure 3.9

3.4 IRROTATIONAL CONTINUOUS FLOWS

All flows of engineering interest satisfy the equation of mass continuity and therefore may be described using the stream function. Although use of the stream function is not restricted to ideal flows, it is an extremely useful concept when coupled with the existence of the velocity potential.

Stream Function in Irrotational Flows

From Eqs. (3.5) and (3.11), for two-dimensional irrotational flow in the x-y plane (and in all planes parallel to this one),

$$\mathbf{\nabla} \times \mathbf{q} = 0 = \frac{\partial v}{\partial x} - \frac{\partial x}{\partial y} = \frac{\partial}{\partial x}\left(-\frac{\partial \psi}{\partial x}\right) - \frac{\partial}{\partial y}\left(\frac{\partial \psi}{\partial y}\right)$$

We see that the stream function satisfies the Laplace equation:

$$\nabla^2 \psi = 0 \qquad\qquad (3.13)$$

Velocity Potential in Continuous Flows

By *continuous* we mean here that the flow satisfies continuity and, if it is also incompressible,

$$\mathbf{\nabla} \cdot \mathbf{q} = 0$$

When this is combined with the velocity potential, Eq. (3.6), (which carries with it the *requirement* that the flow be irrotational),

$$\mathbf{\nabla} \cdot \mathbf{\nabla}\phi = \nabla^2 \phi = 0 \qquad\qquad (3.14)$$

and it is discovered that the velocity potential also satisfies the Laplace equation. Satisfaction of the Laplace equation has far-reaching consequences, as we shall see below.

Functions That Satisfy the Laplace Equation

The number of solutions to the Laplace equation is virtually infinite, the only limit being that of one's imagination in prescribing the boundary conditions. For our present purposes, however, it will suffice to note that these so-called harmonic functions have a few special properties. Suppose, first of all, that we are able to define stream functions, ψ_1 and ψ_2, for each of two flows. If both flows are irrotational, both of these functions are harmonic and

$$\nabla^2 \psi_1 = 0 \qquad \text{and} \qquad \nabla^2 \psi_2 = 0$$

But since *the Laplacian is a linear operator*—and this is a key point—the above expression may also be written as

$$\nabla^2(\psi_1 + \psi_2) = 0$$

Now this expression defines a third flow, also irrotational, whose stream function consists of the sum of the individual stream functions of the two flows. In other words, if two flows are irrotational and continuous, then the flow obtained by adding the stream functions of the two flows will also be irrotational and continuous. The same result is true for the velocity potential so that the analysis of ideal flows is greatly simplified—extremely complicated flows may be "pieced together" by combining very simple flows.

Orthogonality of Stream Function and Velocity Potential

As we have done with the stream function, let us consider lines of constant velocity potential in a two-dimensional flow:

$$d\phi = \frac{\partial \phi}{\partial x} dx + \frac{\partial \phi}{\partial y} dy = -u \, dx - v \, dy = 0$$

From this result we obtain

$$\left(\frac{dy}{dx}\right)_{d\phi=0} = -\frac{u}{v} \qquad \text{(isopotential lines)}$$

But we have already seen that the slopes of lines of constant stream function (i.e., streamlines) are given by

$$\left(\frac{dy}{dx}\right)_{d\psi=0} = \frac{v}{u} \qquad \text{(streamlines)}$$

The slopes of one set of lines are the negative reciprocals of the slopes of the other set. Such lines are perpendicular, and we have shown that streamlines are orthogonal to isopotential lines.

3.5 SUMMARY

To this point we have defined ideal flow from both the kinetic and kinematic points of view. The former defines the nature of the forces that can (and cannot) act, whereas the latter describes the motion that results. In conjunction with these definitions we have been able to describe a number of concepts and functions such as vorticity, irrotationality, velocity potential, and stream function. In addition, we have developed a relationship—the Bernoulli equation—that, when properly applied, allows us to relate differences in velocity to differences in pressure in a field of flow. It is vital to remember that we have been describing ideal fluid flow. All effects that lead to nonideal behavior, mainly those caused by the action of fluid viscosity, must be negligible for us to apply these results legitimately. We have also confined ourselves to two-dimensional flows in either a Cartesian or a polar-cylindrical frame of reference. In many (probably most) applications it is more convenient to work with the cylindrical system of coordinates. Table 3.1 is provided to summarize the important relationships for both systems.

Table 3.1 Useful relationships for two-dimensional flows

General: $(D/Dt)[\ ] = (\partial/\partial t)[\ ] + (\mathbf{q}\cdot\boldsymbol{\nabla})[\ ]$

Cartesian coordinates $(x, y: u, v)$	Cylindrical coordinates $(r, \theta; v_r, v_\theta)$

$$\boldsymbol{\nabla} = \hat{\imath}\frac{\partial}{\partial x} + \hat{\jmath}\frac{\partial}{\partial y} \qquad\qquad \boldsymbol{\nabla} = \hat{\mathbf{i}}_r\frac{\partial}{\partial r} + \hat{\mathbf{i}}_\theta\frac{1}{r}\frac{\partial}{\partial \theta}$$

Continuity

$$\frac{D}{Dt} + \rho(\boldsymbol{\nabla}\cdot\mathbf{q}) = 0$$

$$\frac{D\rho}{Dt} + \rho\left(\frac{\partial u}{\partial x} + \frac{\partial v}{\partial y}\right) = 0 \qquad\qquad \frac{D\rho}{Dt} + \frac{\rho}{r}\left[\frac{\partial}{\partial r}(rv_r) + \frac{\partial v_\theta}{\partial r}\right] = 0$$

Stream function

$$u = \frac{\partial \psi}{\partial y},\; v = -\frac{\partial \psi}{\partial x} \qquad\qquad v_r = \frac{1}{r}\frac{\partial \psi}{\partial \theta},\; v_\theta = -\frac{\partial \psi}{\partial r}$$

Vorticity

$$\zeta = \boldsymbol{\nabla}\times\mathbf{q}$$

$$\zeta_z = \frac{\partial v}{\partial x} - \frac{\partial u}{\partial y} \qquad\qquad \zeta_z = \frac{1}{r}\left[\frac{\partial}{\partial r}(rv_\theta) - \frac{\partial v_r}{\partial \theta}\right]$$

Velocity potential $(\zeta = 0)$

$$u = -\frac{\partial \phi}{\partial x},\; v = -\frac{\partial \phi}{\partial y} \qquad\qquad v_r = -\frac{\partial \phi}{\partial r},\; v_\theta = -\frac{1}{r}\frac{\partial \phi}{\partial \theta}$$

Example Consider the flow given by

$$\phi = \frac{kx}{x^2 + y^2}$$

(Note that for dimensional consistency the constant k must have appropriate dimensions.) The velocity components may be found from

$$u = -\frac{\partial \phi}{\partial x} = k\frac{x^2 - y^2}{(x^2 + y^2)^2}$$

$$v = -\frac{\partial \phi}{\partial y} = k\frac{2xy}{(x^2 + y^2)^2}$$

That the flow is continuous may be checked by verifying that

$$\frac{\partial u}{\partial x} + \frac{\partial v}{\partial y} = 0$$

That the flow is irrotational may be checked by verifying that

$$\zeta_z = \frac{\partial v}{\partial x} - \frac{\partial u}{\partial y} = 0$$

Both of the previous checks could have been accomplished by finding that $\nabla^2 \phi = 0$.

The stream function for this flow may be found from the velocity components, using the procedures outlined in Chapter 1. Thus,

$$\frac{\partial \psi}{\partial x} = -v = -k\frac{2xy}{(x^2 + y^2)^2}$$

Integrating with y held constant,

$$\psi = -k\int \frac{2xy}{(x^2 + y^2)^2}\,dx + f(y) = \frac{ky}{x^2 + y^2} + f(y)$$

Differentiating with respect to y,

$$\frac{\partial \psi}{\partial y} = k\frac{(x^2 - y^2)}{(x^2 + y^2)^2} + f'(y)$$

But $\partial \psi/\partial y = u$ and has the same value as the first term on the right-hand side of this expression. We must conclude, therefore, that $f'(y)$ is zero and $f(y)$ is a constant so that

$$\psi = \frac{ky}{x^2 + y^2} + \text{constant}$$

The value of the constant is determined by the value of ψ on the line $y = 0$ (the x axis), and we may arbitrarily select a convenient value. Zero is the usual value of convenience, and we make this selection here. (Because the *changes* in ψ are the issues of importance in a fluid flow field, the value of the constant is of little concern.) The procedure to obtain ψ would have yielded identical results if we had begun by integrating u with respect to y with a function $f(x)$ to be determined.

Streamline trajectories may be determined by assigning various constant values for ψ. For instance, $\psi = 0$ yields $y = 0$ so that the x axis is the streamline assigned to this value of ψ. A check of the velocity components will indicate that $u = k/x^2$ and $v = 0$ along this line: The flow along the x axis is from left to right, decreasing in velocity with distance from the origin. This particular problem is more easily solved using polar-cylindrical coordinates with the substitution $x = r \cos \theta$ and $y = r \sin \theta$. Thus, in this coordinate system,

$$\phi = \frac{k}{r} \cos \theta \qquad \text{and} \qquad \psi = \frac{k}{r} \sin \theta$$

The lines $\psi = $ constant give $\sin \theta / r = $ constant, which is the equation for circles tangent to the x axis at the origin. The flow is that due to a *doublet,* which will be discussed in more detail in the next chapter.

REFERENCES

Li, W., and S. Lam: *Principles of Fluid Mechanics,* Addison-Wesley, Reading, Mass., 1976.
Robertson, J. H.: *Hydrodynamics in Theory and Application,* Prentice-Hall, Englewood Cliffs, N.J., 1965.
Rouse, H.: *Advanced Fluid Mechanics,* John Wiley & Sons, New York, 1959.
Schey, E. M.: *Div, Grad, Curl, and All That,* W. W. Norton, New York, 1973.

PROBLEMS

3.1 Consider the scalar function F, given by

$$F = Uy \left[1 - \frac{1}{3} \left(\frac{y}{b} \right)^2 \right]$$

(a) Compute the gradient of F.
(b) Compute the divergence of the gradient of F found in part (a).
(c) Compute the curl of the gradient of F found in part (a). Show that your result is general, that is, that $\nabla \times \nabla F = 0$ for any F.

3.2 Suppose that the function given in problem 3.1 is a velocity potential. Find the velocity components for the flow, and check to see if continuity is satisfied. Describe the stream function.

3.3 Suppose that the function given in problem 3.1 is a stream function. Find the velocity components for the flow, and check to see if it is irrotational. Describe the velocity potential.

3.4 Consider the velocity components given in problem 2.4. Find the components of vorticity for this flow. Is this an irrotational flow? Does it satisfy continuity?

3.5 For a flow with velocity

$$\mathbf{q} = \hat{\imath}(x + y) + \hat{\jmath}(y + z) + \hat{k}(x^2 + y^2 + z^2)$$

find the components of vorticity at (2, 2, 2). Does this flow satisfy continuity?

3.6 A velocity potential for a two-dimensional flow is given as

$$\phi = y + x^2 - y^2$$

Find the stream function for this flow. Is this an irrotational flow?

3.7 The stream function for a two-dimensional flow is given by

$$\psi = 9 + 6x - 4y + 7xy$$

Does the velocity potential exist for this flow? If so, find it. Repeat with the constant changed from 9 to 900.

3.8 For a two-dimensional flow that is purely radial, $v_\theta = 0$. If the flow is also steady and incompressible, how must v_r vary to satisfy continuity (see problem 2.7)? Describe the stream function for this flow. What is the equation for streamlines?

3.9 Given the conditions of problem 3.8, how must v_r vary if the flow is to be irrotational? How must v_r vary if the flow is both continuous and irrotational? Describe the stream function and velocity potential for this latter case.

3.10 Describe the flow given by $\psi = -k \ln/r$, where k is a constant. That is, find v_r and v_θ and the velocity potential (if it exists).

3.11 Find the circulation for a flow given by $v_r = 0$ and $v_\theta = kr$.

3.12 Show that if u_1, v_1 and u_2, v_2 are the velocity components of two flows with stream functions ψ_1 and ψ_2, then for a flow with stream function

$$\psi = \psi_1 + \psi_2$$

the velocity components are given by

$$u = u_1 + u_2 \qquad \text{and} \qquad v = v_1 + v_2$$

Repeat in terms of velocity potentials.

3.13 Consider a flow with the following velocity components:

$$u = A\left[1 - \left(\frac{y}{a}\right)^2\right] \qquad v = 0$$

A and a are positive real constants and $y \leq a$. Find the stream function and the vorticity for this flow. Is continuity satisfied? Is the flow irrotational? Can a velocity potential be found for this flow?

FOUR

ANALYSIS OF IDEAL FLUID FLOWS

Our discussion of ideal fluid flows has shown that there are important kinetic and kinematic constraints on their existence. Despite these limitations, however, there are numerous instances in which the application of ideal fluid flow analysis leads to important practical results. In this chapter we illustrate some of the rudimentary ideal flows that find widespread use in such applications. In addition to the general "rules of the road" applicable to ideal flows (principal among these is irrotational flow and the concomitant existence of the velocity potential), we shall be concerned mainly with steady, incompressible flows in two dimensions.

4.1 SOME EXTREMELY USEFUL FLOWS

Uniform Flow

If flow proceeds uniformly in the direction of the positive x axis, as in Fig. 4.1, the velocity components are given by

$$u = U \text{ (a constant)} \qquad v = 0$$

From the definition of the velocity potential, Eq. (3.5), $\partial\phi/\partial x = -U$ so that $\phi = -Ux + f(y)$. Upon differentiation with respect to y, we find that $\partial\phi/\partial y = f'(y)$. But since $\partial\phi/\partial y = -v = 0$, we conclude that $f'(y) = 0$ and the function is a constant, which we shall set equal to zero (thereby making the y axis a line of zero velocity potential). Note again that, as in the case of the stream function ψ, we are almost never interested in the value of ϕ except inasmuch as it is different from values at other points in the flow. It is this difference that

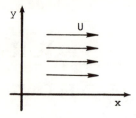

Figure 4.1 Uniform flow in the $+x$ direction.

promotes the flow and defines the velocity so that additive constants are of little physical significance. The result for the velocity potential is

$$\phi = -Ux \qquad \text{(uniform flow)} \tag{4.1}$$

The corresponding equation for the stream function for uniform flow in the x direction is easily obtained by the same procedures involving ψ, with the result

$$\psi = Uy \qquad \text{(uniform flow)} \tag{4.2}$$

Exercise Consider a velocity potential given by

$$\phi = -q(x \cos \alpha + y \sin \alpha)$$

where q is a constant. Differentiation with respect to x and y gives $u = q \cos \alpha$ and $v = q \sin \alpha$—the given velocity potential is for flow with velocity q at an angle α inclined to the x axis. From the integral of v with respect to x, holding y constant, we obtain $\psi = -qx \sin \alpha + f(y)$. Differentiating with respect to y gives $\partial \psi / \partial y = f'(y)$. But $\partial \psi / \partial y = u = q \cos \alpha$ so that $f'(y) = q \cos \alpha$ and $f(y) = qy \cos \alpha$ plus a constant set equal to zero. Thus,

$$\psi = -q(x \sin \alpha - y \cos \alpha)$$

From this we have $d\psi = -q(\sin \alpha \, dx - \cos \alpha \, dy)$ and along streamlines ($d\psi = 0$) $dy/dx = \tan \alpha$—the streamlines are straight lines with slope α.

Source (or Sink) at the Origin

The defined velocity distribution for this flow is radial—that is, with no circumferential component (Fig. 4.2):

$$v_\theta = 0 \qquad \text{and} \qquad v_r = \frac{Q'}{2\pi r}$$

From the definition of v_r it may be deduced that Q' is the volume flow rate per unit depth normal to the plane of the flow. From Table 3.1 we have $rv_\theta = -\partial \phi / \partial \theta = 0$, which gives $\phi = f(r)$ when integrated. It follows that $v_r = -\partial \phi / \partial r = -f'(r) = Q'/2\pi r$. Integration of this result gives $f(r) = -(Q'/2\pi)(\ln r)$ plus a constant of integration that is set equal to zero. Using the notation that $\mu_s = Q'/2\pi$, the result for the velocity potential of a source *at the origin* is

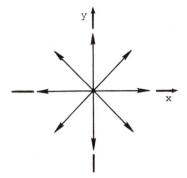

Figure 4.2 Source at the origin.

$$\phi = -\mu_s \ln r \qquad \text{(source at origin)} \qquad (4.3)$$

Repeating these procedures using the definition of the stream function in terms of the given velocity distribution leads to the following for the source at the origin:

$$\psi = \mu_s \theta \qquad \text{(source at origin)} \qquad (4.4)$$

A *sink* is defined in the same way as a source, except that the only velocity component is radially *inward* toward its location (the origin in this case). Development of the velocity potential and stream function for the sink is accomplished simply by taking the negative of the corresponding relationships for the source [Eqs. (4.3) and (4.4)].

Vortex at the Origin

This flow is geometrically orthogonal to that for the source. It might be expected, therefore, that streamlines for the vortex (shown in Fig. 4.3) are potential lines for the source and vice versa. Proceeding from the velocity components as before,

$$v_r = 0 \qquad \text{and} \qquad v_\theta = \frac{\Gamma}{2\pi r}$$

The symbol Γ represents the circulation in the flow and is a constant that is

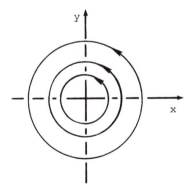

Figure 4.3 Vortex at the origin.

related to the vortex in much the same way that the volume flow rate Q' is related to the source—circulation is the measure of the strength of the vortex. Note that since the flow is in circular paths, the line integral of the component of the velocity *tangent* to one of these paths is given by

$$\int_c \mathbf{q} \cdot ds = \int_0^{2\pi} v_\theta(r \, d\theta) = \Gamma$$

which is in accordance with the definition of circulation, Eq. (3.4).

Rather than beginning with the velocity potential, we shall first determine the stream function for the vortex at the origin. (Of course, the results are independent of the order in which they are determined.) Because $rv_r = \partial\psi/\partial\theta = 0$, we obtain $\psi = f(r)$ through integration with respect to θ. Thus $\partial\psi/\partial r = f'(r)$ and, with the specified circumferential velocity for the vortex, $f'(r) = -\Gamma/2\pi r$. From this, with the constant of integration set equal to zero, $f(r) = -(\Gamma/2\pi) \ln r$. With the notation that $\mu_v = \Gamma/2\pi$, the stream function for a vortex at the origin is obtained:

$$\psi = -\mu_v \ln r \qquad \text{(vortex at origin)} \tag{4.5}$$

To obtain the velocity potential, the procedure is again applied, beginning with one or the other of the velocity components.

$$\phi = -\mu_v\theta \qquad \text{(vortex at origin)} \tag{4.6}$$

These expressions should be compared with their counterparts for the source, Eqs. (4.3) and (4.4). The sense of rotation of a vortex flow is such that positive values of Γ correspond to counterclockwise (positive) rotation.

Recall from Table 3.1 that in polar coordinates the vorticity ζ is given by

$$\zeta_z = \frac{1}{r}\left[\frac{\partial}{\partial r}(rv_\theta) - \frac{\partial v_r}{\partial\theta}\right]$$

so that for the vortex considered here the vorticity is zero everywhere except at the origin (or, more generally, the point of location of the vortex center). All the vorticity of this device is concentrated at its origin, where it is infinite. Elsewhere, remarkably enough, the vorticity is zero. The *influence* of the vortex is felt everywhere, however, with a component of swirl velocity decreasing inversely with distance from the vortex center.

Such a vortex is often termed a *free vortex* to distinguish it from one whose vortical motion is influenced by viscous effects. This latter "forced" vortex is associated with rotation and a distributed vorticity. One such circumferential flow pattern is included as part of the Rankine combined vortex with a velocity distribution given by

$$v_\theta = \frac{\Gamma}{2\pi}\frac{r}{R^2} \qquad \text{for } r \leq R$$

and

$$v_\theta = \frac{\Gamma}{2\pi r} \qquad \text{for } r > R$$

Near the origin, where the free vortex would give infinite values of vorticity and velocity (and, hence, zero pressure), the flow is modeled as one with a constant vorticity out to a radial distance R. This distance defines the extent of the rotational viscous core and is not easily determined. In cases where it is relatively small, such as in the tip vortices emanating from wings, a major portion of the flow field may be modeled as irrotational and having a velocity potential.

Doublet at the Origin Facing in the $+x$ Direction

This irrotational flow device is really a hybrid composed of a source and a sink of equal strength superimposed upon each other at the origin. The doublet "faces" in the direction of its source side and, in the following definition, the "positive" doublet has its source along the positive x axis (Fig. 4.4).

The process of forming the doublet involves taking the limit as the source and sink approach each other. Until this limit is evaluated, we shall assign the symbol m to the strengths so that for the source we have $\phi_1 = -m \ln r_1$ and for the sink $\phi_2 = m \ln r_2$. It is vital to note that in these expressions the radii r_1 and r_2 are measured from the *locations* of the source and sink, respectively. This is the necessary interpretation of r as used in the previous definitions of the velocity potentials and stream functions for sources, sinks, and vortices—it is the distance from the location of the singularity to an arbitrary point in the flow. In the previous definitions the locations of the singularities were at the origin, so there was no necessity to distinguish between local coordinates and those centered at the origin.

If the two flows are combined, the velocity potential for such a combination will be

$$\phi = \phi_1 + \phi_2 = m \ln \frac{r_2}{r_1}$$

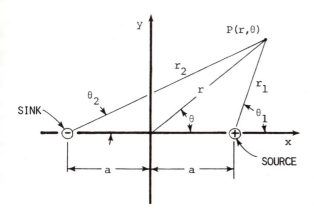

Figure 4.4 Coordinate definitions for the constituents of the doublet.

The utility of this expression is limited because it is not written in terms of the general coordinates r and θ. To accomplish this the cosine law for the geometry shown in Fig. 4.4 is written

$$r_1^2 = r^2 + a^2 - 2ar \cos \theta$$
$$r_2^2 = r^2 + a^2 + 2ar \cos \theta$$

and these give, for the combined velocity potential,

$$\phi = \frac{m}{2} \ln \frac{1 + (a/r)^2 + (2a/r) \cos \theta}{1 + (a/r)^2 - (2a/r) \cos \theta} \tag{4.7}$$

This expression is valid for the flow obtained from the source and sink arranged as in Fig. 4.4. For the *doublet,* however, we must superimpose these two components at the origin. If this were the only thing to be done, the source and sink would simply cancel each other, as can be seen by setting $a = 0$ in Eq. (4.7). Consequently, a limiting process must be evaluated as the distance a becomes vanishingly small. In the limit the strengths of the source and sink are required to increase in such a way that the product of the strength times separation distance is some constant value. This product is a mathematical abstraction that represents the "strength" of the doublet. Before evaluating the limit just described, let us recall the following series expansion:

$$\ln(1 + x) = x - \frac{x^2}{2} + \frac{x^3}{3} - \cdots \qquad (-1 \le x \le 1)$$

and, for convenience, define the intermediate term

$$T = \frac{2ar \cos \theta}{r^2 + a^2}$$

Equation (4.7) may now be written

$$\phi = \frac{m}{2} \ln\left(\frac{1 + T}{1 - T}\right) = \frac{m}{2} [\ln(1 + T) - \ln(1 - T)]$$

and the series expansion gives

$$\phi = \frac{m}{2}\left[\left(T - \frac{T^2}{2} + \frac{T^3}{3} - \cdots\right) - \left(-T - \frac{T^2}{2} - \frac{T^3}{3} - \cdots\right)\right]$$
$$= \frac{m}{2}\left(2T + \frac{2T^3}{3} + \frac{2T^5}{5} + \cdots\right) = mT\left(1 + \frac{T^2}{3} + \frac{T^4}{5} + \cdots\right)$$
$$= 2am \frac{r \cos \theta}{r^2 + a^2}\left(1 + \frac{T^2}{3} + \frac{T^4}{5} + \cdots\right)$$

The velocity potential is now in a form in which it is easy to evaluate the limit as $a \to 0$ with the product $2am$ reaching some constant value. This constant value is defined as

$$\mu_d = \lim_{\substack{a \to 0 \\ m \to \infty}} (2am)$$

The term T vanishes in this limit and the only thing surviving the limiting process is the final result:

$$\phi = \mu_d \frac{\cos \theta}{r} \qquad \text{(positive-facing doublet at the origin)} \qquad (4.8)$$

The velocity components for the doublet may be found from the appropriate derivatives given in Table 3.1:

$$v_r = \mu_d \frac{\cos \theta}{r^2} \qquad v_\theta = \mu_d \frac{\sin \theta}{r^2}$$

From the usual process of taking the derivative and integrating, the stream function will be found to be

$$\psi = \mu_d \frac{\sin \theta}{r} \qquad \text{(positive-facing doublet at the origin)} \qquad (4.9)$$

Exercise Streamline trajectories are obtained, as always, by considering various constant values of ψ. In this case, setting ψ equal to a constant leads to the relationship $r/\sin \theta = \mu_d/\psi = \text{constant}$ on a streamline. Note that the strength of the doublet has units $\mu_d [=] L^3/t$ and the stream function has units $\psi [=] L^2/t$. Therefore, the constant in this relationship has the units of length. The geometric interpretation of this length, D, say, is illustrated in Fig. 4.5. For each set of values of r and θ on the streamline, the distance D is found by erecting the perpendicular to r as shown in Fig. 4.5: $D = r/\sin \theta = \text{constant}$ on a streamline. The locus of points traced out by maintaining this constant while varying θ (or r) is a circle of diameter D tangent to the x axis at the origin. The particular values for $\psi = 0$ correspond to $\sin \theta = 0$ and

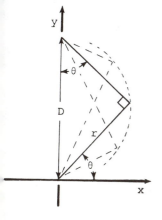

Figure 4.5

give the positive and negative x axes; the corresponding "circle" has a diameter that is infinite in extent.

Tops of the circles are at $\sin \theta = 1$, and a check of the velocity components shows that v_θ is positive for $\theta = \pi/2$ and negative for $\theta = 3\pi/2$. This illustrates that the flow is counterclockwise above the x axis and clockwise below this line: the flow curves in circular paths from the source to the sink of the doublet (see Fig. 4.6).

The reader may wish to verify that lines of constant velocity potential, orthogonal to those of constant stream function, are also circles, but these are tangent to the y axis at the origin.

We have developed four irrotational flows of the most basic kind (uniform flow, source/sink, vortex, doublet). The doublet was constructed, in fact, from the source/sink expressions, and one might imagine that there is an infinity of flow patterns that could be constructed by examining the flows obtained from various combinations of these singularities (so called because they possess mathematical poles at their points of location). By considering various strengths and distributions of these devices it is, in fact, possible to model a wide variety of ideal fluid flows. This has been done and is continuing in the application of the theory of ideal fluid flows; several examples of such studies will be discussed later. For the present purposes, however, we shall consider a simple and very practical geometry that may be created by these methods—the circle (or right-circular cylinder) in a uniform flow. Other flows of a practical nature can be obtained from this "shopping list," summarized in Table 4.1.

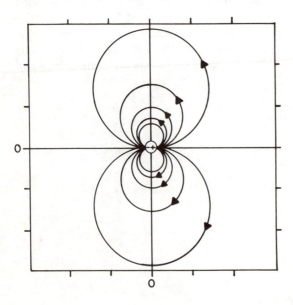

Figure 4.6 Flow patterns for the doublet.

Table 4.1 Simple irrotational flows: a shopping list

Basic flow	Velocity potential	Stream function
Uniform flow in $+x$ direction	$\phi = -Ux$	$\psi = Uy$
Source at origin	$\phi = -\mu_s \ln r$	$\psi = \mu_s \theta$
Vortex at origin	$\phi = -\mu_v \theta$	$\psi = -\mu_v \ln r$
Doublet at origin	$\phi = \mu_d \dfrac{\cos \theta}{r}$	$\psi = \mu_d \dfrac{\sin \theta}{r}$

Note: $\mu_s = Q'/2\pi$, $\mu_v = \Gamma/2\pi$.

4.2 CIRCULAR CYLINDER IN UNIFORM FLOW

The flow pattern we seek is a uniform flow that is required to "go around" an obstacle that is circular in shape. Because of the superposition feature of ideal flows, we are at liberty to add them together in an effort to create the desired pattern. The radial flow emanating from a source would deflect an oncoming uniform flow (recall from problem 3.12 that the velocities add vectorially), but this outward deflection would continue downstream of the source. There would be no recurving of the streamlines to follow a circular path. A sink located downstream would accomplish this recurvature, and we are thus led to consider a combination of a uniform flow with an upstream source and a downstream sink. Although a large family of body shapes can be achieved by such combinations, the particular pattern we seek is obtained in the limiting case when the source and sink are superimposed to form the doublet. With a "positive" uniform flow, that is, in the direction of the $+x$ axis, we must orient the doublet so that the source side faces downstream into the oncoming flow, as shown in Fig. 4.7. (The reader may wish to examine flow patterns obtained with the sink side of the doublet facing upstream into the flow.) For the doublet, therefore,

$$\psi_d = -\mu_d \frac{\sin \theta}{r} \quad \text{and} \quad \phi_d = -\mu_d \frac{\cos \theta}{r}$$

When the uniform flow is added, the stream function and velocity potential for the combined flow are

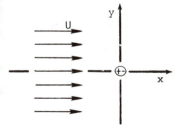

Figure 4.7 Negative doublet facing a uniform flow.

$$\psi = Ur \sin \theta \left(1 - \frac{\mu_d}{Ur^2} \right)$$

$$\phi = -Ur \cos \theta \left(1 + \frac{\mu_d}{Ur^2} \right)$$

The streamline trajectory given by the value of $\psi = 0$ is of particular interest. Setting ψ to this value gives $\theta = 0$, π (the positive and negative x axes) and, in addition to these lines, we have $r^2 = \mu_d/U$ for $\psi = 0$. We shall assign this particular value of r the symbol a; since it is a constant, it describes a circle of that radius, $a = \sqrt{\mu_d/U}$. This allows us to scale the problem in terms of the radius a of the $\psi = 0$ streamline, and we need not deal with the abstract notion of the doublet strength. Nevertheless, we find that strong doublets and/or weak uniform flows lead to larger circles, in accordance with our intuition. Rewriting ψ and ϕ accordingly,

$$\psi = Ur \sin \theta \left[1 - \left(\frac{a}{r} \right)^2 \right] \tag{4.10}$$

$$\phi = -Ur \cos \theta \left[1 + \left(\frac{a}{r} \right)^2 \right] \tag{4.11}$$

Note that at this point we have obtained the desired streamline pattern: a circle centered at the origin in a uniform flow. That the flow is uniform at a distance from the origin is verified by noting that, as r becomes large, the two functions above approach those for the uniform flow. Any number of streamlines may now be drawn by assigning various constant values to ψ in Eq. (4.10). The flow pattern is that sketched in Fig. 4.8a.

Velocities

Differentiation of either ϕ or ψ leads to the following velocity components:

$$v_r = U \cos \theta \left[1 - \left(\frac{a}{r} \right)^2 \right] \tag{4.12}$$

$$v_\theta = -U \sin \theta \left[1 + \left(\frac{a}{r} \right)^2 \right] \tag{4.13}$$

It is useful to note some of the velocities at key points in the flow. On the circle, for instance, where $r = a$, the radial component is zero for all values of θ. This is a necessary condition for that circle to be a streamline. The circumferential component, however, varies from zero at $\theta = 0$, π to a maximum at $\theta = \pi/2$, $3\pi/2$ on the circle. This maximum is $-2U$ at the top of the circle and $+2U$ at the bottom—both are in the direction of the uniform flow, according to the definition of v_θ (component of the velocity in the direction of θ). On the line $\theta = 0$, the tangential velocity is zero everywhere (except at the origin, where it is in-

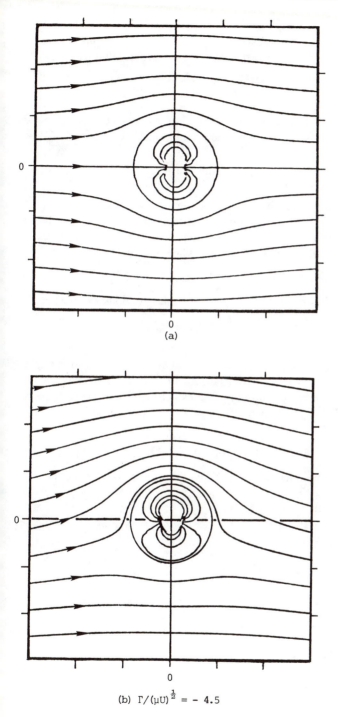

(a)

(b) $\Gamma/(\mu U)^{\frac{1}{2}} = -4.5$

Figure 4.8 Uniform flow past a cylinder: (a) without circulation; (b) with circulation.

determinate), and the radial component is inward (toward the sink) for $r < a$, decreases to zero at $r = a$, and increases to U at distances far from the origin. Similar observations may be made for the velocities along the line $\theta = \pi$.

Pressures

According to the Bernoulli equation, taking the flow to be incompressible (we have already defined the flow to be steady and irrotational), we have

$$\frac{q^2}{2} + \frac{P}{\rho} + gh = \text{constant} \tag{3.8}$$

We are particularly interested in the pressure variation on the line $r = a$ (i.e., along the surface of the right-circular cylinder defined by $\psi = 0$). For this line $v_r = 0$ and $v_\theta = -2U \sin \theta$ so that the magnitude of the velocity vector is $q = -2U \sin \theta$, and the Bernoulli expression reads, when solved for pressure,

$$P = -2\rho U^2 \sin^2 \theta - \rho g h + C \tag{4.14}$$

where C is a constant (density times the constant in the previous expression).

Forces

To determine the force acting on the cylinder of radius a in a uniform flow, we must integrate the pressure around the surface of the cylinder. Because of our assumption of ideal flow, there are no other forces acting except the body force. The pressure force acting on an element of the surface, dA, has components (Fig. 4.9)

$$dF_x = -P \, dA \cos \theta \qquad dF_y = -P \, dA \sin \theta$$

For unit length along the cylinder, $dA = a \, d\theta$, and the force components per unit length are found from

$$F_x' = -a \int_0^{2\pi} P \cos \theta \, d\theta \qquad F_y' = -a \int_0^{2\pi} P \sin \theta \, d\theta \tag{4.15}$$

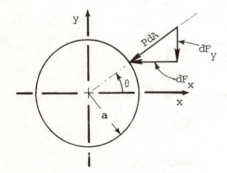

Figure 4.9

When one or the other of these is combined with the Bernoulli equation, Eq. (4.14), the corresponding force component is obtained. For example,

$$-\frac{F_y'}{a} = \int_0^{2\pi} (-2\rho U^2 \sin^2 \theta - \rho g h + C) \sin \theta \, d\theta$$

But note that the distance h will change along the path of integration and we must now be more precise about what we mean by "up." If we take the acceleration of gravity to be directed parallel to the negative y direction (as is the fashion), then for h we may write $h = h_0 + y$, where the constant h_0 locates the origin of our coordinate system relative to the $h = 0$ elevation. The path of integration along the surface of the cylinder gives $y = a \sin \theta$, so in the above integral we must express h as $h_0 + a \sin \theta$.

$$-\frac{F_y'}{a} = \int_0^{2\pi} (-2\rho U^2 \sin^2 \theta - \rho g a \sin \theta + C_1) \sin \theta \, d\theta \qquad (4.16)$$

where $C_1 = C - \rho g h_0$, another constant.

The integral in Eq. (4.16) involves the variation of $\sin \theta$, $\sin^2 \theta$, and $\sin^3 \theta$ over the range $0 \leq \theta \leq 2\pi$. Only the integral of $\sin^2 \theta$ accumulates a nonzero value over this range, as sketches of these functions will show. Evaluating the integral, then,

$$-\frac{F_y'}{a} = \int_0^{2\pi} -\rho g a \sin^2 \theta \, d\theta = -\rho g a(\pi)$$

and the final result for the y direction is

$$F_y' = \pi \rho g a^2 \qquad (4.17)$$

Note, however, that the volume of the cylinder of unit length is $V = \pi a^2$ and the quantity $\pi \rho a^2$ is simply the mass of the volume of fluid (of density ρ) displaced by the cylinder. We have been brought to the conclusion that the *only* force acting in the y direction is upward and equal to the weight of the fluid displaced by the cylinder. This is further evidence that the laws of fluid statics are embedded in those of fluid dynamics. There are no *net* velocity-dependent forces acting in the y direction and, following these same procedures, it will be found that this also is true for the x direction. We would have predicted this result if we had examined the distribution of the static pressure acting normal to the surface of the cylinder. Exclusive of hydrostatic effects due to gravity, this distribution is perfectly symmetrical about both the x and y axes, and for every pressure acting at a given point on the cylinder, a pressure that is equal but opposite in direction may be found at some other point.

In irrotational motion the absence of any net force attributable to the flow is a direct result of neglecting forces stemming from viscous action. Such a result is correct inasmuch as the neglect of viscous forces is justified; but in the mid-eighteenth century, when it was first discovered by d'Alembert, it was thought to be a mathematical error: hence, *d'Alembert's paradox*.

Cylinder in Uniform Flow with Circulation

Because the null result obtained above stems directly from the symmetry of the pressure distribution around the cylinder, it is apparent that a distortion of this distribution will lead to a net force on the cylinder. Such distortion may be introduced into the flow, *without departing from ideality,* if we add another irrotational component to the flow. One way to do this is to include a free vortex at the origin. Since such a component adds only velocities in the circumferential direction, the flow along the circular cylinder will be altered in magnitude only, not in direction, so that the surface of the cylinder is preserved as a streamline. In addition, since the velocity component due to the vortex decreases as $1/r$, the uniform flow at a distance from the origin is preserved. The vortex is thus seen to be a logical addition.

Incorporating a vortex at the origin requires addition of the following terms to ψ and ϕ (Table 4.1):

$$\psi_{\text{vortex}} = -\frac{\Gamma}{2\pi} \ln r \qquad \phi_{\text{vortex}} = \frac{\Gamma\theta}{2\pi}$$

When this component is added to the flow previously constructed, Eq. (4.10) for the stream function becomes

$$\psi = Ur \sin \theta \left[1 - \left(\frac{a}{r}\right)^2 \right] - \frac{\Gamma}{2\pi} \ln r$$

and we see that, as desired, ψ is a constant (albeit a different constant) on $r = a$. As predicted, the flow is uniform as r becomes large relative to a. Note, however, that with the addition of the vortex the lines $\theta = 0, \pi$ are no longer streamlines (Fig. 4.8b).

The velocity contributed by the vortex is $v_\theta = \Gamma/2\pi r$, and differentiation of the stream function given above will verify this. On the surface of the cylinder there is still no velocity in the radial direction. In the circumferential direction, however, the velocity is altered by the vortex and, for the surface of the cylinder,

$$q = v_\theta = -2U \sin \theta + \frac{\Gamma}{2\pi a}$$

When this expression is squared for insertion into the Bernoulli equation, two new terms are obtained. The last term is a constant, however, so only the cross product term will add a function of θ in the force integral. The new expression for F'_y becomes

$$-\frac{F'_y}{a} = \int_0^{2\pi} \left(-2\rho U^2 \sin^2 \theta - \rho ga \sin \theta + \frac{\rho U\Gamma}{\pi a} \sin \theta + C_2 \right) \sin \theta \, d\theta$$

where $C_2 = C_1 - \rho/2(\Gamma/2\pi a)^2$.

Now it can be seen that there are two terms involving $\sin^2 \theta$ in the integral: one due to gravity and the other due to the circulation of the vortex at the origin.

Integration of the latter term gives

$$\left(-\frac{F_y'}{a}\right)_{\text{vortex}} = \frac{\rho U \Gamma}{\pi a} \int_0^{2\pi} \sin^2 \theta \, d\theta = \frac{\rho U \Gamma}{a}$$

The reader may verify that *the addition of the vortex has no effect on the null result for the force in the x direction.* In summary, the cylinder in uniform, steady, incompressible, ideal flow with circulation has force components per unit length of

$$F_y' = -\rho U \Gamma + W_f'$$

$$F_x' = 0 \qquad\qquad (4.18)$$

where W_f' denotes the weight of the displaced fluid per unit length. This result is known as the Kutta-Joukowsky law and dates back to the early twentieth century; it is a very famous result and has much wider application than that of a simple circular cylinder.

The effect of a negative circulation (clockwise—Γ has a negative value) in uniform flow from left to right gives a positive or "lift" force. If the flow pattern is considered, the clockwise swirl components due to the vortex oppose the uniform flow in areas underneath the cylinder and complement the flow around the upper regions. The resulting lower velocities below and higher velocities above the cylinder lead to higher pressures below and lower pressures above. The result is a net lift force.

Exercise We have seen that, without circulation, the velocity (both components) is zero at the intersection of the circle $r = a$ with the x axes. Such points of zero velocity are *stagnation points,* and, as the Bernoulli equation shows, they are points of maximum pressure in the flow. When circulation is included, stagnation points on $r = a$ are obtained from

$$q = -2U \sin \theta + \frac{\Gamma}{2\pi a} = 0$$

The result is that

$$\sin \theta = \frac{\Gamma}{4\pi U a} \qquad r = a \qquad \text{(stagnation points)}$$

With clockwise circulation (negative Γ) $\sin \theta$ must be negative; in this case we observe that the stagnation points are located at points on the lower part of the cylinder, $\pi < \theta < 2\pi$. This further indicates that the circulation has led to relatively high pressures on the lower surface of the cylinder. (At large values of negative circulation, when $\Gamma/4\pi U a < 1$, stagnation points are found to lie off of the cylinder on the line $\theta = 3\pi/2$, $r = -\Gamma/4\pi U$.)

In the preceding developments, circulation has been used to obtain a resolution of d'Alembert's paradox. The root cause of this paradox is the absence of viscous effects in ideal fluid flow and, since circulation provides an "out," it is apparent that viscous effects must in some way be related to the circulation and vice versa. This is in fact the case, and much theory has been developed to connect circulation with viscous effects. Spinning bodies, for example, induce a swirl in the surrounding fluid through the influence of viscosity. The developed circulation leads to a force and to the so-called Magnus effect that is observed to influence the trajectory of spinning bodies—projectiles, tennis balls, baseballs, and (unfortunately) sliced golf balls. In the flow field associated with an airfoil, the influence of viscosity gives rise to a general circulatory motion from the underside of the foil to the upper surface. The resulting high pressure underneath and low pressure above lead to a lift that is well-modeled by the Kutta-Joukowsky law. Unfortunately, the other important force, the drag, cannot be determined from models based on ideal flow considerations. We have had an exposure to this in the null results obtained above, even with circulation, for the force aligned with the oncoming flow.

PROBLEMS

4.1 Verify Eqs. (4.2), (4.4), and (4.6).

4.2 Find the vorticity of the Rankine combined vortex in the regions $r \le R$ and $r > R$.

4.3 A two-dimensional source is located at $(1, 0)$ and another one of the same strength is at $(-1, 0)$. Draw to scale the velocity vectors at the following locations: $(0, 0)$, $(0, 1)$, $(0, -1)$, $(0, -2)$, and $(1, 1)$. Use a scale such that 1 in. is the magnitude of the velocity produced by either source at a distance of 1 in. from its location. Use the results of problem 3.12.

4.4 Determine the velocity potential and stream function for a source located at $(1, 0)$. Write the equation for the velocity potential for the source system of problem 4.3.

4.5 Determine the equation for the velocity on the surface $x = 0$ in the flow given in problem 4.3. Find an equation that gives the pressure along this surface relative to the pressure P_∞ at a great distance from the sources, where the fluid is stationary. Neglecting hydrostatic effects, what is the force on one side of the surface due only to the flow? (Assume that the surface is infinite in extent in the y direction and of width b in the z direction.)

4.6 Find the point of maximum velocity on the surface of problem 4.5. What can be said about the relative magnitude of the pressure at this point?

4.7 In two-dimensional flow, what is the nature of the flow given by $\phi = 7x + 2 \ln r$?

4.8 Verify Eqs. (4.12) and (4.13).

4.9 Select the strength of a doublet needed to model a flow of 20 m/s past a cylinder of radius 2 m. If the cylinder is in water and subject to a clockwise circulation of 3 m^2/s, what must its density be to experience no net vertical force?

4.10 Consider a flow field created by a vortex at $(0, 2m)$ with strength $\mu_v = 25$ m^2/s, a sink at the origin with strength $\mu_s = 20$ m^2/s, and a uniform flow with speed $U = 10$ m/s. Use Table 4.1 to calculate the flow velocity (magnitude and direction) at the point $(2m, -2m)$. Verify your results with a sketch drawn to scale.

4.11 Consider the flow resulting from a source on the x axis at $x = a$ and a uniform flow of velocity U in the positive x direction.

    (a) What are the stream function and velocity potential for this flow?

    (b) Find the velocity at the origin and find the relationship between U, μ_s, and a such that this point is a stagnation point. Find the value of the stream function at the origin and sketch the streamlines through this point.

4.12 The flow in the vicinity of a hurricane (but away from the highly viscous core) is similar to that developed from a sink-vortex combination.

(a) Derive an expression for the stream function of such a flow, assuming counterclockwise circulation.

(b) Describe the trajectory of the streamlines.

(c) If at a point 20 m from the center of the hurricane the wind has components of 45 m/s tangential and 15 m/s inward, what are the strengths of the vortex and sink?

(d) Find the difference in pressure between the point at 20 m and a point at 40 m from the center of the hurricane.

4.13 A uniform flow is inclined at an angle of 30° to the x axis and a source is located at the origin. The uniform flow velocity 10 m/s and $\mu_s = 10$ m^2/s. Give the velocity components for the flow and the location of the stagnation points.

4.14 In a flow with velocity $U = 10$ m/s past a cylinder of radius $a = 105$ mm, the stagnation points are found to be located at 30° up from the centerline aligned with the flow. What is the circulation for this flow? If the fluid is standard sea-level air and the cylinder is 0.5 m long, what is the lift force (magnitude and direction)?

4.15 Consider an incompressible flow for which the velocity potential is given by $\phi = y^2 - x^2$. Does this function satisfy the Laplace equation? Find the stream function for the flow and sketch the line for $\psi = 2$. Indicate the direction of the flow.

4.16 In an incompressible flow, the radial and circumferential components of the velocity are given by

$$v_r = \frac{A}{r} - B \cos \theta \qquad v_\theta = B \sin \theta$$

where A and B are positive real constants. Does a velocity potential exist for this flow? Does the flow satisfy continuity? Find the stream function and give the pressure variation along the line $\theta = 0$ in terms of A, B, and r. The pressure at infinity is constant.

4.17 (Problems 4.18 and 4.19 are extensions.) Develop the stream function and velocity potential for the flow formed by a source at $(-a, 0)$, a sink of equal strength at $(a, 0)$, and a uniform flow of velocity U in the $+x$ direction. Find the velocity components in Cartesian $(x, y; u, v)$ or polar-cylindrical $(r, \theta; v_r, v_\theta)$ coordinates.

4.18 For the flow in problem 4.17, show that the stagnation points are located at

$$\theta_s = 0, \pi \qquad \frac{r_s}{a} = \sqrt{1 + n}$$

where $n = 2\mu_s/Ua$.

4.19 Find the value of the stream function passing through the stagnation points of problem 4.18 and show that the equation for the corresponding streamline is

$$\left(\frac{x_b}{a}\right)^2 = 1 - \left(\frac{y_b}{a}\right)^2 + \left(\frac{2y_b}{a}\right) \cot\left(\frac{2y_b}{na}\right)$$

The subscript b above indicates body coordinates; plot the body contour for $n = 0.1$. Determine the length and thickness of the body (nondimensionalized by a) for this value of n.

FIVE

SOME EXTENSIONS TO THE METHODS OF ANALYSIS OF IDEAL FLOWS

Previous chapters have laid some theoretical groundwork and presented a few examples of important cases of ideal flow. The analysis of such flows represents a major subfield of fluid mechanics and, as such, is a well-developed discipline. This development is characterized by a number of mathematical methods for solving the Laplace equation; in this chapter we introduce a few of these as indications of the analytical power that may be brought to bear upon such problems. We also provide a few examples of results obtained for relatively complex flows. It should be remembered that in what follows we are retaining the basic ideal flow ground rules so that the implications of irrotational motion are retained. The main upshot of these is the emergence of the Laplace equation as a governing relationship and the ability to superpose its solutions to form flows that satisfy practical boundary conditions.

5.1 APPLICATIONS OF THE THEORY OF COMPLEX VARIABLES

In this section we shall only begin to scratch the surface of the extensive range of applications of complex variables. Besides the hydrodynamics literature, much of the mathematical basis for the theory is available in texts such as Churchill (1960). It has been found that in ideal flow the kinematic descriptions for irrotational motion, together with the conservation of mass, reduce to

$$\nabla^2 \phi = 0 \quad \text{and} \quad \nabla^2 \psi = 0$$

which arise from the existence of the velocity potential and stream function defined such that

$$-\frac{\partial \psi}{\partial y} = \frac{\partial \phi}{\partial x} \quad \text{and} \quad \frac{\partial \psi}{\partial x} = \frac{\partial \phi}{\partial y} \tag{5.1}$$

for two-dimensional, incompressible, irrotational flow. The equations in (5.1) are known in the theory of complex variables as the Cauchy-Riemann conditions, and the functions ϕ and ψ satisfying such relationships are called *harmonic*. Whether we choose to obtain solutions for ϕ or ψ, their introduction into the analysis of fluid flows reduces the number of dependent variables (velocity components, u and v in two-dimensional flows) to a single quantity (ϕ or ψ). The simplification can be carried further if we can combine the independent space variables (x, y) into a single variable. This is easily accomplished in two dimensions, and the operation is symbolized by introducing the complex variable $z = x + iy$.

The quantity i is like a unit vector designating the y component of z. Suppose that we have a real quantity R. If we define "real" as having no y component, then we can visualize R as a vector aligned with the x axis. If we multiply R by $-k$, we have $-kR$, which will be located along the negative x axis. The multiplication by $-k$ has in effect rotated the vector π radians and stretched it in proportion to k. If we were to multiply R by -1 and then do it again, we would cause a 2π rotation (see Fig. 5.1). That is multiplication by -1 causes rotation in increments of π. Why can't we define an operation that causes rotation in increments of $\pi/2$? We can, it turns out, and we call the thing that does it the *unit imaginary number*, i. If multiplication by i causes a rotation by $\pi/2$, then doing this twice must result in a rotation by π, or $i^2 = -1$ and $i = \sqrt{-1}$. One should try to feel that i is no more "imaginary" than -1. They are both contrivances to help us keep our numbers sorted out.

The quantity $z = x + iy = r(\cos \theta + i \sin \theta)$ is called the *complex variable* and is basically a bookkeeping device to keep track of x and y with one symbol. The complex variable has a "real" part and an "imaginary" part, x and y, and we sometimes write $z = R(z) + i\,I(z)$. If we consider functions of the complex variable, $F[z(x, y)]$, which have real parts and imaginary parts, we sometimes use the notation

$$F[z(x, y)] = F(z) = R(F) + i\,I(F)$$

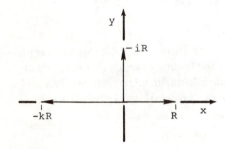

Figure 5.1

Properties of the Complex Variable

In set theorem notation, writing $z = z(x, y)$:

$$(x, y) = (x_1, y_1) \quad \text{iff} \quad x = x_1 \quad \text{and} \quad y = y_1$$

Two complex variables are equal if and only if both of the variables have equal real and equal imaginary parts. In addition,

$$(x, y) + (x_1, y_1) = (x + x_1, y + y_1)$$

$$(x, y) \times (x_1, y_1) = (xx_1 - yy_1, xy_1 + yx_1)$$

z is commutative:

$$z_1 + z_2 = z_2 + z_1$$

z is associative

$$z_1 + (z_2 + z_3) = (z_1 + z_2) + z_3$$

z is distributive:

$$z_1(z_2 + z_3) = z_1 z_2 + z_1 z_3$$

We can evaluate the sums, products, and quotients of two complex variables, $z_1 = x_1 + iy_1$ and $z_2 = x_2 + iy_2$, as follows:

$$z_1 + z_2 = (x_1 + x_2) + i(y_1 + y_2)$$

$$z_1 z_2 = (x_1 + iy_1)(x_2 + iy_2)$$

$$= (x_1 x_2 - y_1 y_2) + i(x_1 y_2 + y_1 x_2)$$

$$\frac{z_1}{z_2} = \left(\frac{x_1 + iy_1}{x_2 + iy_2}\right)\left(\frac{x_2 - iy_2}{x_2 - iy_2}\right)$$

$$= \frac{(x_1 x_2 + y_1 y_2) - i(x_1 y_2 - x_2 y_1)}{x_2^2 + y_2^2}$$

The definitions that follow are especially useful in complex arithmetic. The *conjugate* of a complex variable or number has the same real part but the negative of the imaginary part:

$$z \text{ conjugate} = \bar{z} = \overline{x + iy} = x - iy$$

The *modulus* of a complex variable or number corresponds to the magnitude of a vector with components x and y. Thus,

$$\text{mod } z = |z| = (x^2 + y^2)^{1/2}$$

and note that

$$|z|^2 = x^2 + y^2 = z\bar{z}$$

The *argument* of a complex variable or number is given by

$$\arg z = \tan^{-1} \frac{y}{x}$$

Two Important Theorems

DeMoivre's theorem. We begin by expressing two complex variables in polar notation:

$$z_1 = r_1(\cos \theta_1 + i \sin \theta_1) \qquad z_2 = r_2(\cos \theta_2 + i \sin \theta_2)$$

Multiplying these together gives

$$z_1 z_2 = r_1 r_2[\cos \theta_1 \cos \theta_2 - \sin \theta_1 \sin \theta_2 + i(\sin \theta_1 \cos \theta_2 + \cos \theta_1 \sin \theta_2)]$$
$$= r_1 r_2[\cos (\theta_1 + \theta_2) + i \sin (\theta_1 + \theta_2)]$$

When this is extended to the multiplication of any number of complex variables, we conclude that

$$z_1 z_2 \cdots z_n = r_1 r_2 \cdots r_n[\cos (\theta_1 + \theta_2 + \cdots + \theta_n) + i \sin (\theta_1 + \theta_2 + \cdots + \theta_n)]$$

and in particular, if $z_1 = z_2 = \cdots = z_n = z$, then

$$z^n = r^n(\cos n\theta + i \sin n\theta) \tag{5.2}$$

This is known as *DeMoivre's theorem.*

Euler's theorem. If a function f of a single variable x is evaluated in terms of its value when $x = 0$, the resulting special form of the Taylor series is called the MacLaurin series:

$$f(x) = f(0) + f'(0)x + f''(0)\frac{x^2}{2} + \ldots$$

and, in particular, if $f(x) = e^x$,

$$e^x = 1 + x + \frac{x^2}{2!} + \frac{x^3}{3!} + \ldots$$

Recall also that

$$\sin \theta = \theta - \frac{\theta^3}{3!} + \frac{\theta^5}{5!} - \ldots$$

and

$$\cos \theta = 1 - \frac{\theta^2}{2!} + \frac{\theta^4}{4!} - \ldots$$

If we now let $x = i\theta$,

$$e^{i\theta} = 1 + i\theta - \frac{\theta^2}{2!} - i\frac{\theta^3}{3!} + \frac{\theta^4}{4!} + i\frac{\theta^5}{5!} + \dots$$

$$= \left(1 - \frac{\theta^2}{2!} + \frac{\theta^4}{4!} - \dots\right) + i\left(\theta - \frac{\theta^3}{3!} + \frac{\theta^5}{5!} - \dots\right)$$

and Euler's theorem is the result:

$$e^{i\theta} = \cos\theta + i\sin\theta \tag{5.3}$$

Euler used this result to prove that $e^{i\pi} + 1 = 0$! Because $z = r(\cos\theta + i\sin\theta)$, we may now write the polar notation for z:

$$z = re^{i\theta} \tag{5.4}$$

Using this polar notation we see that

$$\ln z = \ln re^{i\theta} = \ln r + i\theta$$

$$\ln z_1 z_2 = \ln r_1 e^{i\theta_1} r_2 e^{i\theta_2} = \ln r_1 r_2 + i(\theta_1 + \theta_2) \tag{5.5}$$

for example,

$$\ln(-n) = \ln(-1)(n) = \ln n + i\pi$$

Introduction of the complex variable allows the evaluation of logarithms of negative numbers (giving a complex number with real part $\ln n$ and imaginary part π).

5.2 BACK TO FLUID MECHANICS

Complex Velocity Potential

The complex function

$$\Omega = \phi(x, y) - i\psi(x, y) \tag{5.6}$$

is a function of z because $\partial\phi/\partial x = -\partial\psi/\partial y$ and $\partial\phi/\partial y = \partial\psi/\partial x$. These are the Cauchy-Riemann conditions that, in the theory of complex variables, guarantee that Ω may be written as a function of z, that is, $\Omega = \Omega(z)$. This means that even if Ω comes as a complicated collection of x's and y's, some with and some without the unit imaginary number i, the natures of the stream function and velocity potential are such that we can be assured that all the theory applicable to functions of z is applicable to Ω. Appreciation of this capability will come only with application of the theory. We may note further that, with $z = x + iy$,

$$\frac{\partial\Omega}{\partial x} = \frac{d\Omega}{dz}\frac{\partial z}{\partial x} = \frac{d\Omega}{dz}$$

and

$$\frac{\partial \Omega}{\partial y} = \frac{d\Omega}{dz}\frac{\partial z}{\partial y} = i\frac{d\Omega}{dz}$$

thus

$$\frac{d\Omega}{dz} = \frac{\partial \phi}{\partial x} - i\frac{\partial \psi}{\partial x} = -i\frac{\partial \phi}{\partial y} - \frac{\partial \psi}{\partial y}$$

Assigning the symbol $w = d\Omega/dz$, we see from either of the above relationships that

$$w = -u + iv \qquad (5.7)$$

Since Ω is a function whose derivative with respect to z gives velocity components (note, *both* the velocity components), it has come to be called the *complex velocity potential*. The magnitude of the velocity may be found from

$$w\bar{w} = (-u + iv)(-u - iv) = u^2 + v^2 = q^2 \qquad (5.8)$$

The complex velocity potential is a means of collectively expressing ϕ and ψ, and its derivative gives the velocity components as its real and imaginary parts. We shall now examine several flows that may be obtained by simply writing down functions of z for $\Omega(z)$.

Some Basic Two-Dimensional Flows

I. Rectilinear flows. Rectilinear flows are obtained from complex velocity potentials of the form

$$\Omega = Az^n$$

A. With the coefficient A a real constant Using DeMoivre's theorem, Eq. (5.2), Ω may be written

$$\Omega = Ar^n (\cos n\theta + i \sin n\theta)$$

and the velocity potential and stream functions are found directly from Eq. (5.6):

$$\phi = Ar^n \cos n\theta \qquad \text{and} \qquad \psi = -Ar^n \sin n\theta$$

The velocity components are found from

$$w = \frac{d\Omega}{dz} = nAz^{n-1} = nAr^{n-1}[\cos(n-1)\theta + i \sin(n-1)\theta]$$

and, from Eq. (5.7),

$$u = -nAr^{n-1} \cos(n-1)\theta \qquad v = nAr^{n-1} \sin(n-1)\theta$$

In the case of $n = 1$, $\psi = -Ar \sin \theta$ and $\phi = Ar \cos \theta$, or $\psi = -Ay$, $\phi = Ax$. The velocity components are seen to be $u = -A$ and $v = 0$. The flow thus depicted is uniform flow parallel to the x axis with magnitude A and a direction determined by the sign of A.

If $n = 2$, then

$$\psi = -Ar^2 \sin 2\theta \qquad u = -2Ar \cos \theta \qquad v = 2ar \sin \theta$$

Streamlines are obtained by setting ψ equal to a constant:

$$r^2 \sin 2\theta = 2r^2 \sin \theta \cos \theta = \text{constant}$$

In other words, streamlines are given by $xy = \text{constant}$ and are families of hyperbolas that are repeated in the four quadrants of the x-y plane. In particular, the x and y axes are streamlines, and the velocity on the $+y$ axis, for instance, is $u = 0$, $v = 2Ar$. The flow in the first quadrant for a negative value of A is illustrated in Fig. 5.2.

In general, for the flow in a corner of angle α, the complex velocity potential is given by

$$\Omega = Az^{\pi/\alpha}$$

If the exponent of z is a fraction, then the angle of the corner is larger than 180°. For instance, if $n = \frac{2}{3}$ (i.e., if $\alpha = 3\pi/2$), then

$$\psi = -Ar^{2/3} \sin\left(\frac{2}{3}\right)\theta$$

$$u = -\left(\frac{2}{3}\right)Ar^{-1/3} \cos\left(\frac{\theta}{3}\right)$$

$$v = -\left(\frac{2}{3}\right)Ar^{-1/3} \sin\left(\frac{\theta}{3}\right)$$

On the lines $\theta = 0$, $3\pi/2$, the stream function is zero, and this corner represents an obtuse angle of 270° for the flow to turn. On $\theta = 3\pi/2$, for instance, the velocity components are given by

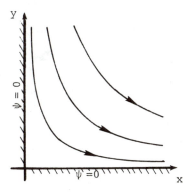

Figure 5.2 Flow in a corner; $n = 2$, $\alpha = \pi/2$.

$$u = 0, \quad v = -\left(\frac{2}{3}\right)Ar^{-1/3} \qquad \left(\theta = \frac{3\pi}{2}\right)$$

If A is negative, then the flow is upward along the negative y axis and decreasing in velocity with distance from the origin. *Note that at the corner, $r = 0$ and the velocity there is infinite.* This departure of ideal flow from reality is a result of neglecting viscous effects. Nevertheless, flows do attain very high velocities in the vicinity of an obtuse corner. The flow pattern for a negative value of A is sketched in Fig. 5.3.

One of the important classes of viscous flows is that past a wedge, and the ideal flow associated with such a boundary may be modeled as that due to a compression turn. The analysis of these types of flows is presented, for instance, in Schlichting (1968, p. 150). The basis of the calculations for the region of viscous flow near the boundary is the pressure distribution in the outer ideal flow region. The ideal flow velocity variation leads to the pressure distribution via the Bernoulli equation, and so the results given above are applicable. Along the wall, following the compression turn, $\theta = 0$ and $r = x$, so the velocity vector is given by $u = -nAx^{n-1}$ and $v = 0$. For a turn of angle β, the exponent n may be replaced by $n = \pi/(\pi - \beta)$. The general form $u = Cx^m$, where C and m depend on the angle β (the half-angle of the effective wedge), is that used in subsequent analysis of the viscous flow region. The flow for a negative value of A is sketched in Fig. 5.4.

B. With the coefficient A complex Replace A with a complex number given by $Ae^{-i\alpha}$. Then, with the previous definition of Ω,

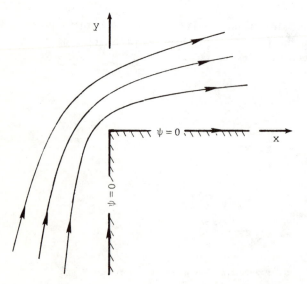

Figure 5.3 Flow past an outside corner; $\alpha = \pi/2$.

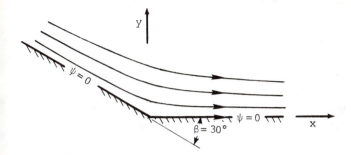

Figure 5.4 Ideal flow model of a wedge.

$$\Omega = Ae^{-i\alpha}z^n = Ae^{-i\alpha}r^n e^{in\theta} = Ar^n e^{i(n\theta - \alpha)}$$

and

$$\Omega = Ar^n[\cos(n\theta - \alpha) + i\sin(n\theta - \alpha)]$$

The stream function is the imaginary part of Ω, so

$$\psi = -Ar^n \sin(n\theta - \alpha)$$

The effect has been to rotate streamlines (values of constant ψ) by the angle α. If $n = 1$, for instance, $\psi = -Ar \sin(\theta - \alpha)$ and $\psi = 0$ on $\theta = \alpha$ and $\alpha + \pi$. This particular streamline is tilted up by the angle α from the x axis. The velocity components for $n = 1$ are found to be $u = -A \cos \alpha$ and $v = -A \sin \alpha$.

II. Radial and swirling flows. For these flow patterns we consider complex velocity potentials of the form $\Omega = A \ln z$, where A may be a real or imaginary constant. The stream function and velocity potential are, by definition, the real and imaginary parts of Ω:

$$\Omega = A \ln re^{i\theta} = A(\ln r + i\theta)$$

A. With the coefficient A real In this case $\phi = A \ln r$ and $\psi = -A\theta$. The velocity components are given by Eq. (5.7) as

$$-u + iv = \frac{A}{z} = \frac{A}{r}e^{-i\theta} = \frac{A}{r}(\cos \theta - i \sin \theta)$$

and

$$u = -\frac{A}{r}\cos \theta \qquad v = -\frac{A}{r}\sin \theta$$

These Cartesian velocity components may be converted to polar notation by noting that

$$v_r = u \cos \theta + v \sin \theta \qquad v_\theta = v \cos \theta - u \sin \theta \qquad (5.9)$$

from which

$$v_r = -\frac{A}{r} \qquad v_\theta = 0$$

(Note that these expressions could also have been obtained from derivatives of ϕ or ψ in accordance with Table 3.1.)

The volume flow rate per unit depth, Q', is given by $Q' = 2\pi r v_r = -2\pi A$. Defining the strength μ_s as before, $\mu_s = Q'/2\pi = -A$, we have

$$v_r = \frac{\mu_s}{r} \qquad v_\theta = 0$$

Thus, we have arrived at the velocity distribution for a source (with μ_s positive) or a sink (with μ_s negative). In terms of μ_s, the complex velocity potential is $\Omega = -\mu_s \ln z$ so that $\phi = -\mu_s \ln r$ and $\psi = \mu_s\theta$. These values should be compared with those listed in Table 4.1.

The complex variable notation is especially useful for defining flows resulting from singularities not at the origin. The complex number z can be thought of as the directed line segment, or vector, connecting the origin with the point (x, y) (see Fig. 5.5). Similarly, the vector z_1 locates the point (x_1, y_1), and the difference $(z - z_1) = (x - x_1) + i(y - y_1)$ has the components of the vector originating at x_1, y_1 and directed to the point x, y. For a singularity located at z_1, the directed distance from that singularity to any point in the x-y plane is $z - z_1$. For the source, the complex velocity potential is simply rewritten with this substitution:

$$\Omega = -\mu_s \ln(z - z_1) \qquad \text{(source at } z_1)$$

B. With the coefficient A imaginary Suppose we let $A = iB$, where B is a real number so that A is a pure imaginary number. Accordingly, $\Omega = iB \ln z$, and we have simply multiplied the previous relationship for the complex velocity potential by the unit imaginary number. But remember that multiplication by i rotates directed line segments by 90°. We can expect that what were radial streamlines for the source will be circumferential streamlines and, likewise, the equipotential lines that were circular for the source will become radial lines. The source is, in effect, rotated 90° into a vortex. Examination of the new complex velocity potential verifies this:

$$\Omega = iB \ln z = iB(\ln r + i\theta) = B(-\theta + i \ln r)$$

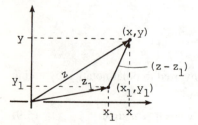

Figure 5.5

from which

$$\phi = -B\theta \qquad \psi = -B \ln r$$

The forms for the equipotential lines and streamlines are reversed, and the source is converted into the vortex through the multiplication by i! (Note that this rotation is about the origin, and conversion of singularities located elsewhere will lead to different results.) The velocity components are given by $d\Omega/dz$, as usual, so that

$$-u + iv = i\frac{B}{z} = i\frac{B}{r}(\cos\theta - i\sin\theta) = \frac{B}{r}(\sin\theta + i\cos\theta)$$

from which $u = -(B/r)\sin\theta$ and $v = (B/r)\cos\theta$, or

$$v_r = 0 \qquad v_\theta = \frac{B}{r}$$

Again, the conversion from radial to rotary flow is noted. The constant B is related to the circulation Γ as follows:

$$\Gamma = \int_C \mathbf{q} \cdot ds = \int_0^{2\pi} \left(\frac{B}{r}\right)(r \, d\theta) = 2\pi B$$

and we may now write for the vortex:

$$\Omega = i\mu_v \ln z \qquad \phi = -\mu_v \theta \qquad \psi = -\mu_v \ln r$$

where $\mu_v = \Gamma/2\pi$ (see Table 4.1). Again, if the vortex center is located elsewhere, at z_1, then this is taken into account by writing $\Omega = i\mu_v \ln(z - z_1)$.

C. Doublets This flow pattern follows from writing $\Omega = A/z$, where A is a real constant. In polar notation,

$$\Omega = \frac{A}{re^{i\theta}} = \frac{A}{r}(\cos\theta - i\sin\theta)$$

and

$$\phi = \frac{A}{r}\cos\theta \qquad \psi = \frac{A}{r}\sin\theta$$

We have seen in the previous chapter that lines of constant ψ yield circles tangent to the x axis. With the substitutions $y = r\sin\theta$ and $r^2 = x^2 + y^2$,

$$\frac{x^2 + y^2}{y} = \frac{A}{\psi}$$

which may be manipulated to form

$$x^2 + \left(y - \frac{A}{2\psi}\right)^2 = \left(\frac{A}{2\psi}\right)^2$$

This is seen to be a family of circles of radii $A/2\psi$ and centers located on the y axis at $A/2\psi$. Positive values of A/ψ will yield circles above the x axis, and negative values of A/ψ will yield circles below that line (see Fig. 4.2). The flow is indeed that of a doublet, and comparison with the results of Chapter 4 shows that for a positive-facing doublet the strength μ_d is identical to the constant A.

III. Simple body shapes. With the methods developed thus far it is quite a simple matter to construct flow patterns from selected combinations of singularities. Some of these are of practical importance, particularly when a uniform flow is included in the field of flow. To illustrate the methods, we demonstrate some classical results in the theory of ideal fluid flows.

A. Rankine half-body Here we consider the flow pattern generated by a source located at the origin in a uniform flow. The complex velocity potential is

$$\Omega = -Uz - \mu_s \ln z$$

so that

$$\phi = -Ur \cos \theta - \mu_s \ln r \qquad \psi = Ur \sin \theta + \mu_s\theta$$

The velocity components are

$$v_r = U \cos \theta + \mu_s/r \qquad v_\theta = -U \sin \theta$$

and a stagnation point is found to lie at $\theta = \pi$ and $r = \mu_s/U$. (Although $v_\theta = 0$ on $\theta = 0$, the solution for $v_r = 0$ on this line leads to a negative value for r and therefore must be rejected.) The value of ψ for the streamline passing through the stagnation point is given by $\psi = \mu_s\pi$, and the expression for all points with this value of ψ is obtained from

$$\mu_s\pi = Ur_b \sin \theta_b + \mu_s\theta_b$$

The subscript b is used here to denote the coordinates of a streamline of particular interest, that is, the "body streamline." When this is sorted out, the body streamline may be drawn according to

$$r_b = \frac{\mu_s}{U} \frac{\pi - \theta_b}{\sin \theta_b} \qquad \text{or} \qquad y_b = \frac{\mu_s}{U}(\pi - \theta_b)$$

From these it may be found that when θ_b is zero, so that r_b must extend to infinity, the asymptotic body thickness, $2y_b$, is $2\pi\mu_s/U$. The shape of this streamline is illustrated in Fig. 5.6.

B. Rankine oval The Rankine oval is obtained if the Rankine half-body is closed by means of a sink of equal strength downstream of the source. A uniform flow is combined with a source at $x = -a$ and a sink of equal strength at $x = a$. The complex velocity potential is

$$\Omega = -Uz - \mu_s \ln(z + a) + \mu_s \ln(z - a)$$

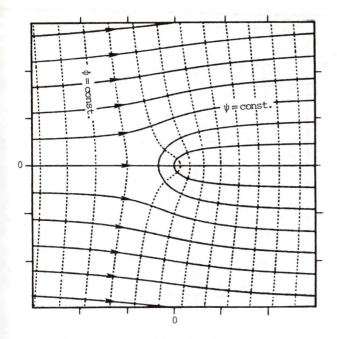

Figure 5.6 Flow past the Rankine half-body.

where μ_s is the strength of the source and sink (the plus sign on the last term accounts for the sign of the sink at $x = a$). The vectors $z + a$ and $z - a$, associated with the source and sink, respectively, may be expressed in polar coordinates relative to their locations (see Fig. 5.7):

$$z + a = r_1 e^{i\theta_1} \qquad z - a = r_2 e^{i\theta_2}$$

so that the complex velocity potential may be written

$$\Omega = -Uz + \mu_s\left[\ln\frac{r_2}{r_1}e^{i(\theta_2 - \theta_1)}\right]$$

$$= -Uz + \mu_s\left[\ln\frac{r_2}{r_1} + i(\theta_2 - \theta_1)\right]$$

The imaginary part of this is the stream function, so

$$\psi = Ur\sin\theta - \mu_s(\theta_2 - \theta_1)$$

The difference in the two angles may be expressed in terms of the original co-ordinates by application of a considerable amount of geometry:

$$\tan(\theta_2 - \theta_1) = \frac{2ar\sin\theta}{r^2 - a^2}$$

Combination of these results and the assignment of various constant values of ψ

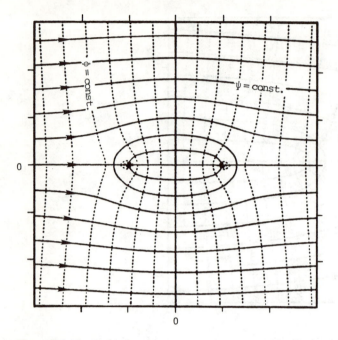

Figure 5.7 The Rankine oval.

will lead to streamlines (see problems 4.17–4.19). Streamlines passing through the stagnation points have the value $\psi = 0$ and will lead to bodies of the sort shown in the Fig. 5.7, with length-to-height ratios depending on the relative strengths of the source and the uniform flow.

C. Kelvin oval The Kelvin oval is obtained from the interesting combination of two vortices in a uniform flow (Fig. 5.8). The vortices are of equal strength but opposite sense and are located equidistant from the origin on the y axis at $y = \pm a$. Because each vortex induces a velocity at the location of its counterpart,

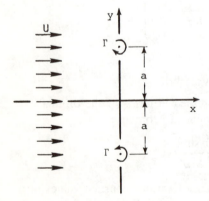

Figure 5.8 Elements of the Kelvin oval.

equal in magnitude to $\Gamma/4\pi a$, the identity of the singularities as fixed vortices can be maintained only if at each vortex location the induced velocity is exactly canceled by the uniform flow. For given values of a and U, this fixes the circulation of the vortices at $\Gamma = 4\pi a U$. The complex velocity potential is given by

$$\Omega = -Uz + i\mu_v \ln(z + ia) - i\mu_v \ln(z - ia)$$

where $\mu_v = 2aU$.

If local polar coordinates (r_1, θ_1) and (r_2, θ_2) are assigned to the vortices at $-a$ and a, respectively, then Ω may be written

$$\Omega = -Uz + i\mu_v \left[\ln\left(\frac{r_2}{r_1}\right) + i(\theta_2 - \theta_1) \right]$$

and from the imaginary part of this we find that

$$\psi = Ur \sin\theta - \mu_v \ln\left(\frac{r_2}{r_1}\right)$$

In terms of the original coordinates

$$\psi = Uy - \frac{\mu_v}{2} \ln\left[\frac{x^2 + (y + a)^2}{x^2 + (y - a)^2}\right]$$

From symmetry considerations it is apparent that the stagnation points will lie on the x axis. The x component of the velocity may be found from the partial derivative of ψ with respect to y, and when this is set equal to zero the stagnation points are found to lie at $x = \pm\sqrt{3}a$ on $y = 0$. It is easily seen from the above expression that the value of the stream function passing through the stagnation points is zero, and the equation for the body streamline is, therefore,

$$\frac{y_b}{a} = \ln\frac{x_b^2 + (y_b + a)^2}{x_b^2 + (y_b - a)^2}$$

The flow pattern is illustrated in Fig. 5.9.

In the preceding sections several approaches were used for finding flow patterns, velocity potentials, stream functions, velocity components, etc. This flexibility (which may have lent an aura of confusion to the presentations) is one of the features of ideal flow analysis. In some cases the so-called direct method has been used in which velocities are specified and integrations lead to the velocity potential and/or stream function. This approach has been used in developing the standard flow elements: uniform flow, source, etc.

Elsewhere, we have simply written down a stream function, velocity potential, or complex velocity potential and examined the resulting flow field. This latter approach is the indirect method, and the fact that the resulting flows have had some interesting features is, in the cases presented here, a matter of prior knowledge. In many cases of practical importance the indirect method leads to a trial-and-error process until the desired flow pattern is found. If the complex ve-

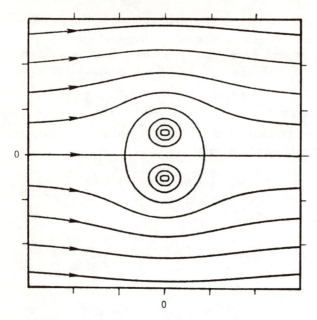

Figure 5.9 Flow past the Kelvin oval.

locity potential is used, the velocity potential and stream function may be found directly from its real and imaginary parts. Velocity components are then found from the partial derivatives of these. Alternatively, the derivative of the complex velocity potential leads directly to the velocity components. Whatever the approach used, the main purpose is to obtain the desired flow field with the velocity vector everywhere determined. The practical goal of all this, which is sometimes lost in the forest of mathematics, is to find the pressure distributions on stream surfaces of particular interest. In cases of irrotational motion, these follow directly from the velocity fields through the Bernoulli equation.

5.3 DISTRIBUTED SINGULARITIES

All of the ideal flows presented so far have involved one or more singularities. Even uniform flow may be constructed by considering a source and a sink at either extreme of the field of flow. Although there is an unlimited variety of patterns that can be created by sprinkling singularities here and there in an otherwise uniform flow, a great deal of progress has been made in developing systematic approaches to the design of flow fields. In this section we provide a brief introduction to the principles underlying these methods.

Indirect Method

We consider, as an example, the expression for the complex velocity potential due to a distribution of singularities along the x axis, as sketched in Fig. 5.10.

Again, we consider two-dimensional plane flows. Analogous methods, developed for two-dimensional axisymmetric flows, are described in Robertson (1965). (Fully three-dimensional flows are analyzed using direct methods of considerable complexity.) The strength per unit length of the proposed source distribution is denoted by μ'. That is, the strength of an increment of source of length ds would be given by $d\mu_s = \mu'\, ds$. The coordinate s locates a point on the singularity distribution that extends from $s = a$ to $s = b$. At a general point $P(z)$ in the x-y plane, the incremental influence of a singularity element of strength $\mu'\, ds$ is, in terms of the complex velocity potential,

$$d\Omega = \mu'\, ds\, f(z - s) \tag{5.10}$$

where $f(z - s)$ expresses the geometric function governing the particular singularity. For a doublet, for example, $f(z - s)$ is $1/(z - s)$. For the entire singularity distribution, therefore,

$$\Omega = \int_a^b \mu'\, f(z - s)\, ds$$

Note that this integration is performed over the singularity distribution with z held constant.

For sources and sinks, in particular, $\mu'\, ds = d\mu_s$ and $f(z - s) = -\ln(z - s)$ with μ_s taken as positive for a source distribution. (A sink distribution is just a negative source distribution, and the sign of μ' may vary along s.) Expressing $(z - s)$ in terms of local coordinates (r_s, θ_s), $f(z - s) = -\ln r_s - i\theta_s$ and

$$\Omega(z) = \int_a^b -\mu'(s)\, (\ln r_s + i\theta_s)\, ds$$

The velocity potential and stream function may be evaluated from this as

$$\phi = -\int_a^b \mu'(s)\ln r_s\, ds \qquad \psi = \int_a^b \mu'(s)\theta_s\, ds \tag{5.11}$$

But $r_s = [y^2 + (x - s)^2]^{1/2}$ and $\theta_s = \tan^{-1}[y/(x - s)]$, so for the stream function, for instance,

$$\psi = \int_a^b \mu'(s)\tan^{-1}\left(\frac{y}{x - s}\right) ds$$

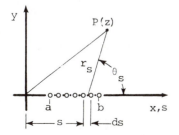

Figure 5.10 Source distribution along the x axis.

General expressions such as these may also be obtained for doublets, vortices, and the corresponding axisymmetric singularities. The distributions need not lie along the x axis; for that matter, they need not lie along straight lines (but inclusion of the latter variation would indicate a marked masochistic tendency).

Final expressions for ϕ and ψ require specification of the variation of $\mu'(s)$ with s. For a source distribution of constant strength, for instance, $\mu'(s)$ is a constant, say C, and

$$\psi = C \int_a^b \theta_s \, ds = C \left[(s - x)\theta_s - y \ln \frac{r_s}{y} \right]_{s=a}^{s=b} \tag{5.12}$$

(The integration above is most easily accomplished by evaluating $\int \cot^{-1} u \, du$, where $u = (s - x)/y$.)

By way of further illustration, let us consider the flow pattern obtained from the combination of a uniform flow with a source at the origin followed by a uniformly distributed sink for the origin to $s = L$. The strength of the sink distribution is selected so that its cumulative flow is equal to that due to the point source. Thus, for the sink distribution,

$$-\mu_s = \int_0^L C \, ds = CL \qquad \text{and} \qquad C = \frac{-\mu_s}{L}$$

Combining Eq. (5.12) with the stream functions for the uniform flow and source at the origin, we have

$$\psi = Uy + \mu_s\theta - \mu_s \left[\theta_L\left(1 - \frac{x}{L}\right) + \theta\frac{x}{L} + \frac{y}{L} \ln \frac{r}{r_L} \right]$$

where r_L and θ_L are the distance and angle measured from the tail of the distribution to any point in the plane. Evaluation of this expression leads to streamlines, and its derivatives give the velocity components in the flow. The $\psi = 0$ streamline is of special interest because it leads to the closed body shape with a blunt upstream contour and a sharp trailing edge (see Fig. 5.11). The distributed singularity used here has in effect replaced the point sink effect in the Rankine oval with a more gradual closing of the downstream body contour.

Clearly, there are a large number of singularity distributions that might be evaluated to determine if they might be of practical interest. The indirect method used here requires considerable insight (and more than a little luck) to create any but the most simple body shapes. As the boundary conditions become more complex, the amount of guesswork involved increases rapidly. In addition, the necessary mathematical manipulations increase in number and difficulty so that the analyst is forced to use computer methods. When this occurs, it is often more efficient to use the direct method, since this is inherently a computer-based approach.

Direct Methods Using Distributed Singularities

We have seen how to write some fairly general expressions for distributed singularities in the hope that some useful streamline patterns can be obtained. The

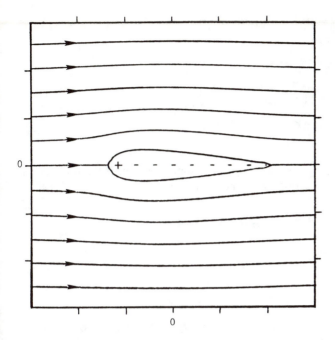

Figure 5.11 Flow pattern obtained from a point source and distributed sink in uniform flow.

direct method is so named because the problem begins with a *specification* of the desired body shape. The problem then is to determine how to distribute singularities to obtain the specified shape, that is, streamline—the flow constituents are "designed" to meet the body shape specification. Although the problem statement is "direct," the mathematical method is more by way of being "indirect" or "implicit." The two most common direct methods involve singularities distributed internal to the body streamline or singularities disposed on the body surface. We shall briefly outline each of these methods, again for plane two-dimensional flows.

Internally distributed singularities. If we continue to consider sources distributed along the x axis, Eq. (5.11) gives, for a particular body streamline,

$$\frac{d\psi_b}{ds} = \mu'(s)\theta_b(x_b, y_b, s)$$

where the subscript b indicates points on the body streamline and θ_b is the angle subtended by a line segment originating on the distributed singularity, as before. In this approach, however, the line segment is not directed to an arbitrary point in the plane but instead ends at a particular point on the prescribed body contour (Fig. 5.12). To find the value of ψ_b due to the total distribution, including the uniform flow, we write

$$\psi_b = Uy_b + \int_a^b \mu'(s)\theta_b(x_b, y_b, s)\, ds$$

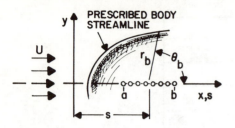

Figure 5.12 Geometry for an internally distributed source.

Since it is known that on the body ψ_b is zero (or some constant if the body is not closed), we make this substitution, with the result that

$$y_b = -\frac{1}{U} \int_a^b \mu'(s)\theta_b(x_b, y_b, s)\, ds$$

But $\theta_b = \tan^{-1}[y_b/(x_b - s)]$, so

$$y_b = -\frac{1}{U} \int_a^b \mu'(s) \tan^{-1}\left(\frac{y_b}{x_b - s}\right) ds \qquad (5.13)$$

Since the coordinates x_b, y_b are known, the only quantity to be resolved is the source distribution $\mu'(s)$. This type of equation is known as a Fredholm integral equation of the first kind, and the function multiplying the quantity to be determined under the integral is called the *kernel*. The kernel in this case is $\tan^{-1}[y_b/(x_b - s)]$, and for every s along the integration path it is a known quantity.

A number of systematic methods are available for solving equations of this type. From a computational point of view, the solution involves discretization of the integral into a summation over a selected number of n increments of s from a to b. For each body point, y_b must be equal to the summation over the distribution of n singularity increments, each with its own constant incremental strength. For n body points, therefore, there are n linear simultaneous algebraic equations implicit in the n incremental strengths. (Among the early advances in this method were those achieved by the famous aerodynamicist Theodore von Karman. Documentation of this work, titled "Calculation of Pressure Distributions on Airship Hulls," is reproduced in NACA TM 574, 1930.) In the "old" days such equations were solved by hand, but with the computer available, solutions are now obtained with much more speed. The accuracy is dependent on the number of points considered, and this often reduces to a matter of who is paying for the computer time.

Surface singularities. In this method singularities are arrayed on the surface of the *specified* body. By doing this we are able to make direct use of the boundary condition on the surface that there is no normal component of velocity there. That is, with q_n denoting the velocity component in the direction of the outward normal,

$$q_n = -\frac{\partial\phi}{\partial n} = 0$$

If the source strength per unit length at a point p on the surface is $\mu'(p)$, then the contribution of this at any other point in the flow field (Fig. 5.13) is

$$d\phi_P = -\mu'(p) \ln r(p, P) ds$$

The velocity potential at point P due to all the singularity elements on the surface is

$$\phi_P = -\int_s \mu'(p) \ln r(p, P) ds$$

If we now consider points on the body, denoted by b, then

$$\left. \frac{\partial \phi}{\partial n} \right|_b = -\int_s \mu'(p) \left\{ \frac{\partial}{\partial n} [\ln r(p, b)] \right\} ds$$

where $r(p, b)$ is the line connecting p (at the singularity element) with b (the general body point). When p is coincident with b, $r(p, b) = 0$ and it is necessary to make a separate evaluation of the integral above. Since at the location of the source element one-half of the fluid flux is in the outward direction, we may write

$$(q_n|_b) ds = \frac{1}{2}[2\pi\mu'(b) ds]$$

or

$$q_n|_b = \pi\mu'(b)$$

This is the negative of the contribution to $\partial\phi/\partial n$ at point b due to the source element $\mu'(b)$ at that point. Expressing this separately in the integral above, we have

$$\left. \frac{\partial \phi}{\partial n} \right|_b = -\pi\mu'(b) - \int_{s'} \mu'(p) \left\{ \frac{\partial}{\partial n} [\ln r(p, b)] \right\} ds' \qquad (5.14)$$

where s' covers the body except for the element at b, the point in question.

If the body is stationary in a uniform flow of velocity U and the slope of the body at b is α, then the component of the uniform flow normal to the body at that point will be inward and of magnitude $U \sin \alpha$. To meet the boundary con-

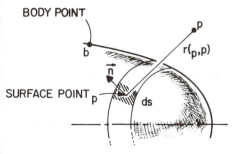

BODY POINT

b

$r(p,p)$

$\vec{n}$

SURFACE POINT p

ds

Figure 5.13 Geometry for surface singularities.

dition, the velocity contribution of the distributed singularities at point b must exactly match this in the outward direction. Thus,

$$-\frac{\partial \phi}{\partial n}\bigg|_b = U \sin \alpha$$

so that we finally obtain

$$\mu'(b) = \frac{U}{\pi} \sin \alpha - \frac{1}{\pi} \int_{s'} \mu'(p) \left\{ \frac{\partial}{\partial n}[\ln r(p, b)] \right\} ds'$$

In this expression, at every point along the singularity distribution, $r(p, b)$ is known relative to a particular body point. In addition, for each point on the prescribed body, α is known. The quantity to be found, for each body point, is $\mu'(b)$, and this is accomplished by repeated evaluation of the integral over the distribution of surface singularities.

The expression above is known as a Fredholm integral of the second kind and, as with internal singularity distributions, there are several approaches to its solution. In discretized form, the equation is written

$$\frac{\mu_i'}{U/\pi} = \sin \alpha - \sum_{j=1}^{n} \frac{\mu_j'}{U} \left\{ \frac{\partial}{\partial n}[\ln r(p, b)] \right\}_{i,j} ds_j$$

This leads to n simultaneous, linear, algebraic equations for n unknown elemental strengths μ_i'.

This concludes our brief exposition of some of the methods of modeling flows around bodies of arbitrary shape. Needless to say, extensions to these beginnings are many, and the computer plays a vital role in such treatments. Many complications are possible, including three-dimensional effects (for instance, with bodies at an angle of attack to the oncoming flow), unsteady flows, and the effects of compressibility. In most approaches to the solution of such problems, the basic elements of analysis are similar to those reviewed here. The reader will do well to remember the basic underpinnings of ideal fluid flows: the absence of viscous effects, irrotational motion, the existence of the velocity potential, the governance of the Laplace equation, etc.

Some Results of Extended Applications

In this chapter we have attempted to present a few of the ideas that form the foundation for extensive application of the principles of ideal fluid flow. This important branch of fluid mechanics has been the source of a vast amount of productivity in predicting the behavior of fluid flows, principally in the form of pressure and velocity distributions. A number of other methods of solution have been developed in addition to those presented here. The applications of methods involving complex variables have been extended to schemes in which known flow solutions may be "mapped" into shapes whose variety (in two dimensions) is virtually limitless. Other analytical methods include separation of variables in which the solution of the governing partial differential equations is transformed into an eigenvalue problem. Further review of methods such as these may be found, for instance, in Robertson (1965).

The extent to which the governing relationships can be solved analytically is basically limited by the complexity of the boundary conditions, that is, the shape of the flow boundaries. For this reason, much of the early work in ideal fluid flow analysis was limited to the treatment of flows in and around shapes that could be relatively simply described. These limitations were lifted with the advent of the high-speed digital computer, and in these days the majority of ideal flow solutions evolve from computer applications. Much of the mathematical elegance and sophistication associated with analytical methods has been forestalled by numerical methodologies that are often innovative, sometimes brutish, but almost always powerful and versatile. In closing this chapter, and this part of the text, we present a few of the practical results that are typical of the advances that have been made via the high-speed digital computer.

Typical of the pioneering work undertaken in computer applications to ideal fluid flows is that of Hess and Smith (1964). The methods used were similar to that of Fig. 5.13 in which a distribution of source density on the surface of a body is determined in order to meet the specific boundary conditions. As we have seen, formulation of the problem in terms of velocity potential leads to a Fredholm integral equation of the second kind, as in Eq. (5.14). Discretization of this equation, for solution by digital computer, is accomplished by simulation of the body surface as a large number of plane quadrilateral elements, over each of which the source strength is assumed constant. These elements are sometimes called *panels*, and the associated computer-based approach to solution is called the *panel method*. By this method, the integral equation is replaced by a set of linear algebraic equations for the values of the source strengths on each of the panels.

Figure 5.14 illustrates the selection of quadrilateral element distribution used by Hess and Smith for calculating the ideal fluid flow about a sonar dome appended to the forward keel of a ship's hull. Also indicated in Fig. 5.14 are the locations of points of pressure measurement used in tests performed by Denny (1963). Calculated and experimental pressure distributions along the bottom of the dome (a line of symmetry) are compared in Fig. 5.15. The agreement is an indication of the usefulness of the method, as well as the relative insensitivity of the pressure distribution to viscous effects. (The tests were at a low Froude number, 0.5, to minimize free surface effects. From a computational point of view, this allows the free surface to be taken as rigid, thereby eliminating wave-making effects that are important at higher Froude numbers.)

Figure 5.16 further illustrates the power of the distributed singularity methods such as that employed by Hess and Smith. The surface singularities were arranged to simulate the flow past a standard hull shape, again assuming a rigid free surface. The results of the calculations are shown in Fig. 5.16, where the predicted longitudinal and azimuthal velocity distributions are shown. Characteristic low-velocity (high-pressure) regions are clearly indicated near the bow and stern regions. The complexity of the velocity variations indicates the capability of such

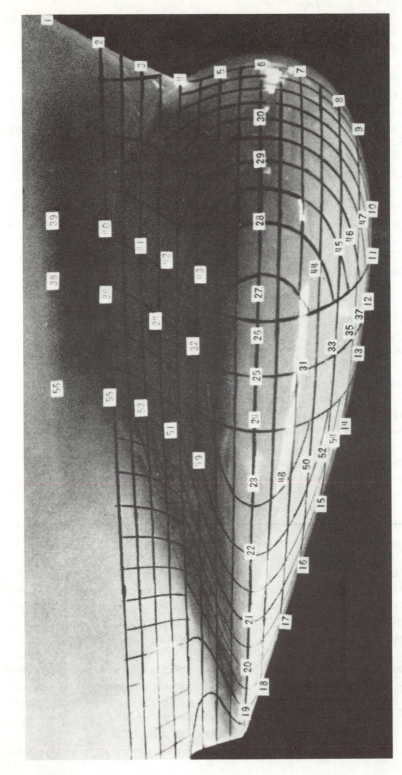

Figure 5.14 Destroyer sonar dome showing element distribution and pitot tube locations. (*From Hess and Smith [1964], by permission of the* Journal of Ship Research.)

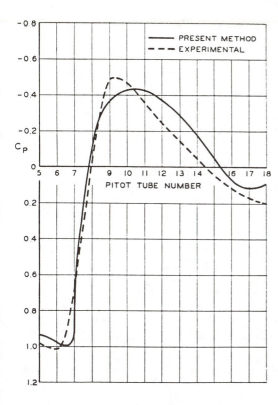

Figure 5.15 Comparison of calculated and experimental pressure distributions along the bottom of a sonar dome. (*Adapted from Hess and Smith [1964], by permission of the* Journal of Ship Research.)

methods. It will be noted, however, that there are broad regions of relatively constant velocity along the hull—a result that supports the usual constant-pressure assumption in estimating the hull frictional resistance, as will be discussed in Part Three.

Use of internally distributed singularities is illustrated in the work of Dalton and Zedan (1984). In this case the so-called inverse problem was considered, in which an integral equation was solved numerically to determine a body shape necessary to give a prescribed surface velocity variation. The line of sources, shown in Fig. 5.17, consisted of *n* elements with strengths represented by a polynomial distribution. Figure 5.18 shows characteristically good results for the predicted velocity distribution, in comparison with the measurements of Joubert et al. (1978). In these studies the aim was to determine if modification of the external velocity distribution would lead to significant drag reductions through body shaping. This required estimation of viscous effects based on the predicted inviscid ideal flow pressure distributions.

As a final example, Figs. 5.19 and 5.20 illustrate the recent work of Wood and Miller (1985) in which a distributed surface singularity method (panel method) was used in predicting high-speed compressible flows. Figure 5.19 shows the arrays of singularity panels used in the estimation of forces and moments on an advanced fighter-type configuration at Mach 1.8 (1.8 times the speed of sound).

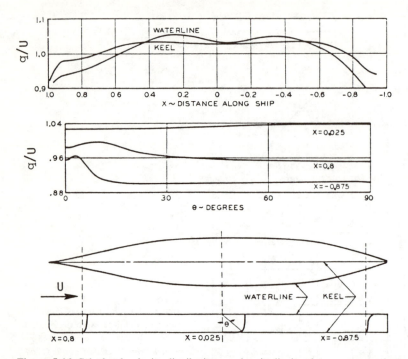

Figure 5.16 Calculated velocity distributions on longitudinal and transverse sections of a standard ship hull. (*Adapted from Hess and Smith [1964], by permission of the* Journal of Ship Research.)

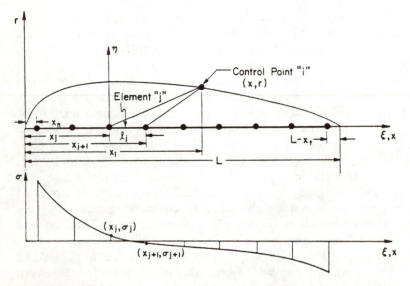

Figure 5.17 Definition sketch of elements and singularity distribution. (*Adapted from Dalton and Zedan [1984], by permission of the* Journal of Hydronautics.)

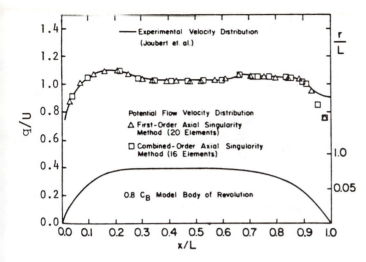

Figure 5.18 Predicted and experimental velocity distribution on a body-of-revolution ship hull. (*Adapted from Dalton and Zedan [1984], by permission of the* Journal of Hydronautics.)

Two methods were used (PAN AIR and SDAS), and comparisons of theory with experiment are shown in Fig. 5.20. (In Fig. 5.20 the symbol δ represents the fuselage incidence angle, C_L is the lift coefficient, and ΔC_D is the incremental change from the minimum value of the drag due to pressure distribution.) The purpose of these studies was to evaluate the effects of fuselage incidence on wing performance and to compare the two computational schemes. The results shown in Fig. 5.20 indicate that both methods are adequate for aerodynamic analysis purposes.

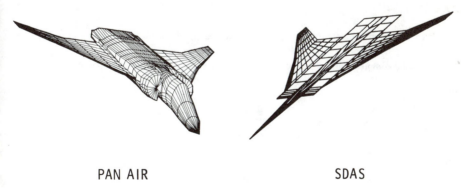

PAN AIR   SDAS

Figure 5.19 Computer graphics drawing of panel models used by Wood and Miller (1985). (*Reproduced with permission of the* Journal of Aircraft.)

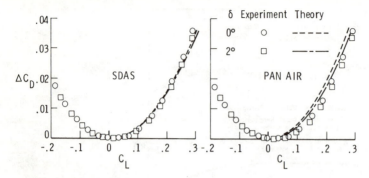

Figure 5.20 Comparison between measured and predicted drag due to lift of the 70/60 deg. crank-wing configuration at Mach 1.8. (*Adapted from Wood and Miller [1985], by permission of the* Journal of Aircraft.)

REFERENCES

Churchill, R. V.: *Complex Variables and Applications*, McGraw-Hill, New York, 1960.

Dalton, C., and M. F. Zedan: Design of Low-Drag Axisymmetric Shapes by the Inverse Method, *J. Hydronautics, 15*, 1–4, pp. 48–54, 1984.

Denny, S. B.: *Applicability of the Douglas Computer Program to Hull Pressure Problems*, David Taylor Model Basin Research Report 1786, October 1963.

Hess, J. L., and A. M. O. Smith: Calculation of Nonlifting Potential Flow About Arbitrary Three-Dimensional Bodies, *J. Ship Research, 8*, 22, pp. 22–44, 1964.

Joubert, P. N., T. J. Sinclair, and P. H. Hoffman: A Further Study of Bodies of Revolution, *J. Ship Research, 22*, 1, pp. 54–63, 1978.

Robertson, J. H.: *Hydrodynamics in Theory and Application*, Prentice-Hall, Englewood Cliffs, N.J., 1965.

Schlichting, H.: *Boundary Layer Theory*, 6th ed., McGraw-Hill, New York, 1968.

Wood, R. M., and D. S. Miller: Impact of Fuselage Incidence on the Supersonic Aerodynamics of Two Fighter Configurations, *J. Aircraft, 22*, 5, pp. 423–428, 1985.

PROBLEMS

5.1 Using DeMoivre's theorem to evaluate z^2, show that $\cos 2\theta = \cos^2 \theta - \sin^2 \theta$ and $\sin 2\theta = 2 \sin \theta \cos \theta$.

5.2 Derive the complex velocity potential for uniform flow past a cylinder. Evaluate $d\Omega/dz$ and find the velocity components. Verify your answer by comparing with the results of Chapter 4.

5.3 Consider the vortex-sink combination located at the origin.

    (a) Write the complex velocity potential.

    (b) Find the magnitude and direction of the velocity vector at any point in the flow.

    (c) Obtain an expression for the difference in pressure between any two points in the flow.

5.4 Write the complex velocity potential for a source located at $z = -a$ combined with a sink of equal strength at $z = a$. Evaluate Ω for large values of a/z. (The result shows that a uniform flow may be constructed from such widely spaced singularities.)

5.5 Develop the complex potential for a flow made up of the following components:

Uniform flow of speed U at an angle α to the x axis

Source of strength μ_1 at $z = z_1$

Source of strength μ_2 at $z = z_2$
Vortex of strength μ_3 at $z = z_3$

Find the complex velocity, $w = d\Omega/dz$. Find the components of the velocity at $z = 2m$ if

$$U = 6 \text{ m/s}, \alpha = 20°$$

$$\mu_1 = 12 \text{ m}^2/\text{s}, z_1 = i \text{ m}$$

$$\mu_2 = -10 \text{ m}^2/\text{s}, z_2 = -i \text{ m}$$

$$\mu_3 = 9 \text{ m}^2/\text{s}, z_3 = 4 \text{ m}$$

5.6 Consider the ideal flow along a surface that is turned 10° into the flow ($\beta = 10°$).
  (a) Write the complex velocity potential for this flow.
  (b) Find the stream function.
  (c) Find the velocity components in both Cartesian and cylindrical coordinates.
  (d) Show that the velocity along the lines $\theta = 0$ and $\theta = 170°$ varies as the 1/17th power of the distance from the point of the turn. How will the pressure vary before and after the turn?

5.7 Find the complex velocity potential for a doublet whose source and sink are on a line through the origin at an angle α with the x axis. That is, the source is at $z_1 = ae^{i\alpha}$ and the sink is at $z_2 = -ae^{i\alpha}$, with a approaching zero in the limit.

5.8 Consider the Rankine half-body. Show that the tangential velocity is zero on $\theta = 0$ and $\theta = \pi$. Is there a stagnation point on either of these lines? If so, verify the location of the stagnation point or points.

5.9 Verify the results found for the stagnation points and body contour of the Kelvin oval.

5.10 A sink is located at $x = a$, near a wall consisting of the y-z plane. Find the complex velocity potential for this flow. Evaluate $d\Omega/dz$ and find the velocity distribution along the wall.

5.11 For problem 5.10, find the velocities at $(0, 0)$ and $(0, \infty)$. Where on the wall is the velocity a maximum?

ONE–DIMENSIONAL
COMPRESSIBLE FLOW

BACKGROUND

As we noted in Table 1.1, the variations possible in the density of a flowing fluid lead to one of the main divisions common in the study of fluid mechanics. In this part we put aside any detailed treatment of other complicating factors such as intricate geometries and/or viscous effects. Limits on geometric variations will be imposed by the one-dimensional constraint and, when viscous effects are taken into account, only cases of constant-area internal flows will be considered. These simplifications will not impede the main goals of Part Two, which are to demonstrate the effects of variable density and to present some of the standard methods for analyzing flows in which density variations play a significant role. We begin with a description of the assumptions that form a foundation for the subsequent developments.

6.1 GROUND RULES

The following constraints apply to our treatment of compressible flows:

1. The important variations in the flow and fluid properties are assumed to occur only in the direction of the flow. This rules out complex flow geometries, such as those with discontinuous boundaries.

In addition, viscous influences can be accommodated in only a crude way, since the effects of variations in velocity across a flow (due to the presence of boundaries) cannot be accurately modeled without adding another dimension to the analysis (see Fig. 6.1). The impact of these assumptions may be minimized by limiting considerations to flows within passages of gradually varying cross sections that are wide relative to the region of viscous flow near the wall. Where

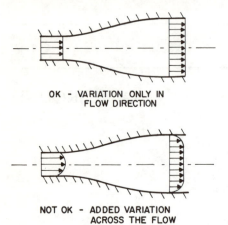

OK - VARIATION ONLY IN
FLOW DIRECTION

NOT OK - ADDED VARIATION
ACROSS THE FLOW

Figure 6.1 Restrictions to one-dimensional flows.

this region is not thin relative to the mainstream flow, consideration is restricted to the variation along the main flow direction of cross-sectional mean values. In the analysis of external compressible flows, the nature of the restrictions is the same, but with only one viscous boundary region to be avoided.

2. The flowing fluid is treated as a perfect gas with constant specific heats. The simplifying effect of this assumption vastly outweighs the importance of the inaccuracies that occur in the analysis of many compressible flows. (If the range of temperature variation is large in the flow, it is a relatively simple matter to account for the variation of specific heats in a piecewise fashion along the flow.)

3. Steady flow and steady thermodynamic state are assumed. Again, the assumption is necessary to constrain the scope of the treatment presented.

4. Shaft work is omitted, and the change in potential energy in the flow direction is assumed to be negligible. Both of these assumptions may be removed without undue complication.

6.2 FUNDAMENTALS AND GOVERNING RELATIONSHIPS

In the treatment of compressible flows, the variation of density requires the introduction of another variable, the temperature. This additional variable leads in turn to the need for an additional governing relationship, which is provided by the flow energy equation. (In ideal fluid flow, energy considerations are implicit in the reversible nature of such flow. The Bernoulli equation, which is, by definition, the integral of the Euler equation, can also be developed by treating the ideal flow as a reversible energy transport process.) In addition to the energy equation, certain other concepts are common to the study of compressible flows, and these will be introduced in this section.

Flow Energy Equation

A familiar concept from thermodynamics is that the rate of change of energy of a system, dE/dt, must be equal to the net rate at which energy crosses the system boundaries: the principle of conservation of energy. Flow of energy across system boundaries may occur in conjunction with fluid flow into and out of the system. In addition, energy transport across system boundaries may be present microscopically, in the form of heat transfer, or macroscopically, as work. [The notion of microscopic and macroscopic forms of energy transport is an extremely useful one. See, for instance, Reynolds and Perkins (1977).] In equation form this energy balance is

$$\frac{dE}{dt} = \dot{Q}_{in} - (\dot{W}_s)_{out} + \dot{m}_{in}\left(h + \frac{v^2}{2} + gz\right)_{in} - \dot{m}_{out}\left(h + \frac{v^2}{2} + gz\right)_{out}$$

where $\dot{Q}$ = heat transfer rate
$\dot{W}_s$ = shaft work rate
$\dot{m}$ = mass flow rate
h = specific enthalpy (internal energy plus flow work)
v = velocity
z = elevation

and the subscripts "in" and "out" refer to flows entering and leaving the system, respectively.

By ground rule 3 above, the mass flow rate into the system must be the same as that leaving, and the rate of energy change of the system, dE/dt, must be zero. Since flow work is included in the enthalpy, ground rule 4 permits the exclusion of $\dot{W}_s$. Finally, neglect of the change in potential energy across the system, relative to the change in kinetic energy, allows removal of the term $(gz)_{out} - (gz)_{in}$.

Exercise Consider a flow that is accelerated from rest to a speed of v. In order to change the potential energy of the flow an equivalent amount, an elevation change of $\Delta z = v^2/2g$ would be required. If the final velocity is $v = 100$ m/s, for instance, the equivalent elevation change is about 500 m. Such an extreme situation is unlikely in most high-speed flows, so the neglect of potential energy effects is almost always a good bet.

With these insights, the flow energy equation may now be written

$$\left(h + \frac{v^2}{2}\right)_{out} - \left(h + \frac{v^2}{2}\right)_{in} = \frac{\dot{Q}_{in}}{\dot{m}} \tag{6.1}$$

With the *stagnation enthalpy* defined as

$$h_0 = h + \frac{v^2}{2} \tag{6.2}$$

the energy equation is reduced to

$$\Delta h_0 = (h_0)_{\text{out}} - (h_0)_{\text{in}} = \frac{\dot{Q}_{\text{in}}}{\dot{m}} \tag{6.3}$$

This is the form of the flow energy valid in all cases for this part of the text (i.e., under the previously discussed ground rules). The stagnation enthalpy of systems under consideration here can only be changed by means of heat transfer! (Shaft work, if present, would play an identical role to that of heat transfer.) Although Eq. (6.3) is extremely compact, it is not all that restrictive, since stagnation enthalpy includes internal energy, flow work (pressure divided by density), and kinetic energy—all of which can vary even though h_0 remains constant. Note further that if the system is *adiabatic,* then stagnation enthalpy is a constant, and

$$\Delta h_0 = 0 \quad \text{(adiabatic)} \tag{6.4}$$

A more complete discussion of the concept of the stagnation state will be presented later.

Mass Flow Effects with Variable Density

In one-dimensional flow the mass flow rate is given by $\dot{m} = \rho A v$, and in steady flow this must be a constant. (The reader is reminded that the area A is the cross-sectional area normal to the flow velocity, v.) The change of flow rate in the direction of the flow is therefore zero, and $d(\rho A v) = 0$. After expanding this triple-product derivative, we may write

$$\frac{d\rho}{\rho} + \frac{dA}{A} + \frac{dv}{v} = 0 \tag{6.5}$$

and, in a form that expresses the velocity variation,

$$\frac{dv}{v} = -\left(\frac{dA}{A} + \frac{d\rho}{\rho}\right) \tag{6.6}$$

In *incompressible flow* $d\rho$ is identically zero, and we have a rather fully developed sense that if the area changes one way, the velocity changes the other—increasing areas lead to decreasing velocities and vice versa. From Eq. (6.6), however, it is easy to see that if the change in density is sufficiently large, and in a sense opposite to that of area, the area-velocity relationships can be reversed. Can density changes in a compressible flow cause the velocity to change in the *same* sense as that of area change? The short answer is "yes," and this reversal of conventional incompressible thinking is one of the very interesting aspects of compressible flow. We know that velocity changes are accompanied by pressure changes in the direction of flow. In a compressible flow these pressure changes can also lead to significant changes in density, which lead in turn to some pretty interesting events.

6.3 SPEED OF PROPAGATION OF A SMALL DISTURBANCE

The speed at which a small disturbance is propagated through a medium (a perfect gas, in our case), has proven to be an extremely useful reference quantity for use in the analysis of compressible flows. For this reason we develop the appropriate expressions here. We set up the problem by imagining the disturbance to be a plane wave traveling through a medium that is at rest, as shown in Fig. 6.2.

Note again that the situation is one-dimensional, in accordance with the ground rules. The problem with the picture as depicted in the sketch is that the passage of the wave creates an unsteady flow problem; fluid that is undisturbed at a given time may be disturbed at some later time—the position of the wave varies with time. In order to make the wave stationary with time so that it stays on the page long enough to be analyzed, we have to imagine that the whole world is moving to the right at the wave velocity, w (see Fig. 3). What had been the undisturbed fluid is now moving at velocity w, and the disturbed fluid is moving to the right at velocity $w - v_2$. In passing through the wave (instead of being passed through *by* the wave), the undisturbed fluid experiences a velocity decrease relative to the wave, just as in the unsteady view in Fig. 6.2.

Analysis of the mass and momentum flux through the control volume surrounding the wave leads to the following results:

Mass:

$$-\rho_1 A w + \rho_2 (w - v_2) A = 0$$

$$w - v_2 = \left(\frac{\rho_1}{\rho_2}\right) w$$

Momentum (vector sense to the right is taken as positive):

$$\Sigma \mathbf{F} = \iint \mathbf{q}(\rho \mathbf{q} \cdot \mathbf{n}) \, dA$$

$$\Sigma F_x = \iint v(\rho \mathbf{q} \cdot \mathbf{n}) \, dA$$

$$P_1 A - P_2 A = w(-\rho_1 w)A + (w - v_2)\rho_2(w - v_2)A$$

Note that the summation of forces here accounts only for the pressure force and, just as in ideal fluid flow, the effects of friction are neglected. (As long as the

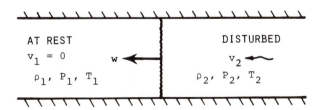

Figure 6.2 Plane wave traveling through a medium at rest.

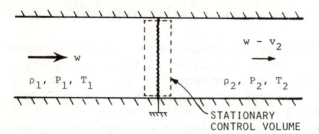

Figure 6.3 Fluid passing through a stationary wave.

disturbance is sufficiently abrupt, the area of the boundary surface over which friction acts will be negligible anyway.) If we now apply the mass continuity result to eliminate the disturbed velocity v_2, we have

$$P_2 - P_1 = \rho_1 w^2 - \rho_2 \left(\frac{\rho_1}{\rho_2}\right)^2 w^2 = w^2 \left(\frac{\rho_1}{\rho_2}\right)(\rho_2 - \rho_1)$$

After rearrangement, the wave speed w is obtained as

$$w^2 = \frac{\rho_2}{\rho_1} \left(\frac{P_2 - P_1}{\rho_2 - \rho_1}\right) \tag{6.7}$$

At this point we may be explicit about what is meant by a "weak" disturbance. To do this we define small perturbation quantities $\delta\rho$ and δP such that

$$\rho_2 = \rho_1 + \delta\rho \qquad \text{and} \qquad P_2 = P_1 + \delta P$$

Upon substitution of these into Eq. (6.7), we have

$$w^2 = \frac{\rho_1 + \delta\rho}{\rho_1} \left(\frac{\delta P}{\delta\rho}\right)$$

This expression is now evaluated in the limit as δP and $\delta\rho$ become vanishingly small. In this limit the ratio $\delta P/\delta\rho$ becomes the derivative of the pressure with respect to the density and, because the passage of the wave is taken to be an adiabatic as well as a frictionless process, the appropriate derivative is the partial derivative at constant entropy. To denote this special value of the speed of disturbance w, we shall use the symbol c—the value obtained in the limiting process. The speed thus deduced is referred to as the "speed of sound"; in other words, sound is a disturbance so weak that it travels at the speed given by

$$c^2 = \left(\frac{\partial P}{\partial \rho}\right)_s \tag{6.8}$$

The relationship between pressure and density, necessary to evaluate Eq. (6.8) further, depends on the particular substance through which the small disturbance travels. The speed c is, in fact, a thermodynamic property of the substance and, as such, may depend on the thermodynamic state. An extremely simple result is obtained for a perfect gas—the substance of interest here. It is known that for a perfect gas with constant specific heats, the pressure and density are related in the following way for an isentropic process:

$$\frac{P}{\rho^k} = \text{constant} \qquad \text{(isentropic)} \qquad (6.9)$$

where $k = c_p/c_v$, the ratio of specific heats of the perfect gas. From this, the result $(\partial P/\partial \rho)_s = kP/\rho = kRT$ is obtained and, from Eq. (6.8),

$$c = \left(\frac{kP}{\rho}\right)^{1/2} = (kRT)^{1/2} \qquad (6.10)$$

Note that, as is the case with enthalpy and internal energy, *the speed of sound in a perfect gas depends only on the temperature* (presuming, of course, that the gas is specified so that R is determined).

For air, the molecular weight is 29 slug/slug mole and the gas constant is $R = 1716$ ft lb/slug R $(= 1716$ ft^2/s^2 R $= 53.34$ ft lb/lb$_m$ R $= 287$ J/kg K). When $k = 1.4$, as for many common gases, such as air, Eq. (6.10) yields $c = 49T^{1/2}$, with c in feet per second and T in degrees Rankine. With standard atmospheric conditions $(T = 518.688$ R at sea level), the speed of sound in air is $c = 1116$ ft/s.

For liquids it is common practice to define the derivative in Eq. (6.8) in terms of the quantity known as the *bulk modulus*, $\beta = \rho(\partial P/\partial \rho)$. The speed of sound in liquids is therefore given by

$$c = \left(\frac{\beta}{\rho}\right)^{1/2}$$

and similar relationships, involving the modulus of elasticity, are available for solids. Pure water has a bulk modulus of about 200,000 psi, giving a speed of sound in water of about 4000 ft/s—one of the main reasons that we do not hear much about supersonic submarines! The term "supersonic" refers to speeds greater than the speed of sound, and is associated with Mach numbers greater than one—the Mach number being the ratio of local speed to the speed of sound:

$$M = \frac{v}{c} = \frac{v}{(kRT)^{1/2}} \qquad (6.11)$$

6.4 STAGNATION STATE

In Eq. (6.2) the stagnation enthalpy was defined as the sum of the static enthalpy and the kinetic energy per unit mass, $v^2/2$. In an adiabatic process the stagnation enthalpy is a constant [Eq. (6.4)], and variations in kinetic energy are accompanied by equal, but opposite, variations in static enthalpy. If a flow is isentropically brought to rest, therefore, the static enthalpy becomes the stagnation enthalpy. In such a situation the flow is said to have achieved its *stagnation state*.

The stagnation state is precisely defined as the state that a flowing substance *would* have *if* the existing state were isentropically transformed to one of zero velocity (i.e., a stagnant condition).

For every flow situation, with the state of the fluid defined by thermodynamic properties P, T, ρ, etc., there is a theoretically achievable *stagnation state* with corresponding thermodynamic properties. These *stagnation properties* are designated by the subscript 0: P_0, T_0, ρ_0, etc. Since attainment of the stagnation state involves an isentropic process, it is a state that can be reached only in a hypothetical (adiabatic and reversible) sense. When a "real" fluid is accelerated from rest, it inevitably experiences some irreversibilities and can never be returned exactly to its initial stagnation state. Nevertheless, the stagnation state is an extremely useful reference condition, as we shall see.

In an isentropic flow the entropy is constant and stagnation conditions are therefore the same everywhere in the flow. If, at a point in a flow, the stagnation conditions are taken to be "reservoir" conditions, this carries with it the implication that the flow is isentropic from the reservoir (where the velocity is zero) to the point in question.

The thermodynamic properties that define the thermodynamic state of a flowing fluid are often referred to as "static" properties—"static pressure," "static temperature," etc. These properties are determined by the molecular activity of the fluid and would be sensed by a measuring device moving with the flow. The term "static" is used, therefore, to denote properties that a fluid possesses in spite of the fact that it is flowing. (Care should be taken to avoid the inference that the use of static properties implies a motionless situation.) The difference between stagnation and static properties is due to the flow, and this *difference* is often (but with some inconsistency) identified by terms such as "dynamic" or "kinetic." Some authors will use the term "total" interchangeably with "stagnation." Thus, one often encounters statements such as "the total pressure is the sum of the static and dynamic pressures."

The utility of the stagnation state may be seen by recognizing that for a perfect gas $h = c_P T$, and $h_0 = c_P T_0$. Dividing Eq. (6.2) by $c_P T$, the ratio of stagnation temperature to static temperature is found to be

$$\frac{T_0}{T} = 1 + \frac{v^2}{2c_P T}$$

But $c_P T = (c_P/kR)kRT = (c_P/kR)c^2$, and $c_P/kR = 1/(k-1)$, so

$$\frac{T_0}{T} = 1 + \frac{k-1}{2}\frac{v^2}{c^2}$$

or

$$\frac{T_0}{T} = 1 + \frac{k-1}{2}M^2 \tag{6.12}$$

This simple result illustrates that the static temperature in an adiabatic flow (in which T_0 is constant) varies in a sense opposite to that of the Mach number. The higher the Mach number, the lower the temperature of an adiabatic flow This is also true in an isentropic flow, of course, since isentropic flows are adiabatic as well as reversible.

6.5 FURTHER IMPLICATIONS OF THE SECOND LAW

We have seen that the stagnation properties do not change in an isentropic flow, and this might lead us to expect that, if the entropy *is* changing, such change would be reflected in a variation in the stagnation properties. This is indeed the case, as may be obtained from the Gibbs equation applied to a perfect gas [see, for instance, Reynolds and Perkins (1977, p. 197)]:

$$T \, ds = du + P \, d\left(\frac{1}{\rho}\right)$$

But $du = c_v \, dT$ and $\rho = P/RT$, so

$$ds = c_p \frac{dT}{T} - R \frac{dP}{P}$$

When this expression is integrated between states 1 and 2 (ground rule 2 is operative):

$$\Delta s = s_2 - s_1 = c_p \ln \frac{T_2}{T_1} - R \ln \frac{P_2}{P_1} \qquad (6.13)$$

In another form, stemming from this,

$$\Delta s = c_v \ln \left[\left(\frac{T_2}{T_1}\right)^k \left(\frac{P_2}{P_1}\right)^{(1-k)} \right] \qquad (6.14)$$

If Eq. (6.14) is now applied to two points in a flow that possess *different* stagnation conditions, then

$$\Delta s = c_v \ln \left[\left(\frac{T_{02}}{T_{01}}\right)^k \left(\frac{P_{02}}{P_{01}}\right)^{(1-k)} \right] \qquad (6.15)$$

This expression reveals that in isentropic flows, in which $\Delta s = 0$, the stagnation conditions are the same from point to point, as we have already postulated. In adiabatic flows, even if there are irreversibilities, the stagnation temperature is constant and

$$\Delta s = c_v \ln \left(\frac{P_{02}}{P_{01}}\right)^{(1-k)} = -R \ln \left(\frac{P_{02}}{P_{01}}\right) \qquad \text{(adiabatic)} \qquad (6.16)$$

The second law of thermodynamics requires that the entropy cannot decrease in an adiabatic process. From Eq. (6.16), therefore, the term on the right-hand side must be either zero (if the process is also reversible) or positive. The irrefutable conclusion is that the stagnation pressure is constant if the flow is isentropic and, if not, the stagnation pressure must decrease. *In adiabatic flow the stagnation pressure cannot increase.*

Entropy Connection to Vorticity

The idealization of compressible flow is related to the constancy of entropy— ideal compressible flows are isentropic. In order to associate this notion with the principles of Part One, it is useful to establish the connecting link between entropy considerations and those associated with what we have termed ideal fluid flow.

We begin by considering two-dimensional flow in streamline coordinates as was done in Chapter 2 (see Fig. 2.5). Two streamlines are illustrated in Fig. 6.4, separated by a small distance δn in the direction the local center of curvature of the streamline at point P. At point P, because it is on the streamline, the normal velocity component v_n is zero. On the nearby streamline, however, there is a small value of normal velocity, δv_n, due to the difference in direction of the two streamlines, $\delta\theta$. This difference in direction so contributes to a growth in the distance between the streamlines when proceeding in the direction s. The incremental change of streamline spacing over a small distance δs (see Fig. 6.4) is $\delta s\,\delta\theta$, so that we may write

$$\left[\frac{\partial}{\partial s}(\delta n)\right]\delta s = \delta s\,\delta\theta = \delta s\left[\frac{\partial\theta}{\partial n}\,\delta n\right]$$

from which

$$\frac{\partial}{\partial s}(\delta n) = \frac{\partial\theta}{\partial n}\,\delta n$$

Since the mass flow rate between the two streamlines must not vary (steady flow is assumed),

$$\frac{\partial}{\partial s}(\rho v\,\delta n) = \rho v\frac{\partial}{\partial s}(\delta n) + \delta n\frac{\partial}{\partial s}(\rho v) = 0$$

with the final result that

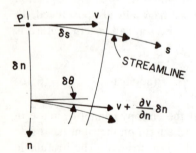

Figure 6.4 Streamline coordinates.

$$\rho v \frac{\partial \theta}{\partial n} + \frac{\partial}{\partial s}(\rho v) = 0 \tag{6.17}$$

Equation (6.17) is the differential form of the continuity equation for steady compressible flow expressed in two-dimensional streamline coordinates.

The expression for vorticity in this coordinate system may be found by evaluating the circulation about the element shown in Fig. 6.4. Thus, with positive circulation taken in the clockwise sense (positive sense being determined by the rotation of s into n in the orientation of Fig. 6.4),

$$\delta\Gamma = v\, \delta s + \delta v_n\, \delta n - (v + \delta v)\, \delta s = \delta v_n\, \delta n - \delta v\, \delta s$$

and, with $\delta A = \delta n\, \delta s$,

$$\zeta = \frac{\delta\Gamma}{\delta A} = \frac{\delta v_n}{\delta s} - \frac{\delta v}{\delta n}$$

In the limit as δA approaches zero,

$$\zeta = \frac{v}{r} - \frac{\partial v}{\partial n} \tag{6.18}$$

where r is the radius of curvature of the streamline at point P.

The connection of vorticity with entropy is found through application of the second law of thermodynamics. For a perfect gas, the Gibbs equation may be written, in terms of the enthalpy, as

$$T\, dS = dh - \frac{dP}{\rho}$$

The variation of entropy with the space variables is expressible, therefore, as

$$T \frac{\partial S}{\partial s} = \frac{\partial h}{\partial s} - \frac{1}{\rho} \frac{\partial P}{\partial s}$$

$$T \frac{\partial S}{\partial n} = \frac{\partial h}{\partial n} - \frac{1}{\rho} \frac{\partial P}{\partial n}$$

(Here we denote specific entropy by capital S because the symbol s has been reserved for the streamline coordinate.) In terms of the stagnation enthalpy, $h_0 = h + v^2/2$,

$$T \frac{\partial S}{\partial s} = \frac{\partial h_0}{\partial s} - \left(v \frac{\partial v}{\partial s} + \frac{1}{\rho} \frac{\partial P}{\partial s} \right)$$

$$T \frac{\partial S}{\partial n} = \frac{\partial h_0}{\partial n} - \left(v \frac{\partial v}{\partial n} + \frac{1}{\rho} \frac{\partial P}{\partial n} \right) \tag{6.19}$$

and from Eq. (6.18),

$$T \frac{\partial S}{\partial n} = \frac{\partial h_0}{\partial n} - \left(\frac{v^2}{r} + \frac{1}{\rho} \frac{\partial P}{\partial n} \right) + v\zeta \tag{6.20}$$

The terms in Eqs. (6.19) and (6.20) involving the stagnation enthalpy express the dependency of entropy change on heat transfer along and across the streamlines. These are zero in adiabatic flow. If we also specify that the flow is frictionless, then, from Eqs. (2.7s) and (2.7n), the parenthetical terms are also zero in Eqs. (6.19) and (6.20). (The hydrostatic pressure variation, if important, may be incorporated within the pressure terms in this analysis.) Equation (6.19) indicates that the entropy is constant along streamlines in adiabatic flow without friction effects. Across streamlines, we see from Eq. (6.20) that, in flow that is isentropic across as well as along streamlines, there can be no vorticity. Isentropic flow is irrotational and may be analyzed accordingly. In particular, the velocity potential exists and may be used as an analytical tool in isentropic flows.

Two important additional relationships may be developed by further consideration of irrotational motion in compressible flow. In such isentropic flows, P/ρ^k = constant and

$$\frac{\partial P}{\partial s} = k \frac{P}{\rho} \frac{\partial \rho}{\partial s} = c^2 \frac{\partial \rho}{\partial s} \tag{6.21}$$

When the streamwise component of the Euler equation, Eq. (2.7s), is considered, we have

$$\rho v \frac{\partial v}{\partial s} = -c^2 \frac{\partial \rho}{\partial s} \qquad \text{or} \qquad \frac{v}{\rho} \frac{\partial \rho}{\partial s} = -M^2 \frac{\partial v}{\partial s}$$

Combining this result with the continuity equation, Eq. (6.17),

$$v \frac{\partial \theta}{\partial n} + (1 - M^2) \frac{\partial v}{\partial s} = 0 \tag{6.22}$$

Consider the implications of Eq. (6.22) for a flow in which the velocity *increases* in the streamline direction; that is, $\partial v / \partial s$ is positive. If the flow is *subsonic*, then $1 - M^2$ is positive and the equality in Eq. (6.22) can hold only if $\partial \theta / \partial n$ is negative—the streamlines must converge, as in incompressible flow. On the other hand, if the flow is *supersonic* and accelerating, the streamlines must diverge—directly opposite the expectation for incompressible flow. If the flow were decelerating along streamlines, the spacing between them would grow in subsonic flow but decrease in supersonic flow. Area-velocity trends are reversed in supersonic flow, an effect that will be treated in more detail in the next chapter.

Equation (6.22) stems from consideration of ideal fluid motion. The expression of zero vorticity may be put in terms of the same dependent variables (θ and v) by substituting $1/r = \partial \theta / \partial s$ for the radius-of-curvature term in Eq. (6.18):

$$v \frac{\partial \theta}{\partial s} - \frac{\partial v}{\partial n} = 0 \tag{6.23}$$

Equations (6.22) and (6.23) constitute two independent relationships with the parameter M for the magnitude and direction of the velocity of two-dimensional ideal compressible flows. They form the basis for the *method of characteristics,* which is a powerful analytical tool for the study of supersonic flows. Several excellent texts are available for further discussion of the method of characteristics and general additional reading. See, for instance, Liepmann and Roshko (1957) and Shapiro (1953).

REFERENCES

Liepmann, H. W., and A. Roshko: *Elements of Gasdynamics,* John Wiley & Sons, New York, 1957.
Reynolds, W. C., and H. C. Perkins: *Engineering Thermodynamics,* McGraw-Hill, New York, 1977.
Shapiro, A. H.: *The Dynamics and Thermodynamics of Compressible Fluid Flow,* Vols. I and II, Ronald Press, New York, 1953.

PROBLEMS

6.1 In an isentropic process, 1 kg of oxygen with a volume of 150 liters at 15°C has its absolute pressure doubled. What is the final temperature?

6.2 A projectile moves through water at 2000 ft/s. Estimate the Mach number of the projectile.

6.3 What will be the Mach number of an airplane traveling at 1500 km/h at sea level, where $P = 101$ kPa and $T = 20°C$? If the plane flies at the same speed at altitude in the stratosphere, where the temperature is $T = -55°C$, what will be its Mach number?

6.4 What is the speed of sound in hydrogen at 80°F?

6.5 Isentropic flow of air occurs at a section of a pipe where $P = 40$ psia, $T = 90°F$, and $v = 537$ ft/s. An object is immersed in the flow that brings the velocity to zero at a point. What are the temperature and pressure at that point?

6.6 What is the Mach number of the flow in problem 6.5?

6.7 Show that the energy equation, as written in Eq. (6.1), may be interpreted in differential form as $v\,dv + d(P/\rho) + du - dQ_{in} = 0$. Under what conditions is this expression the same as the Euler equation?

6.8 Use mass continuity to show that, in a constant-area flow, $dP/\rho = -c^2\,dv/v$.

6.9 Combine the results of problems 6.7 and 6.8 to show that in a constant-area *subsonic* flow the velocity must increase in the direction of the flow. (*Hint:* Incorporate the term $T\,ds - dQ_{in}$, which cannot be negative.)

6.10 Use the Gibbs equation for a perfect gas to derive Eq. (6.9).

6.11 How do the temperature and pressure at the stagnation point in an isentropic flow compare with reservoir conditions?

6.12 Show that Eq. (6.12) may be differentiated to give

$$\frac{dT_0}{T_0} = \frac{(k-1)M^2}{1 + [(k-1)/2]M^2}\frac{dM}{M} + \frac{dT}{T}$$

How are changes in static temperature related to changes in Mach number in adiabatic flow?

6.13 Air flows from a reservoir at 90°C and 7 atm. Assuming isentropic flow, calculate the velocity, temperature, pressure, and density at a section where the Mach number is 0.60.

ISENTROPIC ONE–DIMENSIONAL FLOWS

This chapter examines the dictates of the governing relationships under the constraint that flow is isentropic. The general ground rules listed in Chapter 6 continue to apply and, in addition, the entropy of the flow is constant so that the stagnation conditions are invariant from point to point in the flow. Only internal flows are considered, and changes in the flow conditions along the duct are due solely to changes in cross-sectional area.

7.1 MACH NUMBER RELATIONSHIPS

In isentropic processes involving perfect gases with constant specific heats, the pressure and density changes are simply related to the changes in temperature. For example,

$$\frac{P_0}{P} = \left(\frac{T_0}{T}\right)^{k/(k-1)}$$

and, from Eq. (6.12), we obtain

$$\frac{P_0}{P} = \left(1 + \frac{k-1}{2}M^2\right)^{k/(k-1)} \tag{7.1}$$

For a given Mach number, the static and stagnation pressures at that point are thus simply related. The maximum value of the pressure is the stagnation pressure, and this is realized at a Mach number of zero. The ratio of stagnation to static densities is easily obtained through the equation of state:

$$\frac{\rho_0}{\rho} = \frac{P_0/P}{T_0/T}$$

These Mach number functions are tabulated in Appendix A. As noted, the stagnation values are obtained from these expressions when the Mach number is zero, that is, at the stagnation condition. At all other Mach numbers the ratios are greater than unity, so the stagnation value (of temperature, pressure, etc.) is always equal to or greater than the value of the static property at a point in the flow. If, for instance, we decelerate a flow at Mach number M isentropically to zero velocity, the pressure at that point will increase from P to P_0. Again, such a process is only hypothetical because the necessary reversibility is only approximately attainable.

Relating Conditions at Two Points in Isentropic Flow

If we can approximate a flow as isentropic, then the stagnation values for the fluid are the same at each point in the flow. Denoting two stations in the flow by the subscripts 1 and 2, we may write for the pressure, for instance,

$$\frac{P_2}{P_1} = \frac{P_2}{P_{02}}\frac{P_{01}}{P_1}\frac{P_{02}}{P_{01}} = \frac{P_2}{P_{02}}\frac{P_{01}}{P_1} \quad (1) \qquad \text{(isentropic)} \tag{7.2}$$

Applying Eq. (7.1) successively at points 1 and 2,

$$\frac{P_2}{P_1} = \left\{ \frac{1 + [(k-1)/2]M_1^2}{1 + [(k-1)/2]M_2^2} \right\}^{k/(k-1)} \tag{7.3}$$

Forms such as Eq. (7.3) are seldom used when tabulated values are available since the two ratios necessary in Eq. (7.2) are readily obtained from the tables at the respective Mach numbers. The reader is cautioned always to be alert to the fact that such expressions are valid only if the two points, 1 and 2, are connected by an isentropic process. Manipulations such as the one used in Eq. (7.2) are extremely common in the analysis of compressible flows. Transferring from point to point in a flow is easily accomplished by means of the constant reference values (such as stagnation conditions in isentropic flow) and the Mach number functions.

7.2 DIFFERENTIAL EXPRESSIONS OF CHANGE

Equations such as those presented above may be used to evaluate the differences in flow properties at various points in a isentropic flow. Incremental trends—that is, changes resulting from small departures from a given condition—are equally useful, and these are illustrated by differential relationships governing the flow. The relationships stem from consideration of the differential forms of conservation of momentum, mass, and energy.

From the point of view of momentum, we need only evaluate the form of the Euler equation valid under the present ground rules: steady, one-dimensional, and neglecting hydrostatic effects. Equation (2.7) becomes, with the flow direction designated as x,

$$\frac{Dq}{Dt} = v\frac{dv}{dx} = -\frac{1}{\rho}\frac{dP}{dx} \qquad \text{or} \qquad v\,dv + \frac{dP}{\rho} = 0$$

When the speed of sound, $c^2 = dP/d\rho$, is incorporated,

$$\frac{d\rho}{\rho} = -M^2\frac{dv}{v} \tag{7.4}$$

Mass continuity requires that $\rho Av = $ constant, or $d\rho/\rho + dA/A + dv/v = 0$. Combining this with the momentum result above gives

$$\frac{dv}{v} = \frac{1}{M^2 - 1}\frac{dA}{A} \tag{7.5}$$

Consider Eq. (7.5) in the light that the Mach number M may have values less than or greater than unity. The sign of the coefficient will change accordingly, and we may make the following observation: In *subsonic* flow, velocity varies in a sense that is *opposite* to area variations. In *supersonic flow*, velocity and area variations have the *same* sense. What happens when $M = 1$? We address this important issue later. For now, the sonic ($M = 1$) condition is seen to divide the regions of velocity-area trends.

Energy considerations may be brought to bear by imposing the isentropic requirement in the form of $P/\rho^k = $ constant so that $dP/P = k\,d\rho/\rho$. Combining this with Eq. (7.4), we have

$$\frac{dP}{P} = -kM^2\frac{dv}{v} \tag{7.6}$$

From the equation of state, $dT/T = dP/P - d\rho/\rho$ and

$$\frac{dT}{T} = -(k-1)M^2\frac{dv}{v} \tag{7.7}$$

From Eqs. (7.4), (7.6), and (7.7) we may state that in isentropic flow the density, pressure, and temperature *all* vary in a sense that is opposite to the velocity variation. The velocity, in turn, will vary in one way or the other with area, according to whether the flow is subsonic or supersonic [Eq. (7.5)].

Finally, we consider the variation of Mach number with area. Since $M = v/c = v/(krT)^{1/2}$, we write that $dM/M = dv/v - \frac{1}{2}(dT/T)$ and

$$\frac{dM}{M} = \left[1 + \frac{(k-1)}{2}M^2\right]\frac{dv}{v} \tag{7.8}$$

The Mach number varies in the same sense as does velocity. For instance, in a *subsonic flow with decreasing area*, the velocity increases while the pressure, temperature, and density decrease. The Mach number increases until, eventually, it reaches unity if the area continues to decrease in the direction of flow. Further *decrease* in area, beyond the point at which the Mach number becomes unity, cannot cause a further increase in velocity nor, according to Eq. (7.5), can it cause a decrease in velocity. Thus the Mach number stays at unity, and the only adjustment possible is a decrease in the rate of mass flow through the duct. We discuss this limiting case in some detail later.

If the mass flow rate is to remain the same in the above scenario, the area must begin to increase at the point where the flow reaches a Mach number of unity. Thus, *the sonic point is one of minimum area in isentropic flow,* and dA in Eq. (7.5) must be zero when $M = 1$. Figure 7.1 illustrates the property changes in an isentropic flow for the four possible cases of decreasing or increasing area in subsonic or supersonic flow.

> **Exercise** Equation (7.5) illustrates the sensitivity of velocity to area variations in a compressible flow. In an incompressible flow this is given simply by $dv/v = -dA/A$. The coefficient $1/(1 - M^2)$ is therefore a measure of the effect of compressibility in this regard.
>
> If this coefficient has a value of 1.1, then a 10% deviation from incompressible flow is indicated. This corresponds to a subsonic Mach number of about 0.3—a value that is often used to indicate the relative importance of compressibility effects. In a flow of atmospheric air, for instance, it is often permissible to neglect the effects of compressibility, that is, to assume a constant density, if the flow velocity is less than about 300 ft/s. For water, on the other hand, velocities considerably greater than 1000 ft/s are necessary if compressibility is to be a factor—hence the assumption of constant density in many liquid flows. Note that the fluid as well as the velocity of flow are at issue when making a decision as to whether or not the flow should be treated as compressible.

Flow-Area Relationships in Isentropic Flows

We are well aware that a change in the cross-sectional area of a flow is likely to lead to a change in velocity—an insight that is related to our appreciation of the

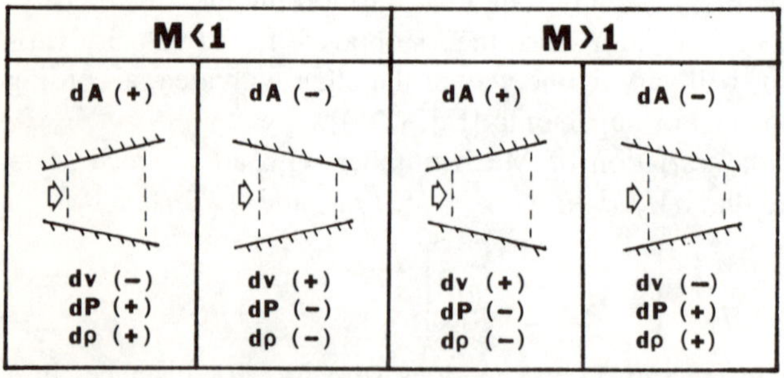

M < 1		M > 1	
dA (+)	**dA (−)**	**dA (+)**	**dA (−)**
dv (−)	dv (+)	dv (+)	dv (−)
dP (+)	dP (−)	dP (−)	dP (+)
dρ (+)	dρ (−)	dρ (−)	dρ (+)

Figure 7.1 Property variations with area in subsonic and supersonic flows.

continuity of mass flow rate: "What goes in, less what comes out, is what stays inside." The fact that the density may vary does not obviate this basic principle, and the approach to compressible flow from this point of view is extremely useful. In one-dimensional flow the mass flow rate per unit area may be written

$$\frac{\dot{m}}{A} = \rho v$$

In our case the flowing medium is a perfect gas, so

$$\rho v = \frac{P}{RT} M(kRT)^{1/2} = \frac{PM}{T^{1/2}} \left(\frac{k}{R}\right)^{1/2}$$

Introducing the stagnation state via the ratios of static to stagnation conditions, $P = P_0(P/P_0)$, etc., the mass flow rate per unit area may be rewritten as

$$\frac{\dot{m}}{A} = \left(\frac{k}{R}\right)^{1/2} \left(\frac{P_0}{T_0^{1/2}}\right) M \left(\frac{P}{P_0}\right) \left(\frac{T_0}{T}\right)^{1/2}$$

Now $(P/P_0) = (T/T_0)^{k/(k-1)}$, and the first two terms on the right-hand side of the expression above are constants in the isentropic flow of a given perfect gas. Expressing this constant as $C_0 = (k/R)^{1/2}(P_0/T_0^{1/2})$, the mass flow rate per unit area is given by

$$\frac{\dot{m}}{A} = C_0 M \left(\frac{T}{T_0}\right)^{(k+1)/2(k-1)} \tag{7.9}$$

The temperature ratio in this expression is only a function of the Mach number in isentropic flow, Eq. (6.12), so *the mass flow rate depends only on the cross-sectional area and the Mach number.*

Consider a flow at Mach number M and imagine that this flow is directed along a duct with changing area such that at some other point the Mach number is unity. We designate this "sonic" point as A^*, and conditions there are also distinguished by the star superscript. Since the two points in the duct (one at Mach number M and the other at $M = 1$) are connected, they must have the same mass flow rate. Applying Eq. (7.9) to the location at which the flow is sonic, we have

$$\frac{\dot{m}}{A} = C_0(1) \left(\frac{T^*}{T_0}\right)^{(k+1)/2(k-1)} \tag{7.10}$$

When the two mass flow rates are equated between Eqs. (7.9) and (7.10), there results a relationship governing the ratio of the two areas—one at sonic conditions, A^*, and the other at the Mach number M.

$$\frac{A}{A^*} = \frac{1}{M} \left[\frac{(T^*/T_0)}{(T/T_0)}\right]^{(k+1)/2(k-1)}$$

From Eq. (6.12) with $M = 1$, $T^*/T_0 = 2/(k + 1)$, and the resulting Mach number function, also tabulated in Appendix A, is

$$\frac{A}{A^*} = \frac{1}{M} \left\{ \frac{2}{k+1} \left[1 + \left(\frac{k-1}{2} \right) M^2 \right] \right\}^{(k+1)/2(k-1)} \tag{7.11}$$

Thus, if the area at Mach number M is known, we may calculate the area change necessary to convert this flow, *isentropically*, to a sonic flow.

If Eq. (7.10) is rewritten with the constant C_0 stated explicitly, an extremely useful relationship is obtained for calculating mass flow rates:

$$\dot{m} = \frac{AP_0}{T_0^{1/2}} \left[\frac{k}{R} \left(\frac{2}{k+1} \right)^{(k+1)/(k-1)} \right]^{1/2} \frac{1}{(A/A^*)} \tag{7.12}$$

Note that in Eq. (7.12), the maximum value of mass flow rate is obtained when A/A^* is a minimum. Inspection of Eq. (7.11), or examination of the tabulated values of A/A^*, reveals that the minimum value is unity and occurs when $M = 1$. *The value of* A/A^* *is greater than one—that is,* A *is greater than* A*—at all Mach numbers—subsonic and supersonic—except M = 1.* We can conclude from these observations that *in isentropic flow through a given area, the maximum possible flow rate obtainable for given stagnation conditions is that which occurs when the flow is sonic at that area.* Under certain circumstances the Mach number at a flow area can be controlled; we return to this principle when discussing such situations.

Since most calculations are done for flows of gases with $k = 1.4$, such as air, it is useful to reduce Eq. (7.12) for such purposes:

$$\dot{m} = 0.685 \frac{AP_0}{(RT_0)^{1/2}} \frac{1}{(A/A^*)}$$

(Note that the area in this expression must have the same units as that used in the pressure, e.g., square feet, pounds per square foot. The dimensions of the gas constant will determine the dimensions of $\dot{m}$.) If the flow is further specified to be that of air, then the gas constant is $R = 1716$ ft lb/slug R and

$$\dot{m} = 0.0165 \frac{AP_0}{T_0^{1/2}} \frac{1}{(A/A^*)}$$

where the constant has the dimension $R^{1/2}$ s/ft and $\dot{m}$ is in lb s/ft (also known as slug/s). Be careful with the dimensions when using these kinds of formulas!

To determine the mass flow rate from Eq. (7.12) and the subsequent expressions, the Mach number must be known at the given area. However, since static conditions are also functions of the Mach number, for given stagnation conditions, any static condition will allow the determination of the Mach number and the subsequent calculation of the mass flow rate. In closing this section, an important insight is drawn from these mass flow rate expressions: In isentropic flow the area A^* is a constant; in adiabatic flow the product $P_0 A^*$ is a constant.

7.3 NOZZLE FLOW AND BACK–PRESSURE EFFECTS

The maximum possible flow rate that can occur through a given cross-sectional area for a given gas in isentropic flow with specified stagnation conditions is that obtained when the flow through the area is sonic—the Mach number at that area is unity. Further appreciation of this fact may be developed through consideration of the events that occur when flow through a duct with varying cross section (called a nozzle) is started from rest. Figure 7.2 has been constructed for the analysis of a particular case in which the cross-sectional area of the nozzle decreases continuously to a minimum value at its exit. The flow in the duct originates from rest at stagnation conditions (as in a pressurized tank) and is assumed to be one-dimensional, isentropic, and otherwise in conformity with the ground rules. The rate of flow through the nozzle is controlled by varying the back pressure, here given the symbol P_b. The reader must imagine that the flow exhausts to a downstream chamber where the pressure can be maintained at a prescribed value of P_b.

The example calculations illustrated in Fig. 7.2 are for the following conditions: $P_0 = 100$ psia, $T_0 = 520$ R, and $A_e = 1$ ft². With air the flowing fluid, the mass flow rate expression previously given leads to

$$\dot{m} = 10.42A*$$

with $\dot{m}$ expressed in slugs per second.

The pressure just barely inside the nozzle exit, P_e, will at first decrease along with P_b. In isentropic flow P_0 remains constant, and the resulting value of P_e/P_0

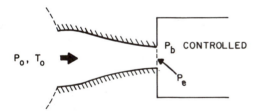

P_b psia	P_e psia	P_e/P_0	M_e	$A_e/A*$	$A*$ ft²	$\dot{m}$ slug/s
100	100	1.00	0	∞	0	0
90	90	0.90	0.39	1.62	0. 62	6. 42
80	90	0.80	0.57	1.23	0.82	8. 49
70	70	0.70	0.73	1.07	0.93	9. 70
60	60	0.60	0.89	1.01	0.99	10. 30
52.8	52.8	0.528	1.00	1.00	1.00	10.42
4 0	40	0.528	1.00	1.00	1.00	10.42
⋮	⋮	⋮	⋮	⋮	⋮	⋮

Figure 7.2 Converging nozzle with varying back pressure.

gives the appropriate Mach number at the exit of the nozzle, M_e, and the corresponding value of A_e/A^*. With this ratio, A^* may be calculated for the given value of A_e, and the mass flow rate is obtained from the expression above. The table of values in Fig. 7.2 gives several sets of values for back pressures decreasing from P_0.

At a back pressure of $P_0 = 52.8$ psia (all values in the figure are approximate), the flow at the nozzle exit reaches unity Mach number. Suppose that the back pressure is further reduced below this value, say to 52.7 psia. Will the exit flow become supersonic? The answer is *no*, for if this were true there would be a *sonic* point *upstream* of the exit with a flow that is increasing in Mach number in a decreasing area. This is in violation of our previous deduction that a supersonic flow decelerates (approaches Mach 1) in a converging nozzle. What actually happens is that when P_b reaches P^* (52.8 psia in air with $P_0 = 100$ psia, as in the example), the nozzle exit flow reaches a velocity equal to the speed of sound. Further decreases in back pressure result in a pressure imbalance that, in subsonic flow, would be transmitted upstream at the speed of sound. But with the nozzle at sonic conditions, the flow leaving the nozzle is traveling at the same speed as the speed with which the signal of pressure imbalance is trying to get upstream into the nozzle. The message never gets there—*back pressures below that necessary to cause sonic exit flow have no effect on the flow within the nozzle*. The nozzle exit pressure stays at $P_e = P^*$, and adjustments to the lower pressure in the surroundings are achieved in a series of complex events that occur *outside the nozzle*. This is an important departure of compressible flow from incompressible flow. In the latter case the speed of sound approaches infinity and, unless the exit velocity approaches infinity, exit pressures are equal to back pressures in liquid flows.

Among those properties that are unaffected by a back pressure lower than the sonic value is the mass flow rate. We say that the nozzle is "choked" under such conditions and that the situation is one of "choked flow"—mass flow rate cannot be increased no matter how hard the "pull" of a lower back pressure. It is important to remember that the mass flow rate *can* be increased in a choked flow by increasing stagnation pressure or exit area or by decreasing the stagnation temperature. In the present example these were held constant with only the back pressure varying. The choking condition expresses only the upper limit on mass flow rate as back pressure is decreased. Choking is a very common occurrence in compressible flows, especially those driven by high stagnation pressures. At stagnation pressures higher than those necessary to choke, for a given back pressure, the flow rate will increase in direct proportion to the stagnation pressure, as indicated in Eq. (7.12) with $A/A^* = 1$.

Exercise Consider the previous converging nozzle, under the given conditions, with a back pressure of $P_b = 70$ psia. What will be the pressure at a location upstream where the area is given by $A_1/A_e = 2.0$? Repeat for a choked nozzle.

With $P_0 = 100$ psia, $P^* = 52.8$ psia and, with the given value of P_b, the nozzle exit flow will be subsonic. Therefore, $P_e = P_b = 70$ psia and $P_e/P_0 = 0.70$. From the tables for isentropic flow (without interpolation), $A_e/A^* = 1.074$ and $M_e = 0.73$ (note, subsonic). At the point where $A_1/A_e = 2.0$, $A_1/A^* = (A_1/A_e)(A_e/A^*) = 2.148$, and, from the tables, $M_1 \approx 0.28$ and $P_1/P_0 \approx 0.947$. The static pressure at this point is therefore 94.7 psia.

When the nozzle is choked ($P_b \leq 52.8$ psia), $M_e = 1.0$, $P_e = P^*$, and $A_e = A^*$. At the cross section where $A_1/A_e = 2.0$, then, $A_1/A^* = 2.0$ and, from the tables (interpolating, this time), $M_1 \approx 0.306$ and $P_1/P_0 = 0.93712$, and $P_1 = 93.7$ psia.

In summary, *for the converging nozzle,* when $P_b > P^*$, the exit pressure equals the back pressure ($P_e = P_b$) and the exit flow is *subsonic.* When $P_b \leq P^*$, the exit pressure equals the sonic pressure ($P_e = P^*$), sonic flow occurs at the exit, and the nozzle is choked at its maximum flow rate for the given area and stagnation conditions. We are now in a position to consider the effects of back pressure on the flow in a converging-diverging nozzle (sometimes referred to as a *C-D nozzle* or a *de-Laval nozzle*).

Establishment of Supersonic Internal Flows

Figure 7.3 is similar to Fig. 7.2 in that an evaluation is made of the effects of back pressure on flow in a duct. In this case, however, the duct consists of a converging section, previously considered, followed by a *diverging* section in which the cross-sectional area increases continuously.

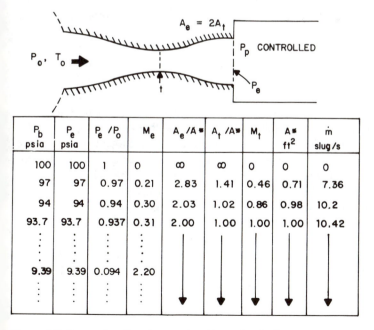

P_b psia	P_e psia	P_e/P_0	M_e	A_e/A^*	A_t/A^*	M_t	A^* ft^2	$\dot{m}$ slug/s
100	100	1	0	∞	∞	0	0	0
97	97	0.97	0.21	2.83	1.41	0.46	0.71	7.36
94	94	0.94	0.30	2.03	1.02	0.86	0.98	10.2
93.7	93.7	0.937	0.31	2.00	1.00	1.00	1.00	10.42
⋮	⋮	⋮	⋮					
9.39	9.39	0.094	2.20					
⋮	⋮	⋮	⋮					

Figure 7.3 Converging-diverging nozzle with varying back pressure.

The maximum flow rate for this duct is also dictated by the stagnation conditions and the minimum cross-sectional area and, if these quantities are the same as in Fig. 7.2, the mass flow rate will be given by the same expression, that is, $\dot{m} = 10.42A^*$. The question remains, as before, concerning the back-pressure conditions necessary to choke this nozzle. That is, at what back pressure, P_b, will the flow at the minimum area (called the "throat" in a C-D nozzle) become sonic?

Flow begins as the back pressure is lowered below the stagnation pressure and, initially, the entire nozzle is subsonic. In such a situation the subsonic flow entering the *diverging* section of the nozzle will decelerate, with an attendant increase in pressure toward the nozzle exit. This behavior is qualitatively the same as if the flow were incompressible. With subsonic flow in the diverging section, the exit pressure is higher than the throat pressure. The pressure at the exit corresponding to sonic pressure at the throat is therefore higher than it would be in the absence of the diverging section. In other words, it is easier to choke the C-D nozzle with reduced back pressure because the diverging section causes the throat pressure to be lower than the exit pressure, *while the nozzle is flowing subsonically*. In Fig. 7.3 for this example we find that when the back pressure is 93.712 psia (interpolating in the tables for a few extra, but dubious, decimal places), the throat pressure is at the sonic value of 52.828 psia. This may be compared with the value of $P_b = 52.8$ necessary to choke the exit without the diverging section, as in Fig. 7.2.

All of this is for isentropic flow, and a number of solutions can be found for back pressures from P_0 down to the value necessary to choke the flow. This latter back pressure we denote by P_c. Figure 7.4 illustrates a few back-pressure conditions where the flow in the C-D nozzle is entirely subsonic (pressures in the range from P_0 to P_c in the figure).

Are there other back-pressure cases of isentropic flow in which the C-D nozzle is choked? The pressure P_c calculated in the example came from letting $A_t = A^*$ (choked throat) so that $A_e/A^* = A_e/A_t = 2.0$. From the *subsonic* branch of the isentropic flow tables it was found that this corresponds to a Mach number at the exit of $M_e = 0.306$. The reader may have observed, in studying the tables, that these give another possibility for $A_e/A^* = 2.0$, and that is when $M_e = 2.1972$. In other words, it is possible, if the back pressure is low enough, we can have isentropic flow throughout the nozzle but with a *supersonic* exit velocity. This case is shown by the line ending at P_j in Fig. 7.4, and it is the main reason why C-D nozzles are of particular interest—they allow expansions to extremely high velocities if the driving pressure ratio is sufficiently large. In the present case the pressure ratio necessary to establish supersonic isentropic flow in the divergent section of the nozzle is $P_j/P_0 = 0.0939$—a back pressure of 9.39 psia. (Incidentally, isentropic flow with supersonic exit conditions could be established with atmospheric back pressure, say 14.7 psia, with a stagnation pressure of about 156 psia—it is the pressure *ratio* that counts.)

The flows obtained with $P_0 > P_b \geq P_c$, and with $P_b = P_j$, are the only ones that satisfy the isentropic constraint presently in effect. In the range of relatively high back pressures, the flow is subsonic throughout the nozzle until the throat

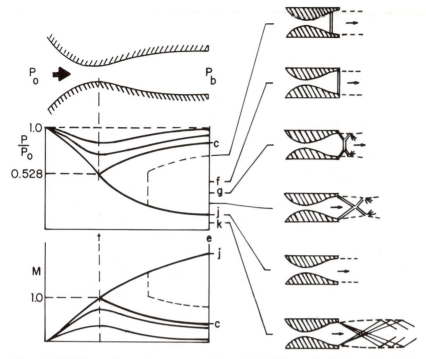

Figure 7.4 Mach number and pressure variations in the C-D nozzle. (*Adapted from Liepmann and Roshko [1957], with permission of John Wiley & Sons.*)

becomes choked. The pressure decreases in the converging section and increases in the diverging section. At a back pressure equal to P_j, the pressure in the nozzle decreases to P^* at the throat and continues to decrease as the flow accelerates in the diverging section of the nozzle. Thus, we have encountered the situation hypothesized earlier: the density variation in the supersonic portion of the nozzle is sufficient to reverse the area-velocity relationship expected in incompressible flow. *In the diverging section the area increases and so does the velocity!*

At back pressures less than P_c but greater than P_j, the throat remains choked (as it will at all back pressures less than or equal to P_c), and there is supersonic flow in at least a portion of the nozzle downstream of the throat. A variety of nonisentropic events occur within the nozzle or at the nozzle exit, leading to exit flows that are subsonic (so that $P_e = P_b$) and then supersonic as P_b is further decreased. These nonisentropic events involve what are known as *normal shock waves,* in which pressures rise according to the back-pressure condition imposed on the nozzle. The isentropic case, without shock waves in the nozzle and with supersonic exit flow, is a rather singular case associated with the specific back pressure P_j. At pressures below P_j, the flow *in the nozzle* is unchanged from fully supersonic isentropic conditions. Adjustments to these low back pressures occur outside of the nozzle in an extremely complex array of shock waves and supersonic expansions. We use the nozzle flow analysis as motivation to enter the next phase of study of compressible flows, concerned with the nature and existence of shock waves.

REFERENCE

Liepmann, H. W., and A. Roshko: *Elements of Gasdynamics,* John Wiley & Sons, New York, 1957.

PROBLEMS

7.1 Oxygen flows from a reservoir, where $P_0 = 100$ psia, $T_0 = 90°F$, through a 6-in.-diameter section where the velocity is 600 ft/s. Calculate the mass rate of flow (isentropic) and the Mach number, pressure, and temperature at the section. Repeat for air.

7.2 A 5-ft^3 tank of compressed air discharges through a converging nozzle of $\frac{1}{2}$-in. diameter. As time goes on the pressure in the tank drops from 150 psia to 50 psia.

(a) If the surroundings are at a pressure of 14.7 psia, how does the Mach number change during the discharge?

(b) If, during the process, the temperature of the air in the tank stays at 70°F, find the time for the tank pressure to reach its final value.

7.3 Air in a reservoir at 250 psia, 290°F, flows through a 2-in.-diameter throat in a C-D nozzle. For sonic conditions at the throat, calculate the pressure, temperature, and density there.

7.4 What must be the velocity, pressure, density, temperature, and diameter at a cross section of the nozzle of problem 7.3 where $M = 2.4$?

7.5 Air is flowing in a supersonic wind tunnel. At a given point the area is 1 ft^2 and the Mach number is 2.5. The reservoir conditions are 2116 lb/ft^2 and 59°F. Find the temperature, pressure, and density at the point where the Mach number is 2.5. Find the area of the throat.

7.6 An altitude control system for a space vehicle uses nitrogen expanding through a nozzle to provide thrust vector control. At maximum thrust nitrogen is used at a rate of 360 lb$_m$ per hour from a reservoir at 500 psia and 0°F. Estimate the nozzle exit velocity and thrust. Calculate the throat area of the nozzle.

7.7 A supersonic nozzle is to be designed for air flow with $M = 3.5$ at the exit section. If the exit is 200 mm in diameter and is to be at a pressure of 7 kPa and a temperature of $-85°C$, calculate the throat area and reservoir conditions. Calculate the mass flow rate at design conditions.

7.8 In problem 7.7, calculate the diameters at locations in the nozzle where the Mach number is 0.5, 1.5, 2.0, and 2.5.

7.9 With reservoir conditions of $P_0 = 180$ psia, $T_0 = 120°F$, air flows through a C-D nozzle with a maximum Mach number of 0.8. The throat diameter is 3 in. Determine the mass rate of flow. Calculate the diameter, pressure, velocity, and temperature at the exit where $M_e = 0.50$.

7.10 Combine Eqs. (6.12) and (7.1) to show that

$$\frac{\rho_0}{\rho} = \left[1 + \left(\frac{k-1}{2} \right) M^2 \right]^{1/(k-1)}$$

7.11 Show that in isentropic flow,

$$\frac{d\rho}{\rho} = - \frac{M \, dM}{1 + [(k-1)/2]M^2}$$

7.12 Show that at small Mach numbers changes in Mach number lead to changes in density given by

$$\frac{d\rho}{\rho} = -\left(\frac{2}{k-1}\right)\frac{dM}{M}$$

What would be the percentage change in density if a Mach number change of 10% occurs in a flow of air at Mach 0.3?

7.13 Show that as the Mach number in a flow becomes very large, changes in density become proportional to the square of the Mach number.

WAVE PROCESSES IN SUPERSONIC FLOWS

We have seen that under certain conditions compressible flow can undergo sudden changes in condition. In the diverging section of a C-D nozzle, for instance, this occurs when the pressure conditions are such that the nozzle cannot sustain an isentropic flow. In order to characterize such an event, we return to the analysis of a sudden disturbance in a one-dimensional flow, as in Chapter 6. In this case, however, we do not call the disturbance small—an assumption that led to the isentropic process expressed in Eq. (6.8). The effect of the disturbance is assumed to take place in an extremely short distance along the flow (instantaneously, in the limit) so that the process may be assumed to occur at constant area, as sketched in Fig. 8.1.

As before, the process is governed by the mass continuity and momentum relationships. Instead of the isentropic assumption, however, the energy relationship will be used in a form that allows for irreversibilities. With conditions upstream and downstream of the disturbance denoted by subscripts x and y, respectively, the equation of mass continuity gives

$$\rho_y v_y = \rho_x v_x$$

Introducing the equation of state for a perfect gas and the Mach number allows this to be written as

$$\frac{P_y}{RT_y} M_y (kRT_y)^{1/2} = \frac{P_x}{RT_x} M_x (kRT_x)^{1/2}$$

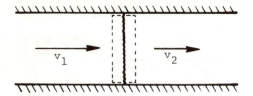

Figure 8.1 Flow through a sudden disturbance.

or

$$\frac{M_y}{M_x} = \frac{P_x}{P_y}\left(\frac{T_y}{T_x}\right)^{1/2}$$ (8.1)

Because the disturbance takes place suddenly, the process is modeled as one in which frictional forces are negligible. If this is true, then the only forces acting on the control volume surrounding the disturbance are those due to pressure differences, and the momentum conservation expression becomes

$$P_y + \rho_y v_y^2 = P_x + \rho_x v_x^2$$

or

$$P_y(1 + kM_y^2) = P_x(1 + kM_x^2)$$ (8.2)

Application of the energy equation entails the assumption that the sudden disturbance occurs adiabatically—justifiable on the basis that there is negligible surface available for heat transfer during the process. (Actually, our working substance, a perfect gas, is inviscid and nonconducting from a rigorous thermochemical point of view. Thus, the above simplifications are preempted by the perfect gas assumption that is already in the ground rules.) For adiabatic processes, Eq. (6.12) applies, so

$$\frac{T_y}{T_x} = \frac{T_{0x}}{T_x}\frac{T_y}{T_{0y}} = \frac{1 + [(k-1)/2]M_x^2}{1 + [(k-1)/2]M_y^2}$$ (8.3)

Equations (8.1)–(8.3) may be combined to eliminate T_y/T_x and P_y/P_x, leaving only the following Mach number relationship:

$$\frac{M_y}{M_x} = \left(\frac{1 + kM_y^2}{1 + kM_x^2}\right)\left\{\frac{1 + [(k-1)/2]M_x^2}{1 + [(k-1)/2]M_y^2}\right\}^{1/2}$$ (8.4)

Before proceeding further, it may be noted that the expression above is satisfied if $M_y = M_x$. This corresponds to a vanishingly small disturbance in which, in effect, nothing happens to the fluid as it passes through the wave—it is vanishingly weak. Here we want to look for another solution for which the disturbance is strong and associated with what is termed a shock wave or, since the wave is oriented at 90° to the oncoming flow, a *normal shock wave*.

8.1 CHANGES ACROSS NORMAL SHOCKS

Equations (8.2) and (8.3) give the pressure and temperature ratios across a normal shock in terms of the two Mach numbers, M_y and M_x. Equation (8.4) expresses one of these Mach numbers in terms of the other, so it only remains to specify one of the four quantities P_y/P_x, T_y/T_x, M_y, or M_x. In practice the most useful choice of independent variable is the Mach number of the flow approaching the shock, M_x. With this information, all other quantities of interest may be calculated. Equation (8.4) may be solved for M_y in terms of M_x. The solution, which requires a fair amount of algebra, is most easily obtained by defining the intermediate variables,

$$A = M_y^2 \quad \text{and} \quad B = M_x^2 \frac{1 + [(k - 1)/2]M_x^2}{(1 + kM_x^2)^2}$$

This leads to a quadratic equation in A, as follows:

$$\left(\frac{k - 1}{2} - k^2 B\right)A^2 + (1 - 2kB)A + B = 0$$

with solution

$$A = M_y^2 = \frac{-(1 + kM_x^4) \pm (1 - M_x^2)(1 + kM_x^2)}{(k - 1) - 2kM_x^2}$$

Selection of the plus sign above leads to the limiting weak wave solution, $M_y = M_x$. For the strong shock, therefore, the negative sign must be chosen, with the result that

$$M_y^2 = \frac{(k - 1)M_x^2 + 2}{2kM_x^2 - (k - 1)} \tag{8.5}$$

Figure 8.2 shows the message conveyed by this expression, and it should be noted that supersonic upstream Mach numbers lead to subsonic downstream Mach numbers. We shall see that normal shocks do not occur in subsonic flows, and the corresponding supersonic values of M_y, given by Eq. (8.5), are not physically realizable.

For a given upstream Mach number M_x, the downstream Mach number M_y is found from Eq. (8.5). The pressure and temperature ratios across the shock may then be found from Eqs. (8.2) and (8.3), and the density ratio may be determined from the equation of state. Because the process is modeled as an adiabatic one, *the stagnation temperature does not change* across the shock. The stagnation pressure does change, however, and may be expressed in terms of the isentropic effects upstream and downstream of the shock. Thus,

$$\frac{P_{0y}}{P_{0x}} = \frac{P_{0y}}{P_y} \frac{P_x}{P_{0x}} \frac{P_y}{P_x}$$

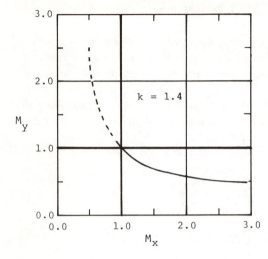

Figure 8.2 Mach numbers before and after a normal shock.

and, with the help of Eqs. (7.1) and (8.2),

$$\frac{P_{0y}}{P_{0x}} = \left\{\frac{1 + [(k-1)/2]M_y^2}{1 + [(k-1)/2]M_x^2}\right\}^{k/(k-1)} \frac{1 + kM_x^2}{1 + kM_y^2} \tag{8.6}$$

From Eq. (6.16) the entropy change across the shock may be calculated from the stagnation pressure ratio given in Eq. (8.6), and, as we have already seen, this ratio must be less than one to satisfy the second law. When $M_x = 1$, Eq. (8.5) gives $M_y = 1$, and it is found that the stagnation pressure as well as the entropy is unchanged in this case. Figure 8.3 shows the entropy variation for other values of M_x, and it is seen from this figure that only supersonic upstream Mach numbers yield increasing entropy in obedience to the second law. Thus we are

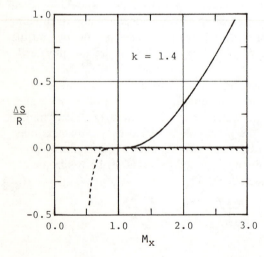

Figure 8.3 Entropy change across a normal shock.

led to the conclusion that *normal shocks occur only in supersonic* flows. Further, we may say that *the stagnation pressure decreases across a normal shock, and the downstream Mach number is always subsonic.*

In this development the shock process has been assumed to be adiabatic and frictionless, but yet there is a change in entropy. The shock wave is an example of an adiabatic and frictionless process that is not isentropic. What happened? In the small print relating the isentropic notion to the frictionless adiabatic process, there is the additional requirement that the process be continuous. The shock wave is a *discontinuous process,* and this is what makes is irreversible, even though it is adiabatic and frictionless.

The Mach number functions mentioned here have been tabulated in Appendix B. The reader should peruse this table and notice the entries and their trends. For instance, there are no entries for subsonic upstream Mach numbers; and the higher the supersonic upstream Mach number, the lower the subsonic downstream Mach number. Static properties increase across the normal shock, whereas stagnation pressure decreases. There is a jump decrease in velocity across a shock wave and, in theory, an infinite rate of deceleration.

Exercise Air flows through a normal shock at $M_x = 2.5$. If the pressure and temperature are 1 atm and 520 R, upstream of the shock, find the Mach number, pressure, temperature, density, and velocity downstream of the shock. Find the change in stagnation pressure and entropy across the shock.

From the normal shock tables, the following values are obtained at $M = 2.5$:

$$M_y = 0.51299 \qquad \frac{P_y}{P_x} = 7.125 \qquad \frac{T_y}{T_x} = 2.1275$$

The downstream pressure and temperature are therefore $P_y = 7.125$ atm and $T_y = 1111.5$ R. The density may be found from the equation of state:

$$\rho_y = \frac{P_y}{RT_y} = \frac{7.125(14.696)(144)}{1716(1111.5)} = 0.0079 \text{ slug/ft}^3$$

The downstream velocity may be found from

$$v_y = M_y(kRT_y)^{1/2} = 838 \text{ ft/s}$$

Also from the normal shock tables we find that, $P_{0y}/P_{0x} = 0.49902$, so

$$\Delta s = -R \ln \frac{P_{0y}}{P_{0x}} = -1716(-0.6951) = 1193 \text{ ft lb/slug R}$$

Note that the upstream stagnation pressure may be determined from the *isentropic* tables for $M_x = 2.5$: $P_x/P_{0x} = 0.05853$, $P_{0x} = 17.08$ atm. Use of the isentropic values is legal here because both P_x and P_{0x} are quantities evaluated at a single point on one side of the shock (upstream)—the entropy change across the shock does not come into play in this evaluation. It would *not* be legal to use the isentropic tables to evaluate a term such as P_x/P_{0y}.

C-D Nozzle with Normal Shocks

As a further illustration of the occurrence of normal shocks, we now calculate more of the back-pressure conditions applicable to the flow in the converging-diverging nozzle. Since the shock can occur only in supersonic flow, it must be in the diverging portion of the nozzle at some back-pressure condition lower than that necessary to choke the nozzle throat. Such a situation is pictured in Fig. 8.4, with the appropriate visualization of the shock process in the temperature-entropy plane. The pressure distribution in a nozzle with an internal normal shock is illustrated in Fig. 7.4 with back pressures in the range from P_c down to P_f.

The pressure P_f, which will cause a normal shock to stand directly at the exit of the nozzle, is particularly easy to calculate. The exit Mach number for isentropic flow just up to the nozzle exit is found from the nozzle area ratio: $M_e = 2.1972$ in the numerical example treated in Chapter 7. If there is a shock at the exit, then this Mach number is M_x for the case under consideration. From the normal shock tables, $M_y = 0.547$, and the pressure ratio across the shock is $P_y/P_x = 5.466$. Since the shock is at the exit, P_y is also the back pressure P_f, and P_x is the exit pressure previously calculated for isentropic flow without the shock. Thus $P_x = P_j = 9.394$ psia, and the back pressure necessary to stand a normal shock just at the nozzle exit is $P_f = 9.394(5.466)$, or $P_f = 51.35$ psia. It can now be stated that in the back-pressure range from P_c to P_f (referring to Fig. 7.4) there will be a normal shock wave standing somewhere in the diverging section of the nozzle.

Calculation for a normal shock in the nozzle. At intermediate pressures the location of the normal shock will move from the nozzle throat (where it occurs at $M_x = 1.0$, the vanishingly weak case) to the nozzle exit as the back pressure

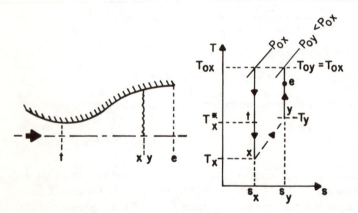

Figure 8.4 Normal shock in a C-D nozzle.

is decreased. Since the flow behind a normal shock is subsonic, the flow in the remaining portion of the diverging section decelerates and increases in pressure to the nozzle exit plane. In this pressure range, therefore, the exit flows are subsonic and the exit pressure meets the existing back pressure. Suppose that the back pressure is in the range $P_c > P_b > P_f$ so that the shock is in the nozzle. For a given nozzle area ratio and upstream stagnation pressure, P_c and P_f are calculated as in Chapter 7. From mass continuity applied to both sides of the shock, Eq. (7.12),

$$P_{0x} A_x^* = P_{0y} A_y^*$$

and if this is divided on both sides by the known quantity $P_b A_e$,

$$\frac{P_{0x}}{P_b} \frac{A_x^*}{A_e} = \frac{P_{0y}}{P_b} \frac{A_y^*}{A_e} \tag{8.7}$$

But the left-hand side of Eq. (8.7) is given because the nozzle geometry is known as well as the pressure ratio. The right-hand side of the expression belongs entirely to the isentrope on the downstream side of the shock and is therefore obtainable from the isentropic tables as a function of the exit Mach number (this function is tabulated in Appendix A). Having solved for M_e, the ratio of stagnation pressures across the shock is given by

$$\frac{P_{0y}}{P_{0x}} = \frac{P_{0y}}{P_b} \frac{P_b}{P_{0x}} \tag{8.8}$$

and this number allows determination of M_x.

Example Using the numerical example of Chapter 7, suppose that $P_b = 70$ psia, in the range that requires a shock in the nozzle. From (Eq. 8.7),

$$\frac{P_{0y}}{P_b} \frac{A_y^*}{A_e} = \frac{100}{70} \frac{1}{2} = 0.714 \qquad \left(= \frac{1}{1.4} \right)$$

and from the isentropic tables (nearest entry), $M_e = 0.41$ and $P_{0y}/P_b = 1/0.89071 = 1.1227$.

$$\frac{P_{0y}}{P_{0x}} = 1.1227 \left(\frac{70}{100} \right) = 0.786$$

and from the normal shock tables, $M_x = 1.86$. From the isentropic tables, at this Mach number we find that the shock stands at a location in the nozzle where $A_x/A_x^* = 1.5$ or, in this case, $A_x = 1.5$ ft^2 since $A_x^* = A_t = 1$ ft^2.

The normal shock wave represents the strongest possible discontinuous disturbance to a supersonic flow. In addition to the nozzle illustrations given here, the normal shock occurs when a supersonic flow is required to turn around a blunt body—the flow must slow to a subsonic velocity in order to negotiate the turn, and this is possible only by means of a normal shock in a supersonic flow. In many instances a less drastic adjustment is required, as in supersonic flow past an oblique (not blunt) body. In the case of the C-D nozzle, when the back pressure is reduced below that necessary for the normal exit shock, there is a range of milder required pressure increases. The events that occur in these cases are associated with the so-called oblique shock.

8.2 OBLIQUE SHOCK WAVES

Oblique shocks are a means by which a supersonic flow can decelerate. Although they are not as brutal as normal shocks, they are just as sudden, in theory, so their analysis again involves a discontinuity in the flow. Consider the supersonic flow past a wall that is deflected into the flow, as in Fig. 8.5. Because the flow is supersonic, it has no advance knowledge of the presence of the deflection and can make no adjustment until arriving at the disturbance. (The effect of viscosity, neglected here, is to reduce the abrupt nature of the deflection in the flow near the wall.) In making the turn, the supersonic flow is decelerated through the oblique shock wave and its pressure is thereby increased. For this reason the flow situation depicted in Fig. 8.5 is sometimes called a *compression corner,* in contrast to flow past a wall that is deflected away from the oncoming stream.

If a portion of the oblique shock is depicted so that only the velocity components normal to the wave are shown, as in Fig. 8.6, then we have a picture that is similar to the normal shock problem—discontinuous, constant area, adiabatic. The only departure is that we are dealing with the normal component of the flow. Rather than repeat the normal shock analysis, we need only write

$$M_{2n} = f_1(M_{1n})$$

where the functional dependency is that given in Eq. (8.5). Further, we may write

$$\frac{P_2}{P_1} = f_2(M_{1n})$$

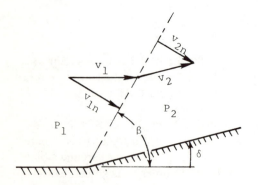

Figure 8.5 Flow past a compression corner.

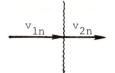

Figure 8.6

and so forth, where previous normal shock relationships are applied to the normal component of the Mach number upstream of the oblique shock. That is, once M_{1n} is known, all other quantities of interest may be obtained from the normal shock tables. Since $M_{1n} = M_1 \sin \beta$, the problem reduces to that of finding the wave angle β.

The geometry of the interaction is further defined in Fig. 8.7. From this we obtain

$$M_{1n} = M_1 \sin \beta \qquad M_{2n} = M_2 \sin(\beta - \delta)$$

$$v_{1n} = v_{1t} \tan \beta \qquad v_{2n} = v_{2t} \tan(\beta - \delta)$$

where the subscript t denotes components tangent to the wave. Although the pressure changes from one side of the wave to the other, the flows on either side are uniform and momentum considerations preclude the existence of accelerations tangent to the wave (the frictionless assumption is still in effect). Therefore $v_{1t} = v_{2t}$. In addition, mass continuity requires that $\rho_1 v_{1n} = \rho_2 v_{2n}$, and we may write

$$\frac{v_{2n}}{v_{1n}} = \frac{\tan(\beta - \delta)}{\tan \beta} = \frac{\rho_1}{\rho_2} \tag{8.9}$$

From the normal shock relationships, after a considerable amount of algebra,

$$\frac{\rho_y}{\rho_x} = \frac{P_y T_x}{P_x T_y} = \frac{(k + 1)M_x^2}{(k - 1)M_x^2 + 2}$$

and for the oblique shock, replacing M_x with $M_1 \sin \beta$,

$$\frac{\rho_2}{\rho_1} = \frac{(k + 1)M_1^2 \sin^2 \beta}{(k - 1)M_1^2 \sin^2 \beta + 2} \tag{8.10}$$

When the density ratio is eliminated between Eqs. (8.9) and (8.10), the geometry

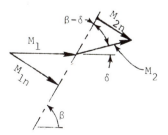

Figure 8.7 Oblique shock geometry.

of the interaction is determined by

$$\frac{\tan(\beta - \delta)}{\tan \beta} = \frac{(k - 1)M_1^2 \sin^2 \beta + 2}{(k + 1)M_1^2 \sin^2 \beta} \tag{8.11}$$

Given upstream Mach number M_1 and the deflection angle δ, Eq. (8.11) specifies the wave angle β. This angle allows the calculation of $M_{1n} = M_1 \sin \beta$, with which the property changes across the shock are found from the normal shock tables. Having M_{2n} from this procedure, the downstream Mach number is obtained from $M_2 = M_{2n}/\sin(\beta - \delta)$.

Although Eq. (8.11) cannot be solved explicitly for the wave angle β, it can be manipulated to give the corresponding deflection angle δ. Thus,

$$\tan \delta = 2 \cot \beta \left[\frac{M_1^2 \sin^2 \beta - 1}{M_1^2(k + \cos 2\beta) + 2} \right] \tag{8.12}$$

This expression is zero for $\beta = \sin^{-1}(1/M_1)$ and for $\beta = \pi/2$. The first of these corresponds to a vanishingly weak (isentropic) disturbance that gives rise to so-called Mach waves in supersonic flows. In the latter case the wave angle $\pi/2$ corresponds to that of a normal shock. The entire spectrum of solutions to Eq. (8.12) is illustrated in Fig. 8.8.

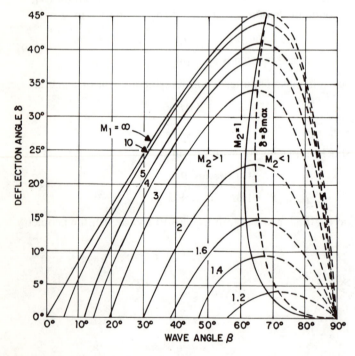

Figure 8.8 Oblique shock solution. (*From Liepmann and Roshko [1957], by permission of John Wiley & Sons.*)

Example Consider flow at Mach 2 past a wedge with a half-angle of 14°, as in Fig. 8.9. Thus, $\delta = 14°$ and M_1 is 2.0. From Fig. 8.8 it is found that $\beta = 44°$ so $M_{1n} = 2 \sin 44° = 1.39$. From the normal shock tables, with this entry, the following values are obtained:

$$M_{2n} = 0.744 \qquad \frac{P_2}{P_1} = 2.089 \qquad \frac{T_2}{T_1} = 1.248$$

$$\frac{P_{02}}{P_{01}} = 0.961$$

and from the geometry, $M_2 = M_{2n}/\sin(\beta - \delta) = 1.48$.

Note that even though M_{2n} is always subsonic, M_2 may be supersonic. Let us continue the example to determine the deflection angle that will cause sonic downstream flow—that is, for $M_2 = 1$. Dealing in terms of temperatures, because the stagnation temperature is constant, we find that

$$\frac{T_2}{T_1} = \frac{T_2}{T_{02}} \frac{T_{01}}{T_1} = \frac{0.8333}{0.5556} = 1.5$$

From the normal shock tables for this temperature ratio we find that $M_{1n} = 1.76$, so $\sin \beta = M_{1n}/M_1 = 1.76/2$, or $\beta = 61.6°$. From Fig. 8.8 an approximate value for the necessary deflection is found to be $\delta = 22.7°$. Figure 8.8 shows the maximum turning angle for an upstream Mach number of 2.0 to be about 23°. Thus, near the maximum turning angle, the downstream Mach number becomes subsonic. The boundary for such points for various upstream Mach numbers is given by the $M_2 = 1.0$ line in Fig. 8.8.

Figure 8.8 also shows a duplicity of solutions corresponding to the same deflection angle δ. Solutions for the smaller wave angles are so-called *weak-wave* solutions in that they give rise to smaller jump changes across the wave. The strong oblique shock solutions, with the larger wave angle for the same deflection, are obtained when pressure conditions are such that they are required, as when a

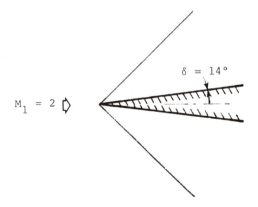

$\delta = 14°$

$M_1 = 2$

Figure 8.9 Supersonic flow past a wedge.

certain range of back pressures are imposed on the C-D nozzle (in the range of P_g in Fig. 7.4). In external flows, as in the example above, deflections beyond the maximum value for a given Mach number lead to detachment of the shock wave from the point of the deflection. Near the axis of the deflection the detached shock wave is very nearly normal—the limiting case of the strong oblique shock. Proceeding away from the origin of the deflection, the wave process spans the range of oblique shock conditions from the normal shock, through the range of strong oblique shocks, through the weak shock range, to the isentropic Mach wave at distances far from the disturbance. Thus, the external shock configuration is curved and detached for deflections larger than the maximum value for the upstream Mach number.

The detached shock configuration is sketched in Fig. 8.10. Until the detached shock reaches a wave angle corresponding to sonic downstream conditions, the flow behind it is subsonic. Thus, the downstream flow, between the detached shock and the body, contains a sonic "throat." The entropy rise across the shock depends on its strength, which varies with the wave angle along a curved shock. Therefore, the flow behind a curved shock cannot be isentropic and there is a considerable amount of frictional shear that acts on the flow downstream of the detached shock.

Figure 8.11 is a series of photographs of supersonic flow past a body mounted in a wind tunnel. The Mach number of the approaching flow is 2.88. The photographs were obtained using a special optical system designed to detect the density discontinuities associated with the various wave processes that occur as the flow adjusts to the presence of the body. [For further discussion and references, see Nunn and Bry (1984).]

There is a cylindrical hole bored in the body and, in Fig. 8.11a, this hole is opened so that flow may pass through the body relatively unobstructed. The pattern shows nearly straight oblique shocks emanating from the sharp leading edges of the body, as well as a series of downstream shocks (including a normal shock wave) in the through-flow emerging from the exit of the body. Figure 8.11b shows

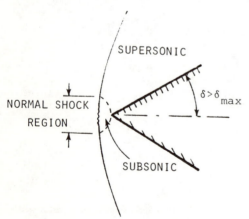

Figure 8.10 Detached "bow" shock configuration.

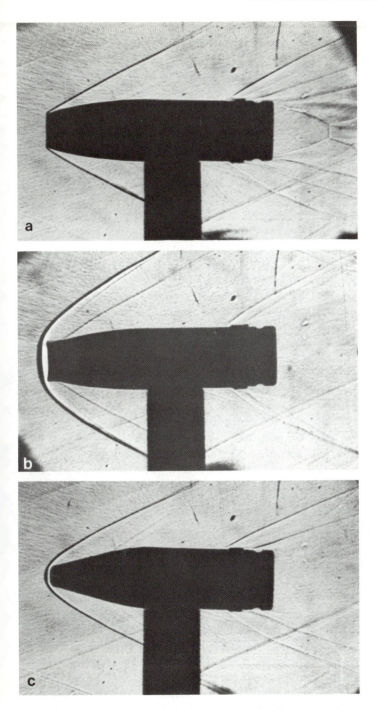

Figure 8.11 Shock wave patterns in flow past a body with varying bluntness. (a) Internal passage open. (b) Internal passage blocked. (c) Modified closed body. (*From Bry [1982].*)

the result of closing the hole in the body. The strong detached shock, together with the curving extension, is clearly illustrated. In Fig. 8.11c the closed body is modified to reduce the frontal area exposed to the normal portion of the detached shock.

8.3 VERY WEAK WAVES (MACH WAVES)

In this chapter we examined so far a range of discontinuous adjustments that occur in a supersonic flow in order to accommodate physical and/or pressure boundary conditions. In terms of wave strength, the range begins with the normal shock and decreases through the spectrum of strong and weak oblique shocks. All of these discontinuities are irreversible, with increasing entropy, and they cause a deceleration and compression of the oncoming flow. At the end of the range it was found that the oblique shock solution reduces to that for a vanishingly weak disturbance [$\delta = 0$ in Eq. (8.12)]. This situation corresponds to an isentropic adjustment and is associated with a wave emanating from the disturbance at an angle of $\beta = \sin^{-1}(1/M_1)$. Such a wave is called a *Mach wave* and is the means by which a supersonic flow may adjust isentropically to one or a series of extremely small disturbances. (For proof that flow through a Mach wave is isentropic, it is sufficient to note that $M_{1n} = M_1 \sin \beta = 1$ and the stagnation pressure is unaffected by a "shock" at this Mach number.)

The occurrence of the Mach wave may be observed in an unsteady framework if we view an infinitesimal point disturbance source traveling through an otherwise stationary fluid. At each point along its trajectory, the source emits a tiny signal that propagates isentropically into the undisturbed medium at the speed of sound, c. If the source moves at a subsonic speed, the distance it moves in time δt is always less than the distance traversed by the signal that was emitted at the beginning of that time interval. This is because $c \, \delta t > v \, \delta t$, and the disturbance signal always precedes its source. The subsonic source "broadcasts" its coming so that the surrounding fluid may make adjustments prior to the arrival of the disturbance.

Figure 8.12 depicts the situation when the source moves supersonically into the surrounding medium. The figure is drawn for $v = 2c$, or a source Mach number of 2, relative to the surroundings. Three time increments are shown in Fig. 8.12, and it is seen that the source precedes its signal as it travels along. In fact, in some regions the fluid remains undisturbed even though the disturbance has gone well past. In the three time increments, the source travels a distance $(3 \, \delta t)v$, while the disturbance emitted at the initial point has traveled only $(3 \, \delta t)c$ into the fluid. From the geometry of the figure it is seen that a line separating the disturbed fluid from that which has not yet gotten the message is oriented at an angle, relative to the path of the disturbance, of $\alpha = \sin^{-1}(3c \, \delta t/3v \, \delta t) = \sin^{-1}(c/v) = \sin^{-1}(1/M_1)$. This line is therefore the Mach wave to which we have alluded. (We chose the symbol α in order to differentiate the Mach wave from the oblique shock wave.) The region of undisturbed fluid is aptly named the *zone of silence*, because sound emitted by the supersonic source does not get to it.

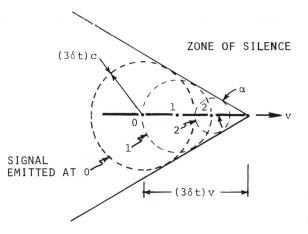

ZONE OF SILENCE

$(3\delta t)c$

α

v

0

SIGNAL
EMITTED AT 0

$\longleftarrow (3\delta t)v \longrightarrow$

Figure 8.12 Disturbance source moving into a stationary fluid.

The steady flow analogy to the moving point disturbance of Fig. 8.12 is that of a flow moving supersonically past a small stationary disturbance. We shall model this as a differential deflection $d\delta$, as illustrated in Fig. 8.13. The geometry is similar to that for the oblique shock analysis except that in this case differential changes are treated. From the geometry of Fig. 8.13 and the constancy of the component of velocity along the wave, v_t, we may write

$$v_t = (v + dv)\cos(\alpha - d\delta) = v \cos \alpha$$

Using the small-angle approximations, $\sin d\delta \approx d\delta$ and $\cos d\delta \approx 1$, this becomes

$$(v + dv)(\cos \alpha + d\delta \sin \alpha) = v \cos \alpha$$

and, retaining only first-order differentials,

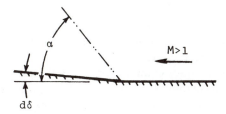

α

$M > 1$

$d\delta$

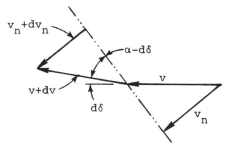

$v_n + dv_n$

$\alpha - d\delta$

v

$v + dv$

$d\delta$

v_n

Figure 8.13 Supersonic flow past a differential deflection.

$$\frac{dv}{d\delta} = -v \tan \alpha$$

But $\alpha = \sin^{-1}(1/M)$, so

$$\frac{dv}{v} = -\frac{d\delta}{(M^2 - 1)^{1/2}} \tag{8.13}$$

Expressing the conservation of momentum normal to the wave,

$$P - (P + dP) = -\rho v_n^2 + (\rho + d\rho)(v_n + dv_n)^2$$

and, from mass continuity,

$$\rho v_n = (\rho + d\rho)(v_n + dv_n)$$

When these two expressions are combined,

$$-dP = -\rho v_n^2 + \rho v_n(v_n + dv_n) = \rho v_n \, dv_n$$

But $v^2 = v_n^2 + v_t^2$ and $dv_t = 0$, so $v_n \, dv_n = v \, dv$ and we obtain

$$dP + \rho v \, dv = 0$$

Thus, we are brought back to the Euler equation for this isentropic (hence, ideal) flow. When ρv is put in terms of Mach number,

$$dP + kPM^2 \frac{dv}{v} = 0$$

and, after incorporating Eq. (8.13), the pressure fraction is given by

$$\frac{dP}{P} = \frac{kM^2}{(M^2 - 1)^{1/2}} \, d\delta \tag{8.14}$$

In Fig. 8.13 we have adopted the convention that $d\delta$ is positive for a flow deflected into itself, and in such a case Eq. (8.15) indicates an increase in pressure. That is, a positive deflection is a compression corner.

Although the results thus far may be used for compression corners, great care must be taken in their application because anything more than a slight compression turn will result in the development of stronger disturbances, that is, shock waves. This is illustrated in Fig. 8.14, which shows successive small deflections that may be thought of as a microscopic view of a finite compression turn. At each deflection a Mach wave is established, and the geometry of the wave pattern is such that at some point from the wall the two waves intersect. The result of this intersection is a coalescence of the weak waves into a stronger one, and for all but the most shallow turns, a shock wave develops.

If the flow is turned *away* from itself, on the other hand, the resulting Mach waves diverge, creating what is known as an *expansion fan*. In such a case the flow remains isentropic and the analysis may proceed under this assumption.

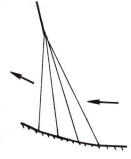

Figure 8.14 Successive compression turns in a supersonic flow.

Prandtl-Meyer Flow

The flow resulting from a series of infinitesimal expansion turns is called *Prandtl-Meyer flow*. Because these flows are isentropic, the results of Chapter 7 apply and, when Eq. (8.14) is combined with Eq. (7.6), we obtain

$$\frac{dv}{v} = -\frac{d\delta}{(M^2 - 1)^{1/2}}$$

Application of Eq. (7.8) allows the differential turn angle to be expressed solely in terms of the Mach number of the oncoming flow:

$$d\delta = -\frac{(M^2 - 1)^{1/2}}{1 + [(k - 1)/2]M^2} \frac{dM}{M} \tag{8.15}$$

Equation (8.15) may be integrated, with the result that

$$-\delta = \left(\frac{k + 1}{k - 1}\right)^{1/2} \tan^{-1} \left[\left(\frac{k - 1}{k + 1}\right)(M^2 - 1)\right]^{1/2}$$

$$- \tan^{-1} (M^2 - 1)^{1/2} + \text{constant} \tag{8.16}$$

The integration of Eq. (8.15) over a finite deflection angle must be approached with care since the process of a series of deflections may violate the isentropic assumptions on which Eq. (8.15) is based. As discussed above, the compression turn may lead to such a violation due to coalescence of the Mach waves, but there is no such problem with the expansion turn. The constant of integration in Eq. (8.16) is the angle of deflection associated with a sonic approach flow. If this angle is set to zero, the resulting expression is the deflection required to change the flow from sonic conditions to one with an arbitrary Mach number M. Since only expansion turns are to be considered here, it is useful to consider such deflections as positive (a reversal of the preliminary sign convention), so the minus sign in Eq. (8.16) may be omitted. The final result is

$$\delta = \left(\frac{k + 1}{k - 1}\right)^{1/2} \tan^{-1} \left[\left(\frac{k - 1}{k + 1}\right)(M^2 - 1)\right]^{1/2} - \tan^{-1} (M^2 - 1)^{1/2} \tag{8.17}$$

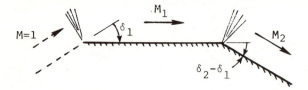

Figure 8.15 Acceleration of a supersonic flow through a Prandtl-Meyer turn.

Equation (8.17) gives the angle through which a sonic flow must be turned to *accelerate* it to the Mach number M. This angle is called the Prandtl-Meyer (P-M) angle, and the Mach number function on the right-hand side also bears the same moniker. For calculation purposes it is useful to tabulate this function in terms of Mach number; the numerical results are presented in Appendix C.

Every supersonic flow has a Prandtl-Meyer function according to its Mach number. The difference between these functions for two flows is the turn angle necessary to accelerate from the lower Mach number to the higher one. This is illustrated in Fig. 8.15. The oncoming flow at M_1 may be thought of as having been accelerated from sonic conditions to that Mach number through the imaginary turn of magnitude δ_1. The flow at M_2 could have been obtained by an acceleration from sonic speed through the turn of magnitude δ_2. The net turning angle to accelerate from M_1 to M_2 is therefore $\delta_2 - \delta_1$. Because the flow is isentropic during this turning process, the changes in properties obey the isentropic relationships. For instance,

$$\frac{P_2}{P_1} = \frac{(P_2/P_0)}{(P_1/P_0)}$$

where the ratios on the right-hand side are found from the isentropic tables for the respective Mach numbers or, for the pressure ratio, Eq. (7.3).

Examples Consider flow at Mach 2 past an expansion corner of 20°. For the approach flow, the Prandtl-Meyer angle is found to be $\delta_1 = 26.38°$. With a turn of 20°, $\delta_2 - \delta_1 = 20°$, so that $\delta_2 = 46.38°$. The Mach number corresponding to this value of the P-M function is $M_2 = 2.83$. The range of the expansion fan may be found from the Mach wave angles corresponding to the two Mach numbers, $\alpha = \sin^{-1}(1/M)$. In this case $\alpha_1 = 30°$ and $\alpha_2 = 20.69°$. Figure 8.16 applies.

As a further example, consider a thin flat plate at an angle of attack of 12.6° to an oncoming flow at Mach 2 (Fig. 8.17). In order to accommodate

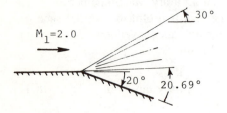

Figure 8.16

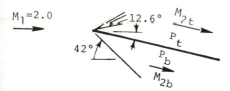

Figure 8.17 Combined compression-expansion turn.

the presence of the plate, the flow must traverse an expansion turn over the top of the plate and a compression turn to flow along the underside. The former adjustment is accomplished through a P-M fan, while the compression turn will give rise to an oblique shock. The subscripts t and b will be used to denote conditions on top of and below the plate. For the compression event we have, using Fig. (8.8) with $M_1 = 2.0$ and $\delta = 12.6°$,

$$\beta = 42° \qquad M_{1n} = 1.34 \qquad \frac{P_b}{P_1} = 1.928$$

$$M_{2n} = 0.7664 \qquad M_{2b} = \frac{0.7664}{\sin(42 - 12.6)} = 1.562$$

For the expansion process over the top of the plate,

$$\delta_1 = 26.38° \qquad \delta_{2t} = 38.98° \qquad M_{2t} = 2.49$$

$$\frac{P_t}{P_1} = \frac{0.05945}{0.1278} = 0.465$$

The ratio of the pressure on the bottom of the plate to that on the top is found to be $P_b/P_t = 1.928/0.465 = 4.15$. In other words, $P_b - P_t = 3.15 P_t$. If the oncoming flow is at standard atmospheric pressure, then $P_t = 6.83$ psia and the pressure difference is about 22 psia. This represents a force of 22 pounds per square inch of plate area in a direction normal to the plate. The force has both lift and drag components, and the drag part is referred to as *wave drag*.

Note that in the combined compression-expansion process the two Mach numbers of the top and bottom flows become widely separated. At the trailing edge of the plate, where the two flows rejoin, there is an extremely complicated interaction due to this mismatch of velocities.

As a final example, consider flow at the exit of a C-D nozzle in which the pressure at the exit is higher than that of the surroundings. This is in the range of pressure conditions denoted by P_k in Fig. 7.4, and is associated with what is termed an underexpanded nozzle flow—the nozzle geometry and pressure ratio are such that flow at the exit has not decreased to a pressure equal to or lower than the back pressure. At the lip of the nozzle exit, therefore, the flow must decrease its pressure to that of the surroundings—an expansion turn is needed. Suppose that the following conditions are given:

$$P_e = 20 \text{ psia} \qquad M_e = 2.0 \qquad P_b = 10 \text{ psia}$$

For the exit Mach number, $P_e/P_0 = 0.1278$, so $P_b/P_0 = (P_b/P_e)(P_e/P_0) = 0.0639$. This corresponds to a Mach number after the expansion of $M_b = 2.44$. The associated P-M angles are $\delta_e = 26.38°$ and $\delta_b = 37.71°$, and the difference between these angles is the angle through which the exit flow must turn, $\theta = 11.33°$. The situation is illustrated in Fig. 8.18.

As a practical matter, the expansion fan at the nozzle exit in the previous example is only the beginning of a complex interaction between the nozzle exit flow and the surrounding fluid. Upon completing the expansion turn, a compression turn is necessary so that the flow may align itself with the nozzle axis. This results in an oblique shock, which interacts with a shock emanating from the opposite nozzle lip. A number of alternating expansions and compressions ensue before the exhaust flow comes to final equilibrium with its surroundings. Of course, the neglect of viscous effects eventually becomes unacceptable in this process.

The underexpanded nozzle is an extremely practical problem in space applications. No matter how large the expansion ratio (A_e/A_t) of its nozzle, the space rocket operates in a region where the surroundings are at a pressure lower than that of the flow at the nozzle exit. The resulting expansions in the flow leaving the nozzle lead to the appearance of the enormous plume that is familiar to anyone who has watched a spacecraft climbing through the atmosphere.

REFERENCES

Bry, W. A.: Aerodynamic Loads on a Ball-Obturated Tubular Projectile, MS thesis, Naval Postgraduate School, Monterey, Calif., 1982.

Liepmann, W. A., and Roshko, A: *Elements of Gasdynamics*, John Wiley & Sons, New York, 1957.

Nunn, R. H., and W. A. Bry: Drag of a Tubular Projectile with Internal Blockage, *J. Spacecraft and Rockets, 21*, 2, p. 216, 1984.

PROBLEMS

8.1 A normal shock wave occurs in a duct carrying air where the upstream Mach number is 2.0 and upstream temperature and pressure are 15°C and 20 kPa. Calculate the Mach number, pressure, temperature, and velocity after the shock.

8.2 Find the entropy change across the shock in problem 8.1.

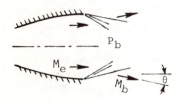

Figure 8.18 Expansion fan at the exit of an underexpanded nozzle.

8.3 Conditions immediately upstream of a normal shock wave in air are P_x = 6 psia, T_x = 100°F, and v_x = 1800 ft/s. Find M_x, M_y, P_y, and T_y.

8.4 If the shock in problem 8.3 occurs at an area of 0.16 ft^2, calculate the rate of entropy rise across the shock in Btu/s R.

8.5 A converging-diverging nozzle is sized so as to operate shock-free with air at an exhaust Mach number of 3. If the ambient pressure is 14.7 psia, find the supply pressures that lead to the following conditions:

    (a) Subsonic flow throughout, except at the throat, which is choked.

    (b) The Mach number at the throat is 0.5.

    (c) A normal shock stands at the nozzle exit.

    (d) Shock-free supersonic flow.

8.6 A normal shock exists in the diverging section of a C-D nozzle. The area of the throat is 1 in.^{2}, the shock is located where the area is 3 in.^{2}, and the area at the nozzle exit is 4 in.^{2}. The working fluid is air, and supply conditions are 500 psia and 3000 R.

    (a) Determine the Mach number, pressure, and temperature immediately downstream of the shock.

    (b) Determine the Mach number, pressure, and temperature at the nozzle exit.

8.7 A C-D nozzle is to provide shock-free supersonic air flow under the following conditions:

Supply (stagnation) pressure, 100 psia
Supply (stagnation) temperature, 60° F
Nozzle back pressure, 14.7 psia
Nozzle throat area, 1.0 in.^{2}

    (a) What is the exit area necessary to meet these conditions? What is the mass flow rate? What is the exit Mach number?

    (b) In starting the flow, as the supply pressure is increased, at what pressure will the nozzle first become choked? What is the mass flow rate at this condition?

    (c) At what supply pressure will a normal shock wave stand at the nozzle exit? What is the mass flow rate at this condition?

8.8 Air flows in a C-D nozzle with area ratio $A_t/A_e = \pi/10$.

    (a) Find the back-pressure ratio, P_b/P_{0x}, for states c, f, and j in Fig. 7.4.

    (b) In which of the cases of part (a) must the nozzle exit pressure match the back pressure, P_b?

    (c) Suppose that P_b/P_{0x} = 0.5. What is the entropy change in the flow through the nozzle?

8.9 A wedge of half-angle δ = 15° is used to measure the Mach number of a supersonic flow. If the observed wave angle is β = 38°, find M_1, P_2/P_1, T_2/T_1, M_2, and the entropy change across the wave.

8.10 In problem 8.9, suppose that the pressure ratio P_2/P_1 is measured rather than β. Describe a solution method to determine M_1 and the remaining unknowns.

8.11 For an approach Mach number of 2.0, it is desired to find the Mach number at an intermediate point in the Prandtl-Meyer fan originating at an expansion corner (see Fig. 8.19). If this location is at τ_x = 10.8°, find the Mach number there. (*Hint:* First show that $\delta_x - \alpha_x = \delta_1 - \tau_x$, where δ_x and α_x are the P-M function and the Mach wave angle corresponding to the unknown Mach number, M_x.)

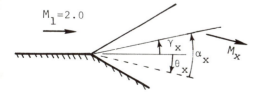

Figure 8.19 Problem 8.11.

8.12 Consider a C-D nozzle with an area ratio of $A_e/A_t = 3.0$. Find the Mach number immediately upstream of a normal shock standing at the exit of this nozzle. Find the back-pressure ratio, P_b/P_{01}, necessary to establish this condition.

8.13 In the nozzle of problem 8.12, suppose that the back-pressure ratio is $P_b/P_{01} = 0.1$. Find the angle of the oblique shock at the nozzle exit and the Mach number downstream of this shock. What is the angle of the flow downstream of the shock. (A trial-and-error solution will be necessary.)

8.14 Repeat problem 8.13 but with a back-pressure ratio of 0.04. (This is in the range that leads to an expansion fan at the exit.)

CONTINUOUS FLOWS WITH ENTROPY CHANGES

We have examined in some detail the effects of area changes on a compressible flow, and in doing so we have invoked the isentropic idealization. When it was impossible to obtain realistic results under the constraints of this idealization, such as in the case of the C-D nozzle with partially supersonic flow, a model was developed for entropy change by means of a discontinuity—the shock wave. In this chapter we look at other common cases in which entropy changes, but, as has been our habit, we remove one complication to make room for another. In this chapter, for the study of compressible flows with entropy change, we restrict our discussion to cases in which *the flow cross-sectional area is constant*. (Recall that in cases involving shock waves, the constant-area approximation was also appropriate.) When computer methods are brought into play, it is a relatively simple matter to relax the constant-area constraint and compute the effects of area change from the principles outlined in Chapter 7.

9.1 GOVERNING DIFFERENTIAL EXPRESSIONS

The overall ground rules, together with the additional constraint of constant area, allow the expression of differential changes in mass flow rate, energy, and entropy, respectively, as follows:

$$\frac{d\rho}{\rho} + \frac{dv}{v} = 0 \tag{9.1}$$

$$dh + v\,dv = dh_0 = dQ \tag{9.2}$$

$$ds = c_v \frac{dT}{T} - R\frac{d\rho}{\rho} = c_v \frac{dh}{h} - R\frac{d\rho}{\rho} \tag{9.3}$$

We will manipulate these relationships so that we are able to pay particular attention to the changes in entropy associated with enthalpy changes. The velocity increment dv is eliminated by combining the first two of the above expressions, and then the terms involving the density may be eliminated. From Eqs. (9.1) and (9.2),

$$dh - v^2 \frac{d\rho}{\rho} = dQ \qquad \text{or} \qquad \frac{d\rho}{\rho} = \frac{dh - dQ}{v^2}$$

But $v^2 = kRTM^2 = (k-1)hM^2$, so that with Eq. (9.3) we have

$$ds = c_v \frac{dh}{h} - R\frac{dh - dQ}{kRTM^2} = c_v \frac{dh}{h} - \frac{c_v}{M^2}\frac{dh - dQ}{h}$$

The final form sought here is

$$\frac{ds}{c_v} = \frac{dh}{h}\left(1 - \frac{1}{M^2}\right) + \frac{1}{M^2}\frac{dQ}{h} \tag{9.4}$$

We have generated a relationship between entropy and enthalpy (which, for a perfect gas, is the temperature times the constant c_p) that includes the effects of heat transfer, dQ. We use this relationship first to examine the adiabatic case.

Constant-Area Adiabatic Flow (Fanno Flow)

In the case of adiabatic flow, $dQ = 0$ and

$$\frac{ds}{c_v} = \frac{dh}{h}\left(\frac{M^2 - 1}{M^2}\right) \tag{9.5}$$

Since $dh/h = dT/T$ and the flow is adiabatic, Eqs. (7.7) and (7.8) may be combined to give

$$\frac{dh}{h} = -\left\{\frac{(k-1)M^2}{1 + [(k-1)/2]M^2}\right\}\frac{dM}{M}$$

and Eq. (9.5) may be written entirely in terms of the Mach number as

$$\frac{ds}{R} = \frac{1 - M^2}{1 + [(k-1)/2]M^2}\frac{dM}{M} \tag{9.5a}$$

Referring again to the second law of thermodynamics, we can deduce from Eq. (9.5) that since entropy cannot decrease in an adiabatic process:

If $M < 1$, then $dh < 0$ and $dM > 0$;
If $M > 1$, then $dh > 0$ and $dM < 0$; and
If $M = 1$, then $ds/dh = 0$.

In adiabatic constant-area flow, irreversibilities drive the flow toward sonic conditions.

These results may be interpreted graphically on the h-s plane as shown in Fig. 9.1. Note that the sonic condition represents a limiting situation. Irreversibilities can cause a subsonic flow to accelerate to, but not beyond, Mach 1. They can also cause a supersonic flow to decelerate toward but not below Mach 1. These types of flows—constant area, adiabatic, with friction—are called *Fanno flows,* and the h-s trajectory is called a *Fanno line.* (The T-s line will, of course, have the same characteristics.) It is important to realize that since these are internal flows, the mass flow is constant along such lines.

Constant-Area Reversible Flow with Heat Transfer (Rayleigh Flow)

Let us now consider the heat transfer term in Eq. (9.4) to be not equal to zero. Instead, we let the entropy term be determined by the second law for *reversible flow: $ds = dQ/T$.* The result, after some rearrangement, is

$$\frac{ds}{c_p} = \frac{M^2 - 1}{kM^2 - 1} \frac{dh}{h} \tag{9.6}$$

Since it is known that the change in entropy must have the same sign as the heat transfer, we may say that for *heat addition to the flow ($dQ > 0$),* for example:

If $M > 1$, then $dh > 0$;

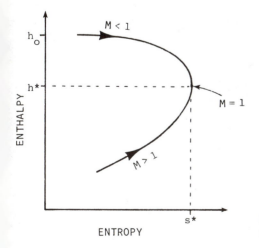

Figure 9.1 Enthalpy-entropy variations in adiabatic constant-area flow—the Fanno line.

If $M < 1/k^{1/2}$, then $dh > 0$; and
If $1/k^{1/2} < M < 1$, then $dh < 0$.

In addition:

If $M = 1/k^{1/2}$, then $dh/ds = 0$; and
If $M = 1$, then $ds/dh = 0$.

Figure 9.2 illustrates a curve with such variations in the *h-s* plane; such a curve is called a *Rayleigh line*. The Mach number range slightly below the limiting sonic value is particularly interesting, since it shows a decreasing enthalpy (and, hence, temperature) with the addition of heat. In this very narrow range, the temperature effect of heating is dominated by that of the very rapid expansion that occurs near Mach 1. It is important to bear in mind that the previous implications were deduced for heat addition. To predict the effect of heat withdrawal from a compressible flow, the signs of dQ and ds [in Eq. (9.6)] must be reversed, with concomitant reversals in all of the trends.

Note further that the sonic condition is again a significant reference point. *In constant-area reversible flow a subsonic flow may be accelerated to but not beyond Mach 1 by heating, and a supersonic flow may be decelerated to but not below Mach 1 by heating.* Heat removal decelerates a subsonic flow and accelerates a supersonic flow, with the sonic condition being the limiting state. In addition, it may be said that heating or cooling a subsonic flow leads to a maximum temperature at $M = 1/k^{1/2}$.

We now proceed to develop the detailed relationships applicable to these flows. The additional case of isothermal constant-area flow can be treated with Eq. (9.4), with the result that $ds/dQ = 1/(kM^2T)$. In this case the flow must involve both irreversibilities and heat transfer, and its treatment leads to a combination of the

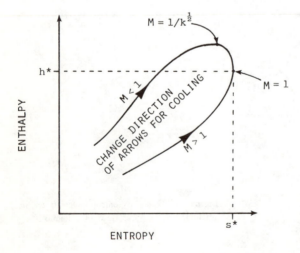

Figure 9.2 Enthalpy-entropy variations in reversible constant-area flow with heat addition—the Rayleigh line.

methods described below. Such treatment may be found, for instance, in John (1969), Rotty (1962), or Benedict (1983).

9.2 ANALYSIS OF FANNO FLOWS

It has been seen that irreversibilities can change the properties of a compressible substance in an adiabatic flow through a constant-area duct. The source of the irreversibilities is assumed to be friction in the following treatment, and the modeling of this effect is achieved through application of the momentum equation. Before this is done, however, the information at hand will be developed further. From Eq. (6.12) we may write, for adiabatic flow (constant stagnation temperature),

$$\frac{dT}{T} = -\frac{d\{1 + [(k - 1)/2]M^2\}}{1 + [(k - 1)/2]M^2} \tag{9.7}$$

This may be integrated from a point in the flow where the Mach number is M to any other point at an arbitrary Mach number M_1 such that

$$\frac{T}{T_1} = \frac{1 + [(k - 1)/2]M_1^2}{1 + [(k - 1)/2]M^2}$$

Since the sonic condition has been found to be a case of particular interest, we assign $M_1 = 1$ with the result that

$$\frac{T}{T^*} = \frac{(k + 1)/2}{1 + [(k - 1)/2]M^2} \tag{9.8}$$

From the equation for mass continuity, Eq. (9.1), we obtain

$$\frac{dP}{P} = \frac{1}{2}\frac{dT}{T} - \frac{dM}{M} = -\frac{1}{2}\frac{d\{1 + [(k - 1)/2]M^2\}}{1 + [(k - 1)/2]M^2} - \frac{dM}{M} \tag{9.9}$$

or, more simply, $PM/T^{1/2} = $ constant so that

$$\frac{P}{P^*} = \frac{1}{M}\left(\frac{T}{T^*}\right)^{1/2} \tag{9.10}$$

with the temperature ratio given by Eq. (9.8). The ratio of stagnation pressures may be found from a combination of Eq. (9.10) and the isentropic relationships at the two states:

$$\frac{P_0}{P_0^*} = \frac{P_0}{P}\frac{P^*}{P_0^*}\frac{P}{P^*}$$

The reader may wish to verify that this boils down to

$$\frac{P_0}{P_0^*} = \frac{1}{M} \frac{2\{1 + [(k-1)/2]M^2\}^{(k+1)/2(k-1)}}{k+1} \tag{9.11}$$

Equations (9.8), (9.10), and (9.11) are tabulated in Appendix D. In the isentropic process the basic causative factor for Mach number change was a change in area. In this case, however, with the area constant, we must evaluate the effect of friction in changing the Mach number from point to point in the flow. We begin with the momentum relationship written for one-dimensional flow with friction.

Accounting for Friction

The effect of friction originates as a shear stress acting at the flow boundaries. Through the momentum transport property of viscosity, this effect is felt in adjacent fluid layers and promotes a general retardation of the fluid motion. Because we are constrained to a one-dimensional treatment here, the velocity and other properties that are to be considered must be taken as average values across the flow cross section. In Part Three the effects of friction are treated in more detail. Figure 9.3 illustrates the flow through a duct of constant cross section A and perimeter p. Accounting for the momentum flux through the control volume of differential length dx gives

$$[P - (P + dP)]A - \tau_w p \, dx = -\rho v^2 A + (\rho + d\rho)(v + dv)^2 A$$

where τ_w is the shear stress acting on the fluid layer adjacent to the wall. From mass continuity,

$$\rho v = (\rho + d\rho)(v + dv)$$

so that

$$-A \, dP - \tau_w p \, dx = -\rho v^2 A + \rho v(v + dv)A = \rho v A \, dv$$

which may be reduced further to

$$dP + \rho v \, dv + 4\tau_w \frac{dx}{D_h} = 0 \tag{9.12}$$

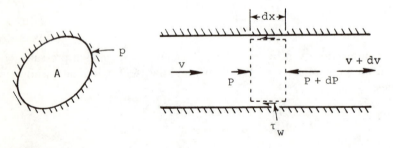

Figure 9.3 Constant-area duct with friction.

where D_h is the hydraulic diameter, defined by $D_h = 4A/p$. The reader should recognize Eq. (9.12) as a modification to the one-dimensional version of the Euler equation, Eq. (2.7), with the addition of a term to account for the friction. This latter term is evaluated further by considering the empirical quantity called the friction factor. Here we use the d'Arcy friction factor f, defined by

$$4\tau_w = f\frac{\rho v^2}{2}$$

Equation (9.12) becomes, after some rearrangement,

$$\frac{f\,dx}{D_h} = -\frac{dP}{\rho v^2/2} - 2\frac{dv}{v}$$

The continuity equation, Eq. (9.1), may be written, for a perfect gas, as

$$\frac{dv}{v} = \frac{dT}{T} - \frac{dP}{P}$$

and, since $\rho v^2 = kPM^2$,

$$\frac{f\,dx}{D_h} = -2\left(\frac{1}{kM^2} - 1\right)\frac{dP}{P} - 2\frac{dT}{T}$$

Using Eqs. (9.7) and (9.9), this may be written as

$$\frac{f\,dx}{D_h} = \frac{1}{kM^2}\frac{dN}{N} + \frac{dN}{N} - 2\frac{dM}{M} + \frac{2}{kM^2}\frac{dM}{M} \tag{9.13}$$

where

$$N = 1 + \left(\frac{k-1}{2}\right)M^2$$

Equation (9.13) is now integrated from a point x in the flow where the Mach number is M to another point downstream where the location is denoted by x_1 and the Mach number is M_1. The result is

$$\frac{f(x_1 - x)}{D_h} = \frac{k+1}{2k}\ln\left(\frac{M^2\{1 + [(k-1)/2]M_1^2\}}{M_1^2\{1 + [(k-1)/2]M^2\}}\right) + \frac{1}{k}\left(\frac{1}{M^2} - \frac{1}{M_1^2}\right) \tag{9.14}$$

But previous expressions have already been expressed with the Mach number of unity used as a reference. In addition, Fig. 9.1 illustrates that this is the Mach number that represents the maximum length of the duct. We therefore write Eq. (9.14) with the downstream state of $M_1 = 1$, and the distance between this point and the upstream point at Mach M denoted by L_{max}. The final working formula is

$$\frac{fL_{max}}{D_h} = \frac{k+1}{2k}\ln\left(\frac{(k+1)M^2}{2\{1 + [(k-1)/2]M^2\}}\right) + \frac{1}{k}\left(\frac{1}{M^2} - 1\right) \tag{9.15}$$

The notion of a maximum length refers to that length of duct that will have a choked sonic exit flow when, *with given upstream entering conditions,* that flow is acted upon by the specified friction effect. In subsonic flow, constant-area ducts longer than L_{max} are not possible without adjusting the upstream entrance conditions. In supersonic flows the possibility of internal shock waves allows another means of adjustment.

Equation (9.15), tabulated in Appendix D, is the necessary relationship between the action of friction and the change of Mach number along a constant-area duct. It must be emphasized that in the integration of Eq. (9.13) the friction factor was assumed to be constant. In general, the friction factor depends on Reynolds number and this, in turn, depends on velocity as well as density and viscosity. The use of these formulas is therefore limited to regions of the flow where the Reynolds number dependency is relatively small—fully developed flow in the completely turbulent regime. Again, the use of the computer greatly alleviates these restrictions.

Illustrative Examples, Subsonic Fanno Flow

Consider a friction duct for which $f = 0.02$, $D_h = 3$ in., and with a length of 100 ft. The entrance conditions, assigned the subscript 1, are given as $P_1 = 100$ psia, $T_1 = 500$ R, and $M_1 = 0.25$—a subsonic entrance. Let us first find out if the length of the pipe is too much for the given entrance conditions and friction factor. From the Fanno tables for $M_1 = 0.25$,

$$\frac{T_1}{T^*} = 1.1852 \qquad \frac{P_1}{P^*} = 4.3456 \qquad \frac{P_{01}}{P_0^*} = 2.4027 \qquad \frac{fL_{max}}{D_h} = 8.4834$$

From the last of these, $L_{max} = 106.0$ ft and, since the actual pipe length is only 100 ft, there is 6 ft of additional pipe length possible until the sonic exit is reached with the given upstream Mach number and friction factor. The situation is shown graphically in Fig. 9.4.

Designating the fictitious length remaining to the sonic point as $L_{2max} = L_{max} - L = 6$ ft, we find $fL_{2max}/D_h = 0.48$. The entry in the Fanno tables corresponding to this value indicates that the Mach number at the exit of the pipe must be

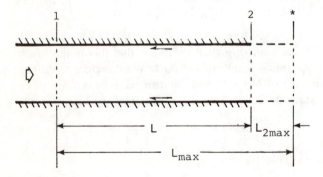

Figure 9.4

$M_2 = 0.608$. Note that most of the range of Mach number change within a pipe choked by friction occurs near the end of the pipe: in this case an extension of 6 ft on the 100-ft pipe would cause the exit Mach number to increase from 0.608 to 1.0.

The reader may verify that for the pipe exit conditions the following ratios are obtained:

$$\frac{T_2}{T^*} = 1.119 \qquad \frac{P_2}{P^*} = 1.755 \qquad \frac{P_{02}}{P_0^*} = 1.185$$

The exit temperature is, therefore,

$$T_2 = \frac{T_2}{T^*}\frac{T^*}{T_1}T_1 = \frac{1.119}{1.185}(500) = 472 \text{ R}$$

By similar methods it can be found that $P_2 = 40.3$ psia, $P_{02} = 51.5$ psia, and $T_{02} = T_{01} = 506$ R. Note that in the subsonic acceleration caused by friction, the pressure, temperature, and stagnation pressure all decrease. Because the flow is adiabatic, the change in entropy may be calculated from the ratio of stagnation pressures, Eq. (6.16), as $\Delta s = 1211$ ft lb/slug R.

The mass flow rate may be calculated in a number of ways. Since state 1 at the entrance is fully specified,

$$\dot{m} = \rho_1 A v_1 = \frac{(k/R)^{1/2}P_1 A M_1}{T_1^{1/2}}$$
$$= \frac{1}{35}\frac{(100)(144)(\pi/4)(1/4)^2(0.25)}{500^{1/2}} = 0.225 \text{ slug/s}$$

This result may be verified by using Eq. (7.12) to calculate the mass flow rate in terms of the stagnation conditions. (For state 1, for instance, it would be necessary to find the upstream stagnation values from the static conditions and Mach number there. The stagnation temperature has been found above, and the stagnation pressure will be found to be $P_{01} = 104.4$ psia.)

Suppose that the entrance conditions in the example are maintained but that a length greater than the maximum, say 150 ft, is specified. What "gives" is dependent on the point of view one takes in posing the problem. Physically, if we were simply to stretch the pipe, the entry conditions at state 1 would have to adjust accordingly, with the exit remaining choked. Remember that, from the h-s diagram, it is not possible to accelerate to supersonic speeds from a subsonic condition via friction. For the suggested case, with a 150-ft choked pipe, $L_{max} = L = 150$ ft, so $fL_{max}/D = 12.0$, and from the Fanno tables $M_1 = 0.217$. With the pipe extended beyond the maximum length for a given Mach number, that Mach number must be decreased accordingly. This is necessary to allow for the greater Mach number increase in the extended length. If the inlet to the pipe were preceded by an isentropic flow, the stagnation conditions would remain the same and the static temperature and pressure would have to increase at the duct entrance according to the decreased value of M_1 there. The reader may verify that for the 150-ft pipe with the same inlet *stagnation* conditions, the following results are obtained: $P_1 = 101.03$ psia, $T_2 = 501.5$ R, and $\dot{m} = 0.197$ slug/s.

In the usual case the duct will be choked because of a supply pressure that is more than adequate under existing friction conditions. This brings up the question of the effects of back pressure on the duct flow; and we will see that this factor imposes an additional control on the flow, in much the same way as in isentropic flows with varying area.

Back-pressure effects. Consider the case illustrated in Fig. 9.5, in which isentropic flow in a converging nozzle feeds into a constant-area friction duct. As the back pressure is reduced, pressure at all points in the nozzle and duct decreases. Along the duct the pressure decreases to the exit plane because of the action of friction in the subsonic flow. The Mach number along the duct increases. Thus the minimum pressure point in this system is at the duct exit, and this is also the point of maximum Mach number. When the exit pressure is sufficiently low to cause the exit flow to reach sonic velocity, further decreases in back pressure can have no effect on the flow within the nozzle or in the duct. In contrast with choking of the converging nozzle, discussed in Chapter 7, the effect of friction moves the choke point to the exit of the duct. At back pressures below P^*, the flow through the system remains the same and adjustments to the surrounding pressure occur in the exhaust flow.

Let us consider again the 100-ft duct described previously. We have found that with the given entrance conditions, the exit Mach number is $M_2 = 0.608$ and

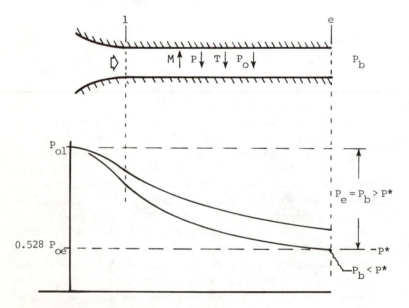

Figure 9.5 Friction duct with subsonic entrance.

the exit pressure is $P_2 = 40.3$ psia. Since this is a subsonic exit condition, it can occur only if the back pressure, P_b, is also 40.3 psia. Any other back pressure above the sonic value would require some other value for the exit Mach number and, therefore, the duct entrance condition would have to be changed accordingly. To find the back pressure necessary to choke the duct, set $P_b = P^* = P_2$ and $L = L_{max}$, so that $fL_{max}/D_h = 8.0$. From the corresponding entry in the Fanno tables, the Mach number at the *entrance* to the duct must be $M_1 = 0.256$ (only a slight increase over the value of 0.25 used in the initial problem). For this value of initial Mach number, $P_1/P^* = 4.25$ and, with $P_1 = 100$ psia (as given), $P_b = P^* = 23.5$ psia. This is to be compared with the higher value that gave the subsonic exit flow. Note that without the friction duct in place, a back pressure of $0.528P_{01}$ would have choked the flow—the additional drop due to friction must be overcome. At lower back pressures, after the duct is choked, conditions in the surroundings can have no influence on the flow within the duct because, as in isentropic choking, these signals cannot be felt upstream of the duct exit. The flow external to the duct must approach these lower pressures through a series of adjustments, beginning with an expansion fan at the duct exit.

Illustrative Examples, Supersonic Fanno Flow

With supersonic flow entering a friction duct, the situation becomes somewhat more complicated because of the possibility of shock processes within the duct. We consider cases for which $L \leq L_{max}$, that is, the duct length is less than or equal to that which will cause the entrance flow to *decelerate* to sonic speed. It is important to realize that if the duct entry flow is supersonic, there must be a sonic throat somewhere upstream of the duct entrance. The mass flow rate is therefore already fixed by this upstream throat and, since the stagnation pressure decreases in the friction duct, a second sonic throat, if it occurs, must be larger than the first in order to pass the same mass flow rate.

Figure 9.6 illustrates the pressure variation in the nozzle-duct system during the start-up process when the back pressure has been lowered a relatively small amount below the upstream stagnation pressure.

The start-up process is similar to the isentropic case, but as the back pressure is reduced and the *nozzle* throat becomes just sonic, the remaining flow in the diverging section of the nozzle and in the duct remains subsonic. In contrast to the friction duct fed by a converging entry nozzle, this system is choked by the throat of the C-D nozzle and not by the friction duct. At some low back pressure the C-D nozzle will be started and flowing supersonically to its exit (which is the entrance to the friction duct). If the pressure drop due to duct friction is also overcome, then the duct will flow supersonically to its exit. If the back pressure is then increased, oblique shocks will first appear at the exit, gradually increasing in strength until the back pressure is such that a normal shock stands at the duct exit. Further increases in back pressure will lead to a normal shock within the duct and, eventually, the normal shock may be made to stand at the exit of the C-D nozzle with supersonic flow eliminated in the friction duct.

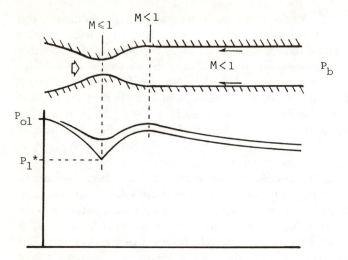

Figure 9.6 Choking of a C-D nozzle feeding a friction duct.

Figure 9.7 illustrates cases for back pressures at or below those necessary to cause a normal shock at the duct exit. Let us use an example to determine the range of back pressures necessary for nozzle choking, a normal shock at the duct entrance, a normal shock at the duct exit, and shock-free supersonic flow. For purposes of illustration, let $P_{01} = 100$ psia, $A_d/A_t = 2.0$, $L/D_h = 10.0$, and $f = 0.02$. (We have retained the same C-D nozzle geometry and stagnation pressure as used in the examples of Chapter 7 in order to avoid repeating the same sorts of calculations performed there.) From the previous work we know that to choke

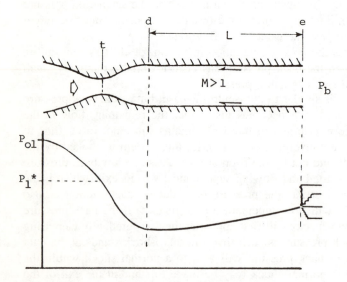

Figure 9.7 Supersonic flow in a friction duct. $L \leq L_{max}$.

this nozzle we must reduce the pressure at the nozzle exit to at least 93.7 psia. At this value of P_d (say, P_{d1}), the nozzle will have an exit Mach number of M_{d1} = 0.306. This is also the entrance Mach number for the duct and, from the Fanno tables, $fL_{\max}/D_h$ = 5.03, or $L_{\max}/L$ = 25.15, and L is indeed less than $L_{\max}$ at this condition. For the exit flow (subscript e), $fL_{e,\max}$ = 5.03 $-$ 0.2 = 4.83, and we see that when the C-D nozzle is just choked, the duct exit Mach number is M_{e1} = 0.31. From the Fanno tables for the two Mach numbers we also find that P_{d1}/P^* = 3.55 and P_{e1}/P^* = 3.49. To find the back pressure for this condition, therefore,

$$P_{b1} = P_{e1} = \left(\frac{P_{e1}}{P^*}\right)\left(\frac{P^*}{P_{d1}}\right)P_{d1}$$

$$= \left(\frac{3.49}{3.55}\right)(9.37) = 92.1 \text{ psia}$$

This pressure is indicated in Fig. 9.8.

Now let us calculate P_{b2}, the back pressure associated with the completely supersonic shock-free flow from the throat of the C-D nozzle to the exit of the friction duct. From the prior calculations we have for this case: M_{d2} = 2.197 and P_{d2} = 9.39 psia. Entering the Fanno tables at this Mach number, we find $fL_{\max}/D_h$ = 0.36 and P_{d2}/P^* = 0.356. Note that under these conditions $L_{\max}/L$ = 1.8.

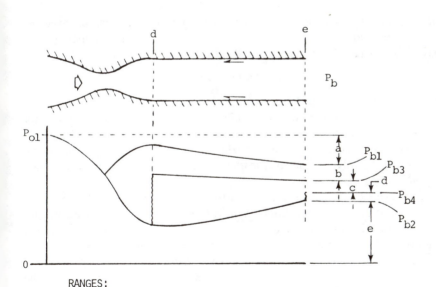

RANGES:

    a. SUBSONIC THROUGHOUT
    b. NORMAL SHOCKS IN C-D NOZZLE
    c. NORMAL SHOCKS IN DUCT
    d. SHOCKS AT DUCT EXIT
    e. EXPANSIONS AT DUCT EXIT

Figure 9.8 Various conditions for the friction duct fed by a C-D nozzle. $L < L_{\max}$.

The exit condition is found from $fL_{e,max}/D_h = 0.36 - 0.2 = 0.16$, and, again from the Fanno tables, $M_{e2} = 1.57$ and $P_{e2}/P^* = 0.573$. With these values we can calculate the corresponding back pressure as

$$P_{b2} = \left(\frac{0.573}{0.356}\right)(9.39) = 15.1 \text{ psia}$$

Note that in this fully supersonic case there is a 5.7 psia pressure *rise* in the friction duct, whereas in the fully subsonic case there was a 1.6 psia pressure *drop* in the duct. (All numbers here are approximate, of course, and where differences are required it would be well to use more significant figures from the tables.)

With a shock at the *entrance* to the duct, the appropriate Mach number upstream of the shock is $M_x = 2.2$ and, as we have already found, the Mach number downstream of the shock is $M_y = 0.547$, and $P_y = P_{d3} = 51.35$ psia. This is the Mach number entering the duct and, from the Fanno tables, $fL_{max}/D_h = 0.746$ and $P_{d3}/P^* = 1.945$. At the duct exit, $fL_{e,max}/D_h = 0.746$ and $M_{e3} = 0.587$ with $P_{e3}/P^* = 1.805$. In a manner similar to that used above for the subsonic duct, the associated back pressure is

$$P_{b3} = \left(\frac{1.805}{1.945}\right)(51.35) = 47.6 \text{ psia}$$

Finally, for a normal shock at the *exit* of the friction duct, we begin at the end of the calculation for P_{b2}, that is, with $M_x = 1.57$. The corresponding jump in pressure across the shock is $P_y/P_x = P_{b4}/P_{b2} = 2.69$, so that

$$P_{b4} = 2.69(15.1) = 40.6 \text{ psia}$$

The four back pressures calculated here are shown in Fig. 9.8. In the range $P_{b4} < P_b < P_{b3}$ there will a normal shock within the friction duct, and solutions for these cases require an interative procedure. In the range $P_{b4} > P_b > P_{b2}$, there will be oblique shocks formed at the nozzle exit. For lower back pressures, the adjustments will involve expansions at the nozzle exit and further downstream interactions.

When the situation is such that $L > L_{max}$ for specified supersonic duct entrance conditions, there can be no reduction in mass flow rate, as was the case in subsonic flows with friction, because the mass flow rate is fixed by the choked throat in the upstream C-D nozzle. For very low back pressures [such as case (b) in Fig. 9.9], the shock takes a position in the duct such that a sonic exit flow occurs at the duct exit and expansion waves are possible in the external flow to reduce the pressure as necessary. For the situation illustrated by curve (a) in Fig. 9.9, the exit plane pressure is equal to the back pressure (the exit flow is just sonic), and as higher back-pressures are imposed, the normal shock moves forward in the duct. Eventually, the shock moves into the diverging section of the nozzle and the duct flow is entirely subsonic. These cases usually require iterative procedures for solution, and they are not treated further here.

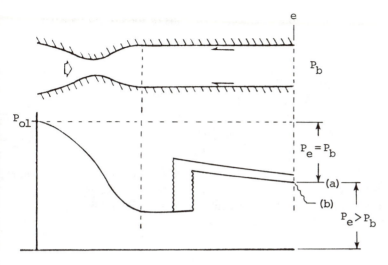

Figure 9.9 Cases for $L > L_{max}$.

9.3 ANALYSIS OF RAYLEIGH FLOWS

In constant-area *reversible* flows with heat transfer, the form of the momentum relationship is simplified by removing the friction term. In fact, it is the same as that used in the normal shock analysis, Eq. (8.2):

$$P(1 + kM^2) = \text{constant}$$

As in Fanno flow, the sonic point is of particular interest, and we may write

$$\frac{P}{P^*} = \frac{k + 1}{1 + kM^2} \tag{9.16}$$

Bringing forward the mass continuity equation for one-dimensional constant-area flow, we may write $PM/T^{1/2} = \text{constant}$, or

$$\frac{T}{T^*} = \left(M\frac{P}{P^*} \right)^2 = \frac{(k + 1)^2 M^2}{(1 + kM^2)^2} \tag{9.17}$$

Stagnation values are easily found from these ratios of static properties, as follows:

$$\frac{T_0}{T_0^*} = \frac{T_0}{T} \frac{T}{T^*} \frac{T^*}{T_0^*} \tag{9.18}$$

$$\frac{P_0}{P_0^*} = \frac{P_0}{P} \frac{P}{P^*} \frac{P^*}{P_0^*} \tag{9.19}$$

In this case the basic change mechanism in the flow is the transfer of heat, and this is related to the Mach number change by means of the energy equation, $dQ = dh_0 = c_p \, dT_0$, or

$$T_0^* - T_0 = \frac{Q_{max}}{c_p} \qquad (9.20)$$

Equation (9.20) may also be written

$$\frac{T_0^*}{T_0} = 1 + \frac{Q_{max}}{c_p T_0} \qquad (9.20a)$$

Where Q_{max} (positive for heating, negative for cooling) represents the maximum amount of heat that can be transferred to or from the flow. In other words, Q_{max} is the heat transfer necessary to cause the flow to become sonic—it is sometimes given the symbol Q^*. The change in entropy of the flowing gas may be found from the formula for perfect gases with constant specific heats, Eq. (6.14). When this is applied to the sonic reference point,

$$\frac{\Delta s^*}{c_v} = \frac{s - s^*}{c_v} = \ln\left[\frac{1}{M^2}\left(\frac{1 + kM^2}{k + 1}\right)\right]^{(k+1)/2} \qquad (9.21)$$

In Rayleigh flow the quantity Q_{max} is analogous in its application to that of L_{max} in Fanno flow: it is a measure of the "distance" that separates a given flow condition from the reference sonic condition. It must be remembered, however, that in Rayleigh flow the process *is* reversible and heat may be extracted from a flow to reverse the effect of the same amount of heat addition. We illustrate the use of these formulas (and the Rayleigh tables in Appendix E) by a few examples.

Illustrative Examples

Consider the addition of heat to a flow of air, such as in the combustion chamber of a jet engine. For given inlet conditions of the air (subscript i), the amount of heat that can be added is limited by the choking point of the Rayleigh line. Suppose that standard sea-level air enters the combustor at a Mach number of 0.25. That is,

$$M_i = 0.25 \qquad P_i = 14.696 \text{ psia} \qquad T_i = 518.69 \text{ R}$$

From the isentropic tables for the inlet Mach number, $P_i/P_{0i} = 0.95745$ and $T_i/T_{0i} = 0.98765$ so that $P_{0i} = 15.349$ psia and $T_{0i} = 525.18$ R. From Appendix E, at the inlet Mach number, $T_{0i}/T_0^* = 0.25684$, and the maximum achievable temperature is found to be $T_0^* = 525.18/0.25684 = 2044.8$ R. The maximum heat that can be added to the flow is given by Eq. (9.20) as $Q_{max} = c_p(T_0^* - T_{0i}) = 0.24(1519.6) = 364.70$ Btu/lb$_m$. If, for example, the fuel has a heating value of 16,000 Btu/lb$_m$ of fuel, then the maximum fuel/air ratio is $f/a = 0.0228$.

This example can be embellished by considering what would happen if, instead of this "maximum" rate of fuel addition, a greater amount is injected into the combustion chamber. Suppose that $f/a = 0.3$, say, some 30% greater than that calculated above. With the given heating value, this corresponds to a heat addition of $Q_{max} = 480$ Btu/lb (assuming that the flow remains choked and that nothing blows up). Retaining the given inlet stagnation conditions, $T_0^* = 525.18 + 480/0.24 = 2525.2$ R, and $T_{0i}/T_0^* = 0.20798$. Designating the revised inlet conditions with ('), this ratio allows us to enter the Rayleigh tables and find that $M_i' = 0.22134$. For this Mach number the isentropic tables give $P_i'/P_{0i} = 0.96644$ and $T_i'/T_{0i} = 0.99029$. From these, $P_i' = 14.834$ psia and $T_i' = 520.07$ R. (In this example we have interpolated between table values in order to retain more significant figures. The reader should realize that the accuracy of these results depends on the validity of the approximations that form their theoretical basis—the number of significant figures may well be illusory.)

With the increase in heat addition, the Mach number at the combustor inlet must decrease, with associated increases in static pressure and temperature there. The concomitant decrease in mass flow rate may be evaluated by noting that $\dot{m} \alpha PM/T^{1/2}$. Thus,

$$\frac{\dot{m}'}{\dot{m}} = \frac{P_i'}{P_i}\left(\frac{T_i}{T_i'}\right)^{1/2}\frac{M_i'}{M_i} = 0.89248$$

The 30% increase in fuel/air ratio leads to an 11% decrease in the air mass flow rate.

As a more or less standard example, consider the addition of heat at the rate of 480 Btu/lb$_m$ to air flowing at $M_1 = 0.2$ with a stagnation temperature of $T_{01} = 1000$ R. In this case $T_{02}/T_{01} = 1 + Q/c_p T_{01} = 3.0$. The Rayleigh tables give $T_{01}/T_0^* = 0.17355$, so $T_{02}/T_0^* = (T_{02}/T_{01})(T_{01}/T_0^*) = 0.52065$. Returning to the Rayleigh tables, we find that $M_2 = 0.395$. With these two Mach numbers and the tables, the following table of ratios is constructed:

	1/*	2/*	2/1
Temperature	0.20661	0.60584	2.9323
Pressure	2.2727	1.9693	0.86650
Stagnation pressure	1.2346	1.1586	0.93844

Let us take a look at the range of heating that reduces the temperature, that is, for $1/k^{1/2} < M_1 < 1$. Consider the flow of air at $M_1 = 0.85$ and $T_1 = 1000$ R. From the Rayleigh tables, $T_1/T^* = 1.02854$ and $T_{01}/T_0^* = 0.98097$. If enough heat is added to cause sonic conditions, then $T_2 = T^* = 1000/1.02854 = 972$ R, a reduction in temperature! The heat transferred is given by $Q_{max} = c_p(T_0^* - T_{01})$ or, in terms of given quantities, $Q_{max} = c_p T_1(T_{01}/T_1)[(T_0^*/T_{01}) - 1]$. This leads to the result that $Q_{max} = 5.33$ Btu/lb$_m$. Addition of heat at this rate chokes the flow and causes the temperature to *decrease* by 28 R or about 3%.

Back-Pressure Considerations

As is always the case, subsonic exhaust flows must reach pressure equilibrium at a duct exit. Supersonic flows, if they do not happen to attain this match, will contain internal or exit shock waves and/or expansion waves at the duct exit. We consider a few examples here, but for more complete treatments the reader is referred to the literature. [A particularly good text for presentations at this level is John (1969).]

Since heat addition to a constant-area subsonic flow increases its Mach number, the Mach number at the duct exit will be the maximum for the flow and the duct exit will choke before any other point, in this case. (But note that for supersonic flow entering the duct this is not necessarily the case.) The pressure variation along the duct will look something like that sketched in Fig. 9.10. With a given heat addition and upstream stagnation temperature, we may calculate the back pressure necessary to choke the duct exit by noting that in such a case the heat added is Q_{max} and $T_0^*/T_{01} = 1 + Q_{max}/(c_p T_{01})$. Calculation of this ratio determines M_1 and the associated ratios tabulated for M_1 in the Rayleigh tables.

For example, consider the addition of heat at the rate of 100 Btu/lb$_m$ to a subsonic flow with stagnation conditions $T_{01} = 500$ R, $P_{01} = 100$ psia. If the duct is to be choked, then the back pressure must be at or below the value of P^* for this flow. From the energy balance, $T_0^*/T_{01} = 1.8333$, and from this we find that the entrance Mach number for the heating duct must be $M_1 = 0.41$, and $P_1/P^* = 1.9428$. For the sonic pressure $P^* = P_{01}(P_1/P_{01})(P^*/P_1) = 100(0.89071/1.94279) = 45.85$ psia. At any back pressure higher than this, the duct exit will be subsonic and the value of M_1 must be recalculated (for the specified values of heat addition and entrance stagnation conditions). At lower back pressures the duct flow is unaffected, and the pressure adjustment begins with an expansion of the sonic flow at the duct exit.

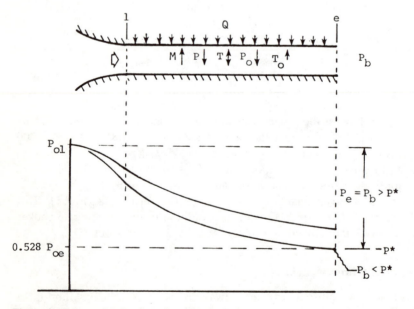

Figure 9.10 Reversible heating of a subsonic flow.

If, for example, the back pressure were 52.8 psia (the value necessary to choke the duct in the absence of heat addition) and the heating rate is kept the same, then $T_{0e}/T_{01} = 1.83333$, as before, but the exit is not choked (that is, T_{0e} is not T_0^*). An iterative solution may be obtained by first guessing M_1, from which P_1/P_{01}, P_1/P^*, and T_{01}/T_0^* are obtained. Then $T_{0e}/T_0^* = (T_{0e}/T_{01})(T_{01}/T_0^*)$, and this ratio determines the exit Mach number, M_e. From M_e, P_e/P^* is obtained and then $P_e/P_{01} = (P_e/P^*)(P^*/P_1)(P_1/P_{01})$ may be calculated. The procedure may be repeated, with new guesses for M_1, until the pressure P_e is suitably close to the given back pressure. For the suggested back pressure (52.8 psia), the solution for M_1 is the same as before ($M_1 = 0.41$), to two significant figures. The reader may wish to verify that for an even higher back pressure—say, 70 psia—the duct entrance Mach number must be about 0.36, which leads to an exit Mach number of 0.62.

An example of reversible heat transfer to a supersonic flow may be given by considering the amount of heat addition necessary to cause a given change in Mach number; for example, if $M_1 = 3.0$, find Q such that $M_2 = 2.0$. From the Rayleigh tables, $T_{01}/T_0^* = 0.65398$ and $T_{02}/T_0^* = 0.79339$. From these, $T_{02}/T_{01} = 1.21317$ and $Q/c_p T_{01} = 0.21317$. With the initial stagnation temperature given, the heat addition may be found. Note that if station 2 is at the end of the duct (subscript e), then there are two possibilities for the exit pressure situation. In the first place, if $P_e < P_b$, there is a range of oblique shocks (the strongest being the normal shock) at $M = 2.0$, by which the pressure may be brought up to that of the surroundings. The second possibility is that $P_e \geq P_b$, in which case there will be expansion waves emanating from the duct exit. Such cases are illustrated in Fig. 9.11, and readers may wish to verify their ability to find the ratios P_f/P_{01} and P_j/P_{01} with data such as those given above (M_1 and M_e).

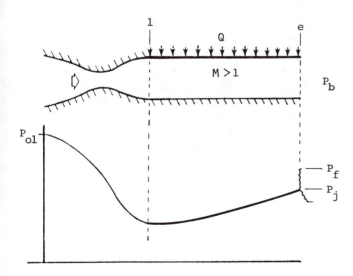

Figure 9.11 Reversible heating of a supersonic flow.

9.4 ISOTHERMAL CONSTANT–AREA FLOWS

For completeness, the third case of continuous flows with entropy change, that of an isothermal flow, is briefly outlined here. Under the constraint of constant temperature, $dv/v = dM/M$, and the continuity equation, Eq. (9.1), may be written

$$\frac{d\rho}{\rho} + \frac{dM}{M} = 0 \tag{9.22}$$

The momentum balance for isothermal flow [see step prior to Eq. (9.13)] gives

$$\frac{f\,dx}{D_h} = -2\left(\frac{1}{kM^2} - 1\right)\frac{dP}{P} \tag{9.23}$$

From the equation of state for a perfect gas at constant temperature, $dP/P = d\rho/\rho$, so that Eq. (9.22) gives

$$\frac{dP}{P} = -\frac{dM}{M} \tag{9.24}$$

and, in combination with Eq. (9.23),

$$\frac{f\,dx}{D_h} = 2\left(\frac{1}{kM^2} - 1\right)\frac{dM}{M} \tag{9.25}$$

Equation (9.25) illustrates that for isothermal flow, a critical value of Mach number occurs at a value of $M = 1/k^{1/2}$. Since the left-hand side of Eq. (9.25) must always be positive, the sign of dM changes at this critical Mach number. When $M < 1/k^{1/2}$, the Mach number will increase in the flow direction, and when $M > 1/k^{1/2}$, the Mach number will decrease. Further insight may be derived from Eq. (9.4) with constant temperature ($dh = 0$): $ds = dQ/kM^2T$. From the second law, for irreversible flow, $ds > dQ/T$. The combination of these two requirements leads to the conclusion that, for isothermal flow, heating is required if $kM^2 < 1$ and cooling is required for the remaining Mach number range, including supersonic flow. Note that the case of $M = 1/k^{1/2}$ is not achievable since this is the case for reversible flow. It is physically impossible to isothermally accelerate a low-speed flow beyond this critical Mach number.

Equation (9.25) may be integrated over a length of duct, $L = x_2 - x_1$, with the result that

$$\frac{fL}{D_h} = \ln\left(\frac{M_1}{M_2}\right)^2 + \frac{1}{kM_1^2} - \frac{1}{kM_2^2} \tag{9.26}$$

Typically, these relationships are used to find the pressure drop in a friction duct where a given gas flows under approximately isothermal conditions. With the duct geometry and friction factor specified, Eq. (9.26) determines the downstream Mach number, M_2, in terms of the upstream Mach number, M_1. Equation (9.24) then gives the pressure change. As has been repeatedly discussed in previous sections, the range of possible flow conditions is constrained by the exit pressure balance. Further discussion and numerical examples may be found in the references cited previously.

REFERENCES

Benedict, R. P.: *Fundamentals of Gas Dynamics,* John Wiley & Sons, New York, 1983.
John, J. E. A.: *Gas Dynamics,* Allyn & Bacon, Boston, 1969.
Rotty, R. M.: *Introduction to Gas Dynamics,* John Wiley & Sons, New York, 1962.

PROBLEMS

9.1 Air enters a friction duct at Mach 0.4 and leaves the duct at Mach 0.6. At what fraction of the duct length will the Mach number be 0.5?

9.2 Air flows adiabatically in a duct with a hydraulic diameter of 110 mm and a friction factor of 0.025. The entry conditions are 200 m/s, 50°C, and 2 atm. What is the maximum length of the duct under these conditions? With the duct exit choked, what are the pressure and temperature there?

9.3 What is the mass rate of flow of air from a reservoir, $T_0 = 15$°C, through 6 m of insulated 25-mm-diameter pipe, $f = 0.02$? The exit pressure is the same as that of the surroundings, 1 atm. Calculate the corresponding reservoir stagnation pressure.

9.4 A constant-area duct is supplied with air at a Mach number of 2.5. The flow is adiabatic, with $f = 0.0296$, and the duct length/diameter ratio is 10. The upstream stagnation pressure is 250 psia, and the stagnation temperature is 500 R. Find the back pressure necessary to establish a normal shock wave at the exit.

9.5 For the duct of problem 9.4, find the Mach number, pressure, temperature, stagnation pressure, and stagnation temperature at a point one-half the distance down the duct.

9.6 Consider a friction duct 50 ft long with an inside diameter of 6 in. The friction factor is $f = 0.02$. The upstream stagnation conditions are 200 psia and 540 R. Find the maximum flow rate achievable in the duct.

9.7 With the duct of problem 9.6 choked, what will be the pressure drop? Find the rate of entropy change in the duct.

9.8 A duct 6 in. in diameter and 30 ft long has a friction factor of $f = 0.03045$. $k = 1.4$. If the surrounding pressure is $P_b = 14.7$ psia and the duct is to be choked, what is the minimum value that the stagnation pressure must be at the duct entrance?

9.9 Given the same duct as in problem 9.8 but with an exit Mach number of 0.5, find the entrance pressure and stagnation pressure.

9.10 In frictionless air flow through a 120-mm duct, 0.15 kg/s enters at 0°C and 7 kPa. How much heat, in kilocalories per kilogram (kcal/kg), can be added without choking the flow?

9.11 How much heat transfer is necessary in a reversible flow of air at 500 m/s to increase its Mach number from 2.0 to 2.8? Is the heat transferred to or from the flow? What would be the effect of reversing the direction of heat transfer?

9.12 Consider a duct supplying air in which heat is added at the rate of 200 Btu/lb_m. The upstream stagnation conditions are 100 psia and 20°F. Find the maximum back pressure that will allow the duct to be choked.

9.13 If the duct of problem 9.12 has a cross-sectional area of 30 in.^{2} and an exit Mach number of 0.5, find the mass flow rate of air.

9.14 Heat extraction is used to increase the Mach number of a supersonic flow in a constant-area duct. The initial conditions are $M_1 = 2.0$, $T_{01} = 2000$ R, $P_{01} = 500$ psia. If the downstream Mach number is to be $M_2 = 3.0$, find the downstream pressure, temperature, stagnation pressure, and stagnation temperature.

9.15 If the duct in problem 9.14 has a diameter of 6 in., find the mass flow rate and the rate of heat removal in kilowatts.

9.16 Air flows in a constant-area heating duct that is attached to the exit of a C-D nozzle. $A_e/A_t = 2.4$. The nozzle stagnation pressure and temperature are 100 psia and 1000 R. The heat is added at a rate of 60 Btu/lb$_m$. What is the most that the back pressure can be if the nozzle throat is choked?

9.17 If the nozzle of problem 9.16 flows supersonically without shocks, what will be the Mach number at the nozzle exit (entrance to the duct)?

9.18 For the nozzle-duct system of problem 9.16, what back pressure is necessary to establish a normal shock at the *nozzle* exit?

9.19 Air at 60°F and 30 psia flows through a 1-ft-diameter duct at a uniform velocity of 200 ft/s. For parts (a) through (h) below, assume isentropic flow.

(a) Determine the mass flow rate, stagnation temperature, and stagnation pressure.

(b) Repeat part (a) with a uniform velocity of 1000 ft/s.

(c) With the speed of sound given by $c = C_1 T^{1/2}$, determine C_1 for air at 520 R.

(d) Determine the Mach number of the flows described in parts (a) and (b).

(e) The duct of part (a) reduces to a diameter of 9 in. Determine the velocity, pressure, temperature, and Mach number at this section.

(f) Determine the minimum diameter of the duct that will still permit the same flow as found in part (a).

(g) Assume that a converging-diverging duct is downstream of the section described in part (a) so that supersonic flow exists in the diverging section. Determine the Mach number, pressure, and temperature at a section of the divergent portion where the diameter is 1 ft.

(h) A normal shock exists at the section in part (g). Determine the Mach number and properties of the fluid downstream of the shock.

(i) Represent the state points indicated in parts (a), (e), (f), (g), and (h) on a temperature-entropy diagram.

(j) If the duct flow described in part (a) has a friction factor of $f = 0.02$, determine the conditions at the exit if the duct is 900 ft long. Determine the length necessary to choke the flow at the given mass flow rate.

(k) A duct 1 ft in diameter is attached to the C-D nozzle of part (g). With $f = 0.02$, determine the length of the duct to choke the flow and the exit conditions when choked.

(l) A duct with the flow described in part (a) at the entrance has heat added such that choked conditions exist at the exit. Determine the heat added and conditions at the duct exit.

(m) A duct with the inlet conditions described in part (g) has heat added such that choked conditions exist at the exit. Determine the heat added and thermodynamic conditions at the exit.

THREE

SIMPLE VISCOUS FLOWS

FUNDAMENTAL CONCEPTS AND GOVERNING RELATIONSHIPS

We now arrive at the third part of our three-part coverage of the subject of intermediate fluid mechanics. (The reader may find it useful at this point to review the remarks in Chapter 1 concerning the organization of this text.) Here we put aside the complicating factor of variable density and consider in relative detail the role of viscosity in fluid flows.

Let us first recall that *viscosity is a fluid property* that is a measure of the extent to which *momentum differences* are "sensed" by a fluid. In this chapter we place much emphasis on this concept, for its appreciation is vital to understanding the physical nature of viscous flows. Viscosity is to shear stress and velocity as thermal conductivity is to heat flux and temperature. Thus Newton's law of viscosity, which expresses shear stress as $\tau_{yx} = \mu \, \partial u / \partial y$, is analogous in form and meaning to Fourier's law of heat conduction, which describes heat flux as $q_y = k \, \partial T / \partial y$. The quantities μ (molecular viscosity) and k (thermal conductivity) may be viewed as constants of proportionality between effects and causes.

We begin the discussion of viscous flows by considering the nature of viscosity. Then we take another look at the force balance on a fluid element and, in doing so, see the need for mathematical description of the shear stresses at work in a viscous fluid. This description is obtained in the form of constitutive relationships (stress-strain laws), which are then used to form the basic equations that govern the motion of a viscous fluid—the Navier-Stokes equations. After development and discussion of the Navier-Stokes equations, a chapter is devoted to their solution for several simple yet practical laminar flow problems. This leads to the boundary layer concept, and the subsequent chapter deals with this important approach to the solution of a large class of viscous flow problems. Finally, the nature of turbulence is discussed, together with some preliminary treatment of the analysis of turbulent flows.

10.1 THE NATURE OF VISCOSITY

Viscosity is a measure of the sensitivity of a fluid to momentum differences that exist in directions normal to the direction of flow. This sensitivity is nicely illustrated by the theory of the molecular behavior of perfect gases when they flow in the presence of transverse velocity gradients. Although the molecular description of real gases—and liquids—is much more complicated than that of a perfect gas, the nature of viscosity as a fluid property is fundamental to all substances that flow. Perfect gas behavior is realized at relatively low densities (high temperatures and/or low pressures), so that the distance between molecules is many times their diameter. In the subsequent development the molecules are taken to be rigid, nonattracting, and spherical, and we make use of the following nomenclature:

$$d = \text{diameter of molecules}$$

$$m = \text{mass of a single molecule}$$

$$n = \text{molecular density (molecules per unit volume)}$$

$$T = \text{temperature of the gas}$$

From the kinetic theory for low-density gases, the following expressions will be needed [see, for instance, Sears (1959, chap. 13)]:

$v = $ average speed of random molecular motion:

$$v = (8kT/\pi m)^{1/2}$$

where the Boltzmann constant is $k = 1.4 \times 10^{-16}$ erg/K

$z = $ frequency at which molecules cross a given planar surface from one side, per unit surface area:

$$z = \tfrac{1}{4}nv$$

$\lambda = $ mean free path (average distance traveled by a molecule between successive collisions):

$$\lambda = \frac{1}{\pi\sqrt{2}\, d^2 n}$$

$a = $ distance from a plane through which, on the average, molecules will travel without suffering a collision. Put another way, molecules reaching a plane will have had their last collision a distance a from that plane (on the average):

$$a = \tfrac{2}{3}\lambda$$

Momentum Transport Due to Molecular Activity

Consider a two-dimensional flow of a perfect gas, with no component or variation in the z direction. The only *flow* velocity component is $u = u(y)$ in the x direction, as shown in Fig. 10.1. This flow is superimposed on the random molecular motion of the gas. We evaluate the momentum of "average" molecules located at a distance a above and below a reference plane located at $y = y_0$. At the level $y_0 + a$, the flow velocity will be $u(y_0 + a) = u_0 + \frac{2}{3}\lambda(du/dy)$, and the x-wise momentum of a molecule at this point will be

$$p^+ = m\left(u_0 + \frac{2}{3}\lambda \frac{du}{dy}\right)$$

Since the rate that molecules cross the plane is $z = \frac{1}{4}nv$, the momentum flow rate per unit area, or *momentum flux*, through the plane due to molecules crossing from above is $p^+/A = \frac{1}{4}nvp^+$, or

$$\frac{p^+}{A} = \frac{1}{4}nvm\left(u_0 + \frac{2}{3}\lambda \frac{du}{dy}\right)$$

Similarly, the molecules arriving at the plane from below bring with them a momentum flux of

$$\frac{p^-}{A} = \frac{1}{4}nvm\left(u_0 - \frac{2}{3}\lambda \frac{du}{dy}\right)$$

Molecules at the plane are traveling at the speed u_0, with momentum flux $p^0/A = \frac{1}{4}nvmu_0$. Relative to molecules from above, the difference in momentum flux at the plane is

$$\frac{\Delta p^+}{A} = \frac{p^+ - p^0}{A} = \frac{1}{4}nvm\left(u_0 + \frac{2}{3}\lambda \frac{du}{dy}\right) - \frac{1}{4}nvmu_0$$

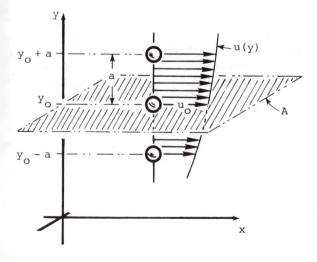

Figure 10.1 Perfect gas molecules in a flow with a velocity gradient.

or

$$\frac{\Delta \dot{p}^+}{A} = \frac{1}{6}nvm \frac{du}{dy}$$

In a similar way, relative momentum flux of molecules crossing the plane from below is found to be

$$\frac{\Delta \dot{p}^-}{A} = -\frac{1}{6}nvm \frac{du}{dy}$$

The total rate of momentum exchange per unit area due to collisions at y_0 between molecules arriving from above and below is, on average,

$$\frac{\dot{p}}{A} = \frac{1}{3}nvm\lambda \frac{du}{dy}$$

This flux of momentum, *due to random molecular motion,* is manifested as an *x*-wise force between adjacent fluid layers per unit area: *a shear stress.* This stress is given by

$$\tau_{yx} = \frac{\dot{p}}{A} = \frac{1}{3}nvm\lambda \frac{du}{dy}$$

The quantity $\frac{1}{3}nvm\lambda$ is known as the *molecular viscosity* of the gas and, as can be seen, it depends not on the flow but on the thermodynamic state of the gas. (It is more often called the absolute viscosity, but the "molecular" handle is more descriptive of its nature.) The molecular viscosity represents the proportionality between the shear stress τ_{yx} and the velocity gradient du/dy. That is,

$$\tau_{yx} = \mu \frac{du}{dy} \tag{10.1}$$

where, for a perfect gas, $\mu = \frac{1}{3}nvm\lambda$.

It should be emphasized that μ, as derived above, is a fluid property and is relevant to *momentum transport due to random molecular motion*. Momentum is often transported by other means, such as turbulent motions of relatively large "lumps" of fluid. In this case, other versions of "viscosity" have been hypothe-cized, such as the "eddy viscosity." These devices are often more dependent on the *flow* than on the fluid. Viscous flows in which the shear stresses are due solely to molecular momentum transport are characterized by the term *laminar,* and fluids that satisfy the shear stress law given in Eq. (10.1) are called Newtonian fluids. Equation (10.1) is often referred to as the Newtonian shear stress law.

An unexpected conclusion from the analysis of a perfect gas presented here is that the molecular viscosity of a perfect gas depends not on pressure but on temperature. Thus, if the relationship between molecular speed v and temperature is used in the above result, we have

$$\mu = \frac{2}{3\pi} \frac{1}{d^2} \left(\frac{mkT}{\pi} \right)^{1/2}$$

This theoretical model has been found to be valid up to pressures as high as 10 atm.

Many common fluids exhibit Newtonian behavior; that is, $\tau \alpha \, du/dy$. Such fluids are analogous to Hookean solids, for which stress is proportional to strain. For Newtonian fluids the stress is seen to be proportional to *strain rate,* a property that features prominently in the development of the equations of viscous motion. Figure 10.2 illustrates the stress-strain rate behavior of some common non-Newtonian fluids.

Newtonian behavior is closely approximated by many single-component fluids: air, water, organic solvents, liquid metals, and low-density gases. The flow of Bingham plastics requires a finite initial shear stress. Such substances include suspensions, pastes, nuclear fuels, and particles in heavy water. A somewhat general model is the Ostwald-deWaele constitutive relationship:

$$\tau \alpha \left| \frac{du}{dy} \right|^{(n-1)} \frac{du}{dy}$$

If the constant n is unity, this model is seen to give Newtonian behavior. For $n > 1$ the effective viscosity increases with $|du/dy|$, exhibiting what is called a dilatant, or *shear thickening* behavior, as in Fig. 10.2. *Shear thinning, or pseudoplastic* effects are obtained with $n < 1$ so that the effective viscosity decreases

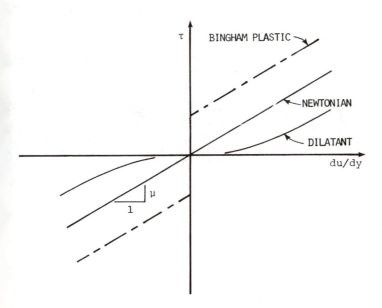

Figure 10.2 Stress-strain rate models of viscous fluids.

with $|du/dy|$. For more on the rheology of fluids, an excellent discussion is given in Bird et al. (1960, chap. 1).

10.2 SURFACE FORCES ACTING ON A FLUID ELEMENT

The surface forces acting on a fluid element are best expressed in terms of per-unit-area values or stresses. Figure 10.3 shows the method by which we express these stresses in terms of their values acting at a central point. We have seen a figure such as this in Chapter 2 (Fig. 2.3), but in that case we considered only the normal stresses—for example, τ_{xx}—acting on the element. In doing so we arrived at what was called the Euler equation of motion, Eq. (2.7). The shear stresses were conspicuous in their absence in this relationship, and it is the main goal of this part of the text to remove this simplification.

In Fig. 10.3 the standard notation for stresses is used: a subscript (ji) refers to the stress acting in the i direction on a surface oriented normal to the j direction. Thus, τ_{xz} is a z component of stress on a surface normal to the x direction. From this figure we may obtain the total surface force acting on a fluid element, *including shear forces,* by summing all the terms appropriate for each of the coordinate directions. In the z direction, for instance,

$$\Sigma F_z = \left[\frac{\partial}{\partial z}(\tau_{zz})\,\delta z\right]\delta x\,\delta y + \left[\frac{\partial}{\partial x}(\tau_{xz})\,\delta x\right]\delta y\,\delta z + \left[\frac{\partial}{\partial y}(\tau_{yz})\,\delta y\right]\delta x\,\delta z$$

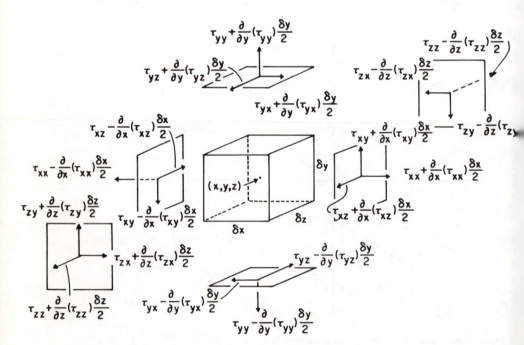

Figure 10.3 Stresses acting on the surface of a fluid element.

or, since the volume of the element is $\delta V = \delta x\,\delta y\,\delta z$,

$$\frac{\Sigma F_z}{\delta V} = \frac{\partial}{\partial z}(\tau_{zz}) + \frac{\partial}{\partial x}(\tau_{xz}) + \frac{\partial}{\partial y}(\tau_{yz})$$

Note the ordering of the subscripts in this expression. The last subscript on each of the stresses is z, indicating the z direction for which the force is designated. The first subscript on each of the stresses indicates the direction across which a difference is evaluated—hence these subscripts are the same as the directions indicated in the partial derivatives. The organization of terms is significant, and this can be nicely streamlined by using the index notation described in Appendix F. With such a notation, all three Cartesian components of the force acting on an elemental volume are given by

$$\frac{\Sigma F_i}{\delta V} = \frac{\partial}{\partial x_j}(\tau_{ji}) \tag{10.2}$$

Readers not familiar with this notation should refer to Appendix F. From time to time we will expand such expressions in order to retain an appreciation for the form of their components. If we wish to find the net force in the y direction, for instance, Eq. (10.2) is evaluated with $i = 2$ (corresponding to y). The right-hand side is evaluated as the sum of the three terms that result from letting the repeated (or dummy) index j take on each of its possible values: 1, 2, and 3 (corresponding to x, y, and z). Thus,

$$\frac{\Sigma F_y}{\delta V} = \frac{\partial}{\partial x}(\tau_{xy}) + \frac{\partial}{\partial y}(\tau_{yy}) + \frac{\partial}{\partial z}(\tau_{zy})$$

That this is correct may be demonstrated by summing the forces in the y direction in Fig. 10.3.

We now have an expression that according to Newton's second law should be set equal to the mass per unit volume (the density) times the acceleration of a fluid particle. If all we had to worry about was the normal stresses, we could simply multiply Eq. (10.2) by the Kronecker delta, δ_{ij}, and add in the body forces. The result would be the Euler equation. With the viscous stresses not zero, however, it is necessary to develop relationships that link these stresses to the velocities present in the flow field—the so-called constitutive relationships for a flowing viscous fluid. Viscosity does this for us.

10.3 DEVELOPMENT OF CONSTITUTIVE EQUATIONS FOR AN ISOTROPIC FLUID

Although we will not place our complete faith in it, the analogy between the deformation of a solid and the flow of a viscous fluid provides a "quick-and-dirty" path to the constitutive equations for a fluid that obeys the Newtonian viscosity law. To do this, it is logical first to develop the constitutive laws for an elastic isotropic solid (a *Hookean solid*) under a general three-dimensional state of strain. The restriction to an isotropic medium requires that the proportionality constant relating stress and strain does not depend on the direction of the applied stress.

Deformation of a Solid

As sketched in Fig. 10.4, the displacement vector **s** describes the motion of a solid particle as it is strained. Using the Cartesian coordinates, ξ, η, and ζ, for the components of **s**,

$$\mathbf{s} = i\xi + j\eta + k\zeta \tag{10.3}$$

In addition to the displacement, the particle may be strained under the action of shear and normal stresses. In the x-y plane, for example, the distortion of an element will appear as sketched in Fig. 10.5. From this figure the strains may be deduced to have the following forms:

Linear strain:

$$\epsilon_x = \frac{(\partial\xi/\partial x)}{\delta x}\,\delta x = \frac{\partial\xi}{\partial x}$$

Angular strain:

$$\gamma_{xy} = \frac{\partial\xi}{\partial y} + \frac{\partial\eta}{\partial x}$$

In three dimensions, the growth due to linear strain of the element with sides δx, δy, and δz will be

$$\delta V = \epsilon_x\,\delta x(\delta y\,\delta z) + \epsilon_y\,\delta y(\delta x\,\delta z) + \epsilon_z\,\delta z(\delta x\,\delta y)$$

The total change fraction of the volume $V = \delta x\,\delta y\,\delta z$ is defined as $\epsilon = \delta V/V$, which is simply the sum of the linear strains in the three coordinate directions:

$$\epsilon = \epsilon_x + \epsilon_y + \epsilon_z = \text{div } \mathbf{s} \tag{10.4}$$

The complete set of linear and angular strains is

$$\epsilon_x = \frac{\partial\xi}{\partial x} \qquad \epsilon_y = \frac{\partial\eta}{\partial y} \qquad \epsilon_z = \frac{\partial\zeta}{\partial z} \tag{10.5}$$

$$\gamma_{xy} = \frac{\partial\xi}{\partial y} + \frac{\partial\eta}{\partial x} \qquad \gamma_{yz} = \frac{\partial\eta}{\partial z} + \frac{\partial\zeta}{\partial y} \qquad \gamma_{zx} = \frac{\partial\zeta}{\partial x} + \frac{\partial\xi}{\partial z} \tag{10.6}$$

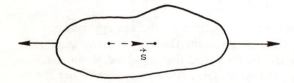

Figure 10.4

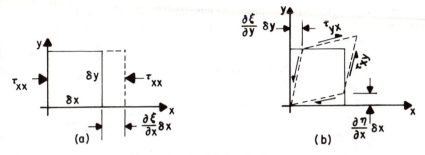

Figure 10.5 Distortion of an element: (a) linear strain; (b) angular strain.

The shear stresses are related to the shear (angular) strains by the shear modulus G,

$$\tau_{xy} = G\gamma_{xy} \qquad \tau_{yz} = G\gamma_{yz} \qquad \tau_{zx} = G\gamma_{zx} \tag{10.7}$$

and the normal stresses are related to the normal (linear) strains through the modulus of elasticity E and Poisson's ratio, υ:

$$\epsilon_x = \frac{\tau_{xx}}{E} - \frac{\upsilon}{E}(\tau_{yy} + \tau_{zz})$$

$$\epsilon_y = \frac{\tau_{yy}}{E} - \frac{\upsilon}{E}(\tau_{xx} + \tau_{zz})$$

$$\epsilon_z = \frac{\tau_{zz}}{E} - \frac{\upsilon}{E}(\tau_{xx} + \tau_{yy}) \tag{10.8}$$

The shear stress expressions, Eqs. (10.6) and (10.7), are already in a form that is suitable for modeling of fluid deformation. The normal stresses and strains require a bit more manipulation, however. Because of the constraint of isotropy, the three constants G, E, and υ have no directional dependency (they are scalars). They are related to each other by

$$G = \frac{E}{2(\upsilon + 1)} \tag{10.9}$$

and if Eqs. (10.8) are added together, we obtain

$$\epsilon = \frac{1 - 2\upsilon}{E}(\tau_{xx} + \tau_{yy} + \tau_{zz})$$

It is also convenient at this point to introduce the mean value of the normal stresses, $\bar{\sigma} = (\tau_{xx} + \tau_{yy} + \tau_{zz})/3$ so that $\epsilon = 3\bar{\sigma}(1 - 2\upsilon)/E$ and, with Eq. (10.9),

$$\epsilon = \frac{3(1 - 2\upsilon)}{2G(\upsilon + 1)}\bar{\sigma} \tag{10.10}$$

We are now in a position to express the normal stresses in a way that involves only the shear modulus G and the strains. For the x direction, for instance, we rearrange Eq. (10.8) such that

$$\tau_{xx} = E\epsilon_x + \upsilon(\tau_{yy} + \tau_{zz} + \tau_{xx}) - \upsilon\tau_{xx}$$

or

$$\tau_{xx} = \frac{E}{1 + \upsilon}\epsilon_x + \frac{3}{1 + \upsilon}\bar{\sigma}$$

The modulus E is eliminated in favor of G by means of Eq. (10.9), and Eq. (10.10) allows the elimination of Poisson's ratio after the devilishly clever manipulation

$$\frac{3\upsilon}{1 + \upsilon} = 1 - \frac{1 - 2\upsilon}{\upsilon + 1}$$

After similar maneuvers for the y and z directions, we have

$$\tau_{xx} = 2G\epsilon_x + \bar{\sigma} - \frac{2}{3}G \text{ div } \mathbf{s} \tag{10.11a}$$

$$\tau_{yy} = 2G\epsilon_y + \bar{\sigma} - \frac{2}{3}G \text{ div } \mathbf{s} \tag{10.11b}$$

$$\tau_{zz} = 2G\epsilon_z + \bar{\sigma} - \frac{2}{3}G \text{ div } \mathbf{s} \tag{10.11c}$$

and, from Eqs. (10.6) and (10.7),

$$\tau_{xy} = G\left(\frac{\partial\xi}{\partial y} + \frac{\partial\eta}{\partial x}\right) \tag{10.11d}$$

$$\tau_{yz} = G\left(\frac{\partial\eta}{\partial z} + \frac{\partial\zeta}{\partial y}\right) \tag{10.11e}$$

$$\tau_{zx} = G\left(\frac{\partial\zeta}{\partial x} + \frac{\partial\xi}{\partial z}\right) \tag{10.11f}$$

Fluid-Solid Analogy

The six stresses (three normal and three shear) are now in a form that involves only the displacements and the property G, which expresses the proportionality between shear stress and strain. In the laminar flow of a Newtonian *fluid*, the only differences are that stresses lead to strain *rates* (for example, $\partial u/\partial y$—units of strain per unit time), and the proportionality constant is the viscosity, μ. Both constitutive laws are linear, expressing the linear behavior of such well-behaved substances. The analogy is thus very strong. In addition, the mean normal stress may be replaced with the negative of the pressure at the point x, y, z in the fluid. $\bar{\sigma} = -P$. The analogy may be summarized as follows:

1. The role of G in a solid is played by μ in a fluid.
2. The roles of displacements in solid motions are played by velocities in fluid motions.
3. The role of compressive stress in a solid is played by pressure in a fluid.

Under this analogy, Eqs. (10.11) become:

$$\tau_{xx} = 2\mu \frac{\partial u}{\partial x} - P - \frac{2}{3}\mu \text{ div } \mathbf{q} \tag{10.12a}$$

$$\tau_{yy} = 2\mu \frac{\partial v}{\partial y} - P - \frac{2}{3}\mu \text{ div } \mathbf{q} \tag{10.12b}$$

$$\tau_{zz} = 2\mu \frac{\partial w}{\partial z} - P - \frac{2}{3}\mu \text{ div } \mathbf{q} \tag{10.12c}$$

$$\tau_{xy} = \mu \left(\frac{\partial u}{\partial y} + \frac{\partial v}{\partial x} \right) \tag{10.12d}$$

$$\tau_{yz} = \mu \left(\frac{\partial v}{\partial z} + \frac{\partial w}{\partial y} \right) \tag{10.12e}$$

$$\tau_{zx} = \mu \left(\frac{\partial w}{\partial x} + \frac{\partial u}{\partial z} \right) \tag{10.12f}$$

With the exception of issues regarding the second coefficient of viscosity, a quantity that describes the viscous resistance to normal strains, the fluid-solid analogy yields the exact form of the constitutive equations for laminar flow in a Newtonian fluid. The second coefficient of viscosity is important only in flow systems where compressibility cannot be neglected, and even in such cases it is often insignificant in its effect. [These relationships and methods receive widespread treatment in the fluid mechanics literature. This presentation follows closely that of Schlichting (1960, chap. 3). Later editions, containing newer presentations, seem to be less lucid. For more rigorous proof based on tensor calculus, see, for instance, Li and Lam (1976, pp. 199–222).]

Equations (10.12) give the nine components of the stress tensor, since it may be shown that in isotropic fluids (and solids) $\tau_{ij} = \tau_{ji}$ [Schlichting (1968, p. 47); problem 10.1 also applies]. Using the compaction of index notation,

$$\tau_{ji} = -\frac{2}{3}\mu\theta\,\delta_{ji} - P\,\delta_{ji} + 2\mu e_{ji} \tag{10.13}$$

where δ_{ji} is the Kronecker delta, $\theta = \text{div } \mathbf{q}$ (sometimes called the dilation), and e_{ji} is the rate of strain tensor:

$$e_{ji} = \frac{1}{2}\left(\frac{\partial u_i}{\partial x_j} + \frac{\partial u_j}{\partial x_i}\right) \tag{10.14}$$

Note that the stresses τ_{ji} are forces per unit area and are directional in nature. In order to formulate the equations of motion, each component of τ_{ji} must be allocated to its appropriate surface and direction of application. When this is done, as in Fig. 10.3, the resulting forces may be equated with the corresponding acceleration per unit mass, in accordance with Newton's second law. The resulting expressions are called the *Navier-Stokes equations*. Before making these substitutions, however, we examine some of the general characteristics of the strain rate tensor.

Velocity Gradient Tensor

By means of the fluid-solid analogy we have found relationships between the elements of the stress tensor and the fluid motion. Looking at a fluid element with its various stresses acting on it, as in Fig. 10.3, we can easily imagine that it is going to be warped and moved by the resulting forces. In a more general search for a match-up between forces and motions, without resorting to the fluid-solid analogy, it is reasonable to try to identify the various kinds of motion with mathematical precision. This will also tend to link the concepts of ideal fluid motion, examined in Part One, with the additional complexities that are introduced by the shearing action of viscous effects.

An examination of general motion begins with expression of the velocity at a point in the flow, u_i, in terms of the conditions at a nearby point. This is done via a Taylor series expansion:

$$u_i = u_{0i} + \left(\frac{\partial u_i}{\partial x_j}\right)_0 \delta x_j + \frac{1}{2!}\left(\frac{\partial^2 u_i}{\partial x_j^2}\right)(\delta x_j)^2 + \cdots$$

or

$$u_i = u_{0i} + \left(\frac{\partial u_i}{\partial x_j}\right)_0 \delta x_j + \text{terms that vanish as } \delta x_j \to 0$$

The tensor $\partial u_i/\partial x_j$ is called the *velocity gradient tensor* and may be decomposed into symmetric and antisymmetric parts as follows:

$$\frac{\partial u_i}{\partial x_j} = \frac{1}{2}\left(\frac{\partial u_i}{\partial x_j} + \frac{\partial u_j}{\partial x_i}\right) + \frac{1}{2}\left(\frac{\partial u_i}{\partial x_j} - \frac{\partial u_j}{\partial x_i}\right) \tag{10.15}$$

In expanded form, Eq. (10.15) appears as follows:

$$\frac{\partial u_i}{\partial x_j} = \frac{1}{2} \left\{ \begin{matrix} \dfrac{\partial u_1}{\partial x_1} + \dfrac{\partial u_1}{\partial x_1} & \dfrac{\partial u_1}{\partial x_2} + \dfrac{\partial u_2}{\partial x_1} & \dfrac{\partial u_1}{\partial x_3} + \dfrac{\partial u_3}{\partial x_1} \\ \dfrac{\partial u_2}{\partial x_1} + \dfrac{\partial u_1}{\partial x_2} & \cdots & \vdots \\ \dfrac{\partial u_3}{\partial x_1} + \dfrac{\partial u_1}{\partial x_3} & \cdots & \vdots \end{matrix} \right\}$$

$$+ \frac{1}{2} \left\{ \begin{matrix} 0 & \dfrac{\partial u_1}{\partial x_2} - \dfrac{\partial u_2}{\partial x_1} & \dfrac{\partial u_1}{\partial x_3} - \dfrac{\partial u_3}{\partial x_1} \\ \dfrac{\partial u_2}{\partial x_1} - \dfrac{\partial u_1}{\partial x_2} & 0 & \dfrac{\partial u_2}{\partial x_3} - \dfrac{\partial u_3}{\partial x_2} \\ \dfrac{\partial u_3}{\partial x_1} - \dfrac{\partial u_1}{\partial x_3} & \dfrac{\partial u_3}{\partial x_2} - \dfrac{\partial u_2}{\partial x_3} & 0 \end{matrix} \right\}$$

It is seen that the first term is a symmetric tensor with elements $a_{ji} = a_{ij}$. The second is antisymmetric with $a_{ji} = -a_{ij}$, and $a_{ji} = 0$ when $i = j$.

To improve our insight regarding the motions represented by this very general expression, we examine three special cases. First, let u_1 be the only nonzero velocity and $\partial/\partial x_1$ be the only gradient. In this case $\partial u_i/\partial x_j = \partial u_1/\partial x_1$, and only the symmetric part of the velocity gradient tensor is nonzero. The situation is illustrated in Fig. 10.6. There is pure linear strain (in the x_1 direction) and no other deformation. This hints that the symmetric part of $\partial u_i/\partial x_j$ is somehow related to strain.

Now let us allow another term to be nonzero in the symmetric tensor, while maintaining a zero antisymmetric part. Let $u_3 = 0$ and $\partial/\partial x_3 = 0$. To keep the antisymmetric part of the tensor zero, it is necessary to require that

$$\frac{\partial u_1}{\partial x_2} = \frac{\partial u_2}{\partial x_1}$$

A fluid element that finds itself in this flow field will find that its boundaries are rotating because of the motion of its fluid particles there. But the relative motions of the two sides of the element are equal but opposite, as illustrated in Fig. 10.7.

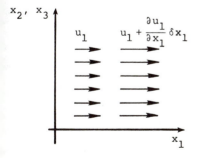

Figure 10.6 Pure linear strain in one direction.

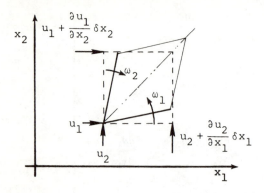

Figure 10.7 Pure angular strain.

If ω_2 and ω_1 are the angular velocities of the two sides, then

$$\omega_2 = -\frac{\partial u_1}{\partial x_2} \quad \text{and} \quad \omega_1 = \frac{\partial u_2}{\partial x_1}$$

But these two gradients are taken as equal in the present case, so $\omega_2 = -\omega_1$. The diagonal of the element *does not rotate,* and neither will the element itself as its size becomes vanishingly small. We have discussed such irrotational motion in Part One; in the special case above, it is called *pure angular strain.*

Note that in the previous cases, the part of Eq. (10.15) with the antisymmetric array was made to be zero. In both cases there was no rotation. Thus only the symmetric part of the velocity gradient tensor represents straining motions. Similarly, only the antisymmetric part contains the terms that describe rotation of a fluid element. This may be illustrated by a third special case that is the same as that above but with

$$\frac{\partial u_1}{\partial x_2} = -\frac{\partial u_2}{\partial x_1}$$

For such a case, from Eq. (10.15),

$$\frac{\partial u_i}{\partial x_j} = \frac{1}{2}\{0\}$$

$$+ \frac{1}{2}\left\{\begin{array}{ccc} 0 & \dfrac{\partial u_1}{\partial x_2} - \dfrac{\partial u_2}{\partial x_1} & 0 \\ \dfrac{\partial u_2}{\partial x_1} - \dfrac{\partial u_1}{\partial x_2} & 0 & 0 \\ 0 & 0 & 0 \end{array}\right\}$$

The motion is as shown in Fig. 10.8, and it is seen that $\omega_2 = \omega_1$. Both sides of the element rotate together and at the same angular velocity. The motion is called *rigid body rotation.*

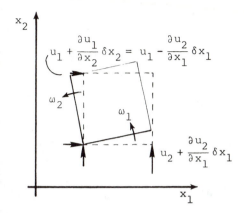

Figure 10.8 Rigid body rotation.

In summary, the following generalizations can be made:

1. If the fluid element is made sufficiently small so that only the linear terms have significant values, the fluid motion can be resolved into translations (u_{0i}), deformations, and rotations.
2. The deformation terms are grouped into the symmetric part of the velocity gradient tensor and the antisymmetric part contains terms that represent rotations without strain.

Because of the arguments presented here, the symmetric part of the velocity gradient tensor is called the *rate-of-strain tensor*. It plays a prominent role in the development of the stress tensor, as we have seen in the results of the fluid-solid analogy, Eq. (10.13). The cases illustrated above are very special, of course, and they seldom occur alone in fluid motion. In the general case the kinematic description of a flow must include all the elements of the velocity gradient tensor. Further manipulations of the velocity gradient tensor allow the rate-of-strain tensor to be expressed in terms of principal stresses. Subsequent derivation of the constitutive relationships follows with relative ease. In addition, tensor algebra may be used to show that the antisymmetric part of Eq. (10.15) is related directly to the vorticity vector—a result that should not be especially surprising to the reader. We now return to the use of the stress tensor in formulating the equations of motion.

10.4 NAVIER–STOKES EQUATIONS

From Newton's second law we may write, in index notation,

$$\frac{1}{\rho} \frac{\Sigma F_i}{\delta V} = a_i = \frac{Du_i}{Dt}$$

and, when the body forces per unit volume are added to Eq. (10.2),

$$\frac{\Sigma F_i}{\delta V} = \frac{\partial \tau_{ji}}{\partial x_j} - \gamma \frac{\partial h}{\partial x_i}$$

Combining these two expressions,

$$\rho \frac{Du_i}{Dt} = \frac{\partial \tau_{ij}}{\partial x_j} - \gamma \frac{\partial h}{\partial x_i} \tag{10.16}$$

The final ingredient is the constitutive relationship given in Eq. (10.13). With this included, the result is the Navier-Stokes equation (or equations, since the three coordinate directions are incorporated):

$$\rho \frac{Du_i}{Dt} = -\gamma \frac{\partial h}{\partial x_i} - \frac{\partial P}{\partial x_i} + \frac{\partial}{\partial x_j}\left[\mu\left(\frac{\partial u_j}{\partial x_i} + \frac{\partial u_i}{\partial x_j}\right)\right] - \frac{2}{3}\frac{\partial}{\partial x_i}\left(\mu \frac{\partial u_m}{\partial x_m}\right) \tag{10.17}$$

This equation forms the basis for the analysis of laminar flows of Newtonian fluids. We shall look at several reductions of Eq. (10.17), for various special cases, but before doing so, let us look at an expansion into one of the coordinate directions. For example, let $i = 1$, corresponding to the x direction:

$$\rho \frac{Du}{Dt} = -\gamma \frac{\partial h}{\partial x} - \frac{\partial P}{\partial x} + \frac{\partial}{\partial x}\left[\mu\left(\frac{\partial u}{\partial x} + \frac{\partial u}{\partial x}\right)\right] + \frac{\partial}{\partial y}\left[\mu\left(\frac{\partial v}{\partial x} + \frac{\partial u}{\partial y}\right)\right]\cdots$$
$$+ \frac{\partial}{\partial z}\left[\mu\left(\frac{\partial w}{\partial x} + \frac{\partial u}{\partial z}\right)\right] - \frac{2}{3}\frac{\partial}{\partial x}\left[\mu\left(\frac{\partial u}{\partial x} + \frac{\partial v}{\partial y} + \frac{\partial w}{\partial z}\right)\right]$$

Various Forms

Principal among the major realistic assumptions that can be applied to Eq. (10.17) are those of constant viscosity and constant density. Here we give these simplified forms, leaving the index notation algebra to the problems at the end of the chapter.

Constant viscosity. Removal of the viscosity from the partial derivatives in Eq. (10.17) results in considerable compaction:

$$\rho \frac{Du_i}{Dt} = -\gamma \frac{\partial h}{\partial x_i} - \frac{\partial P}{\partial x_i} + \mu \nabla^2 u_i + \frac{1}{3}\mu \frac{\partial}{\partial x_i}\left(\frac{\partial u_m}{\partial x_m}\right) \tag{10.18}$$

and with $i = 1$, for example,

$$\rho \frac{Du}{Dt} = -\gamma \frac{\partial h}{\partial x} - \frac{\partial P}{\partial x} + \mu \nabla^2 u + \frac{1}{3}\mu \frac{\partial}{\partial x}\left(\frac{\partial u}{\partial x} + \frac{\partial v}{\partial y} + \frac{\partial w}{\partial z}\right)$$

Constant viscosity and constant density. For a fluid of constant density, the equation of mass continuity gives $\theta = \partial u_m/\partial x_m = 0$, and we have

$$\rho \frac{Du_i}{Dt} = -\gamma \frac{\partial h}{\partial x_i} - \frac{\partial P}{\partial x_i} + \mu \nabla^2 u_i \tag{10.19}$$

Expanding the case for $i = 2$, by way of example:

$$\rho \frac{Dv}{Dt} = -\gamma \frac{\partial h}{\partial y} - \frac{\partial P}{\partial y} + \mu \nabla^2 v$$

In this part of the text we treat only cases in which the assumptions of constant viscosity and constant density are valid, that is, where the motion is governed by Eq. (10.19).

Polar-cylindrical coordinates. In Part One we saw that in many instances the nature of the flow boundaries is such that the Cartesian coordinate system is less than convenient for describing the motion. In polar-cylindrical coordinates, the Navier-Stokes (N-S) equations contain a few additional terms. For cases of constant viscosity and density,

$$\rho \left(\frac{Dv_r}{Dt} - \frac{v_\theta^2}{r} \right) = -\gamma \frac{\partial h}{\partial r} - \frac{\partial P}{\partial r} + \mu \left(\nabla^2 v_r - \frac{v_r}{r^2} - \frac{2}{r^2} \frac{\partial v_\theta}{\partial \theta} \right) \qquad (10.20)$$

$$\rho \left(\frac{Dv_\theta}{Dt} + \frac{v_r v_\theta}{r} \right) = -\frac{\gamma}{r} \left(\frac{\partial h}{\partial \theta} - \frac{\partial P}{\partial \theta} \right) + \mu \left(\nabla^2 v_\theta - \frac{v_\theta}{r^2} + \frac{2}{r^2} \frac{\partial v_r}{\partial \theta} \right) \qquad (10.21)$$

$$\rho \frac{Dv_z}{Dt} = -\gamma \frac{\partial h}{\partial z} - \frac{\partial P}{\partial z} + \mu \nabla^2 v_z \qquad (10.22)$$

where

$$\frac{D}{Dt} = \frac{\partial}{\partial t} + \mathbf{q} \cdot \nabla = \frac{\partial}{\partial t} + v_r \frac{\partial}{\partial r} + \frac{v_\theta}{\gamma} \frac{\partial}{\partial \theta} + v_z \frac{\partial}{\partial z}$$

and

$$\nabla^2 = \frac{\partial^2}{\partial r^2} + \frac{1}{r} \frac{\partial}{\partial r} + \frac{1}{r^2} \frac{\partial^2}{\partial \theta^2} + \frac{\partial^2}{\partial z^2}$$

The corresponding expression of mass continuity is

$$\frac{D\rho}{Dt} + \rho \left[\frac{1}{r} \frac{\partial}{\partial r} (rv_r) + \frac{1}{r} \frac{\partial v_\theta}{\partial \theta} + \frac{\partial v_z}{\partial z} \right] = 0 \qquad (10.23)$$

10.5 OTHER FORMS OF NAVIER–STOKES EQUATIONS

It is important to retain the physical meaning of the various terms in the N-S equations. For instance, the term $\mu \nabla^2 u_i$ in Eq. (10.19) may be reexpanded as

$$\mu \nabla^2 u_i = \mu \frac{\partial}{\partial x_j} \left(\frac{\partial u_i}{\partial x_j} \right) = \frac{\partial}{\partial x_j} \left(\mu \frac{\partial u_i}{\partial x_j} \right) = \frac{\partial}{\partial x_j} (\tau_{ji})$$

This expresses the net (over the $j = 1, 2, 3$ dimensions) unbalance of shear stress in the i direction. Likewise, terms such as $\rho u_j(\partial u_i/\partial x_j)$, appearing in $\rho Du_i/Dt$, represent the net variation of the i component of the momentum flux. To examine these terms further and to connect these concepts with those developed previously, we look now at other ways of expressing the N-S relationship.

Vorticity Transport Form of N-S Equations

Let us restrict our attention to only two flow dimensions and examine the terms on the left-hand and right-hand sides of Eq. (10.19). For the acceleration terms we have

$$\frac{Du}{Dt} = \frac{\partial u}{\partial t} + u\frac{\partial u}{\partial x} + v\frac{\partial u}{\partial y} = \frac{\partial u}{\partial t} + u\frac{\partial u}{\partial x} + v\left(\frac{\partial u}{\partial y} - \frac{\partial v}{\partial x}\right) + v\frac{\partial v}{\partial x}$$

$$\frac{Dv}{Dt} = \frac{\partial v}{\partial t} + u\frac{\partial v}{\partial x} + v\frac{\partial v}{\partial y} = \frac{\partial v}{\partial t} + u\left(\frac{\partial v}{\partial x} - \frac{\partial u}{\partial y}\right) + u\frac{\partial u}{\partial y} + v\frac{\partial v}{\partial y}$$

Recall from Eq. (3.3) that $\zeta_z = \partial v/\partial x - \partial u/\partial y$. The above accelerations may therefore be written as

$$\frac{Du}{Dt} = \frac{\partial u}{\partial t} + \frac{1}{2}\frac{\partial}{\partial x}(u^2 + v^2) - v\zeta_z \tag{10.24}$$

$$\frac{Dv}{Dt} = \frac{\partial v}{\partial t} + \frac{1}{2}\frac{\partial}{\partial y}(u^2 + v^2) + u\zeta_z \tag{10.25}$$

Turning now to the right-hand side of Eq. (10.19) and applying mass continuity, $\partial u/\partial x + \partial v/\partial y = 0$,

$$\nabla^2 u = \frac{\partial^2 u}{\partial x^2} + \frac{\partial^2 u}{\partial y^2} = \frac{\partial}{\partial x}\left(-\frac{\partial v}{\partial y}\right) + \frac{\partial^2 u}{\partial y^2} = \frac{\partial}{\partial y}\left(-\frac{\partial v}{\partial x} + \frac{\partial u}{\partial y}\right)$$

or

$$\nabla^2 u = -\frac{\partial \zeta_z}{\partial y} \tag{10.26}$$

and, similarly,

$$\nabla^2 v = \frac{\partial \zeta_z}{\partial x} \tag{10.27}$$

Combining these results, the two components of the N-S equations may be written as

$$\frac{\partial u}{\partial t} + \frac{\partial}{\partial x}\left(\frac{q^2}{2}\right) - v\zeta_z = -g\frac{\partial h}{\partial x} - \frac{\partial P}{\partial x} - v\frac{\partial \zeta_z}{\partial y} \tag{10.28}$$

$$\frac{\partial v}{\partial t} + \frac{\partial}{\partial y}\left(\frac{q^2}{2}\right) + u\zeta_z = -g\frac{\partial h}{\partial y} - \frac{\partial P}{\partial y} + v\frac{\partial \zeta_z}{\partial x} \tag{10.29}$$

Taking the derivatives of Eqs. (10.28) and (10.29) with respect to y and x, respectively, the two equations may be combined to yield a single expression in terms of the vorticity. This is

$$\frac{\partial \zeta_z}{\partial t} + u\frac{\partial \zeta_z}{\partial x} + v\frac{\partial \zeta_z}{\partial y} = v\,\nabla^2\zeta_z$$

or

$$\frac{D\zeta_z}{Dt} = v\,\nabla^2\zeta_z \tag{10.30}$$

This form of the N-S equations, which also holds in three dimensions, is called the *vorticity transport* equation. The combined local and convective change (i.e., the substantive derivative) of the vorticity in a flow is due to the diffusion of vorticity through the effects of viscosity. In irrotational flow, where the vorticity is zero, Eqs. (10.28) and (10.29) reduce to components of the Euler equation. In the flow of a hypothetical inviscid fluid—that is, when $v = 0$—Eq. (10.30) shows that $D\zeta/Dt = 0$: there is no transport of vorticity. The constancy of vorticity, or lack thereof, is seen to be related directly to the action of viscosity.

Expression in Terms of Stream Function

In Part One the stream function was developed from considerations of mass continuity. Recall [Eq. (3.11)] that $u = \partial\psi/\partial y$ and $v = -\partial\psi/\partial x$. With these, the z component of the vorticity may be written $\zeta_z = -\nabla^2\psi$ (which we have observed to be zero in irrotational flow). Equation (10.30) may now be expressed with the stream function as the sole dependent variable:

$$\frac{D}{Dt}(\nabla^2\psi) = -v\,\nabla^4\psi$$

or

$$\frac{\partial}{\partial t}(\nabla^2\psi) - \frac{\partial\psi}{\partial y}\frac{\partial}{\partial x}(\nabla^2\psi) + \frac{\partial\psi}{\partial x}\frac{\partial}{\partial y}(\nabla^2\psi) = v\,\nabla^4\psi$$

The simplification of the N-S equations to a form that involves a single dependent variable is obtained at the expense of an increase to fourth order. Whatever form of the N-S equations is used, the big hitch in the "git-along" is the inherent nonlinearity arising from the acceleration terms. Since our model of viscous effects is essentially linear, the treatment of these effects simply increases the number of terms in the equations of motion—the mathematical complexity, from the point of view of nonlinearities, is the same as that of the Euler equation.

Slow Flow

In several practical cases the essential nonlinearity present in the inertia terms of the N-S equations can be removed by neglecting these terms relative to the viscous terms. This can be illustrated with the vorticity transport equation, but in order to "scale" the problem, it is first necessary to establish common reference quantities for the several variables. Consider two-dimensional flow in which the vorticity is given by $\zeta = \partial v/\partial x - \partial u/\partial y$. The velocities are nondimensionalized by a reference velocity V, and the length coordinates are given the reference length L. Nondimensional quantities are designated with an asterisk, *, so that

$$u* = \frac{u}{V} \qquad v* = \frac{v}{V} \qquad x* = \frac{x}{L} \qquad y* = \frac{y}{L} \qquad t* = t\left(\frac{V}{L}\right)$$

and

$$\zeta = \frac{V}{L}\left(\frac{\partial v*}{\partial x*} - \frac{\partial u*}{\partial y*}\right)$$

This leads to the definition of nondimensional vorticity, $\zeta* = \zeta(L/V)$.

Equation (10.30) may now be written in terms of the nondimensional quantities:

$$\frac{(V/L)}{(L/V)}\frac{D\zeta*}{Dt*} = v\frac{1}{L^2}\nabla^2\left[\left(\frac{V}{L}\right)\zeta*\right]$$

or

$$\frac{D\zeta*}{Dt*} = \frac{v}{VL}\nabla^2\zeta*$$

where

$$\nabla^2 = [(\partial^2/\partial x*^2) + (\partial^2/\partial y*^2)]$$

The quantity VL/v is the Reynolds number, Re, based on the reference quantities V and L. Thus the vorticity transport equation is

$$\frac{D\zeta*}{Dt*} = \frac{1}{Re}\nabla^2\zeta* \qquad (10.31)$$

"Slow flow" is the term used to designate situations in which the Reynolds number is extremely small. Such flows may not actually be slow, because a small Reynolds number is obtained if the product VL is small. That is, if the characteristic velocity of the flow is very small, or if the characteristic length is very small (such as that of tiny particles in a flow), then the characteristic Reynolds number will be small. For instance, consider flow of a fluid with a kinematic viscosity on the order of 10^{-5} ft^2/s (water, for example). Then Re = $10^5 VL$, and it is seen that either V and/or L must be extremely small to yield a small Reynolds number. A value of Re = 1 is often considered to be the measure of "small," and if we wish to consider the flow about an object with a characteristic length of 1 ft, say, then slow flow would obtain at a speed of about 10^{-5} ft/s or about 10 in./day—basketballs do not experience slow flow!

Slow flow is, in fact, a valid approximation to some practical flow situations. These include particle flow (aerosols) and the flows occurring in the thin gaps between lubricated surfaces. The latter instance gives rise to the hydrodynamic theory of lubrication. The important fact, for our purposes, is that in such cases the viscous forces dominate the inertia forces and, with a sufficiently small Reynolds number, Eq. (10.31) may be approximated by

$$\nabla^2 \zeta^* = 0$$

In terms of the velocities, neglecting the inertia terms leads to

$$\mu \nabla^2 u_i = \gamma \frac{\partial h}{\partial x_i} + \frac{\partial P}{\partial x_i}$$

Slow motion is therefore one way to circumvent the acceleration terms in the N-S equations and their associated nonlinearities. [Slow flows are not discussed further in this text. The interested reader may refer to Schlichting (1968, chap. 6).] In the next chapter we examine other ways to treat linear versions of the N-S equations for several practical cases.

10.6 SUMMARY

The following is provided as a review of the steps leading to the Navier-Stokes equations and an indication of the basic nature of these important relationships.

I. Ingredients
   A. Force Balance
      Body forces
      Surface forces   Assumption: The fluid is
      Accelerations   continuous
   B. Constitutive Relationships
      Stress/strain rate connections   isotropic
                              Assumptions: Newtonian
                                                    laminar
II. Results
   Nonlinear: $u_j(\partial u_i/\partial x_j) \neq 0$ except in special cases
   Second-order: Velocity conditions must be prescribed at two boundaries
   Partial: u, v, w, ρ, P are functions of $x, y, z,$ and t (in general)
III. Auxiliary relationship (continuity)

$$\frac{\partial}{\partial t} + \frac{\partial}{\partial x_i} (\rho u_i) = 0$$

$\cdots$ four equations in u, v, w, ρ, P *sufficient,* if ρ = constant
IV. Essential boundary condition at a solid surface with velocity $(u_w)_i$, $u_i = (u_w)_i$

The last of these is the so-called no-slip condition required of a viscous fluid in the presence of a wall. We will have much more to say about this.

The N-S equations are rather formidable in their complete form, but the reduced expressions for incompressible flow with constant viscosity can be transformed into manageable shape in many cases. These transformations have largely to do with the boundary conditions, which, when they are simple enough, lead to useful solutions (integrations) of the N-S equations. Such solutions are often called "exact solutions" of the N-S equations because the mathematical steps that are involved require no approximation. In fact, these "exact" solutions are only approximate representations of real flows in which the true boundary conditions are more complex. Nevertheless, the solutions lead to significant insights into the effects of viscosity and, perhaps most important, they lead us to the boundary layer concept. This concept allows an approach to the study of viscous flows that is much more simple than treatment of the mathematical bag of worms embodied in Eq. (10.17).

Finally, before proceeding with exact solutions, it should be reemphasized that the form of our governing equations of motion is valid only for laminar flow of Newtonian fluid. Non-Newtonian fluids have more complicated constitutive relationships. A more severe restriction is that of laminar flow. Turbulent flows, which are very common in nature, are only partially understood, and the analytical techniques at our disposal are mostly unproven or limited to special cases. What happens when we are confronted with a turbulent flow situation? What seems to occur is that the problem is so complicated that we must use highly empirical methods and keep our fingers crossed. As often happens, the sophistication of the method of solution becomes inversely proportional to the complexity of the problem. This will become apparent in Chapter 13.

REFERENCES

Bird, R. B., W. E. Stewart, and E. N. Lightfoot: *Transport Phenomena,* John Wiley & Sons, New York, 1960.
Li, W., and S. Lam: *Principles of Fluid Mechanics,* 2nd ed., Addison-Wesley, Reading, Mass., 1976.
Schlichting, H.: *Boundary Layer Theory,* 4th ed., McGraw-Hill, New York, 1960.
————: 6th ed., 1968.
Sears, F. W.: *Thermodynamics,* Addison-Wesley, Reading, Mass., 1959.

PROBLEMS

10.1 Consider air at standard conditions (15°C, 1 atm).

(a) Estimate the average molecular velocity and the molecular density. The number of molecules per gram-mole (Avogadro's number) is $N = 6.022 \times 10^{23}$.

(b) Assume that the viscosity of standard air is 1.762×10^{-4} g/cm s (poise). Estimate the mean free path and the diameter of the molecules.

10.2 Show that the kinetic theory result for the viscosity of a perfect gas is dimensionally consistent with the definition given in Eq. (10.1).

10.3 Show that $\epsilon = $ div **s** as given in Eq. (10.4).

10.4 Figure 10.9 shows a face of a fluid element with sides δx, δy, and δz. Complete the shear stress description for the four sides of the element and sum the moments of the forces due to these stresses about the center point C. Using Newton's second law, show that

$$\tau_{xy} - \tau_{yx} = \rho(\delta r)^2 \alpha$$

where δr is the radius of gyration of the element and α is its angular acceleration. Note that upon contraction of the element the right-hand side of this expression approaches zero, with the result that $\tau_{xy} = \tau_{yx}$.

10.5 Show that the dilation θ [Eq. (10.13)] is zero in incompressible flow. What does this mean regarding the net linear strain of a fluid element in such a flow?

10.6 Write the three components of Du_i/Dt. Show that this may be written

$$\frac{Du_i}{Dt} = \frac{\partial u_i}{\partial t} + u_j \frac{\partial u_i}{\partial x_j}$$

10.7 Expand

$$\frac{\partial}{\partial x_j}\left(\frac{\partial u_j}{\partial x_i} + \frac{\partial u_i}{\partial x_j}\right)$$

and complete the derivation of Eq. (10.18).

10.8 Show that flow with the following velocity components is in rigid body rotation (without angular strain)

$$u = a + by - cz$$
$$v = d - bx + ez$$
$$w = f + cx - ey$$

where a through f are arbitrary constants.

10.9 Describe the following flows in terms of their rates of strain, rotation, and dilation:
(a) $u = cx$, $v = cy$, $w = -2cz$
(b) $u = c$, $v = w = 0$
(c) $u = 2cy$, $v = w = 0$
(d) $u = u(x, y)$, $v = v(x, y)$, $w = 0$

10.10 Determine the components of stress from the results obtained in problem 10.9.

10.11 The following vector gives the velocity field for an incompressible flow:

$$\mathbf{q} = 3x^2\hat{i} - 6xy\hat{j} + 16xy^2\hat{k}$$

Show that the flow satisfies mass continuity. If $\tau_{zz} = 0.100$ lb/ft^2, what are the other normal and shear stresses at the point $x = 20$, $y = 10$, and $z = 0$ (dimensions in feet)? The molecular viscosity is $\mu = 4.5 \times 10^{-5}$ lb s/ft^2.

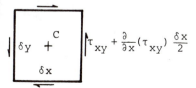

Figure 10.9 Problem 10.4.

ELEVEN

EXACT SOLUTIONS TO THE NAVIER–STOKES EQUATIONS

This chapter presents some classical solutions of the N-S equations; these are of practical significance, even though they represent relatively simple cases of viscous flow. Their study is useful in developing a sense of the influence of viscosity and in identifying the significant features of the various terms in the N-S equations. In addition, it will be found that the influence of viscosity is sometimes confined to a distinct region of the flow, and this insight will lead directly to a very important concept in the analysis of viscous flows.

In all cases we treat problems in which the flow is considered to be laminar, incompressible, and with constant viscosity. It has already been mentioned that our main thrust is to look at flow situations in which the boundary conditions lead to significant simplifications to the N-S equations. In addition to these simplifying boundary conditions, we will consistently apply a fundamental rule of viscous flow—the no-slip boundary condition.

11.1 NO–SLIP BOUNDARY CONDITION

In the previous chapter the influence of viscosity was seen to be manifested in the transport of momentum between adjacent fluid layers. In laminar flows this transport is due to the molecular activity of the flowing fluid. This activity is present everywhere in the flow, including the layers immediately adjacent to bounding surfaces. At such surfaces the existence of intermolecular attractions causes the fluid to adhere to a solid wall, and this gives rise to a shear stress there—the wall shear stress.

This is the *condition of no-slip* that obtains near solid walls, and its existence gives rise to an essential difference between viscous flows and ideal fluid flows. In the latter case it is necessary to consider one of two situations: (1) the field of flow is outside of the influence of the no-slip condition, or (2) the fluid has no viscosity and therefore feels no shear effect, even in the presence of boundaries. In viscous flows, which are sometimes referred to as flows of "real" fluids, the wall region may be thought of as the "birthplace" of viscous effects. In the vast majority of practical cases (excluding so-called free-molecule flows), fluid particles near a solid wall move at the same velocity as the wall. If the wall is stationary, then so are the fluid particles there. The no-slip condition is broadcast throughout the field of flow through the mechanism of viscosity, as illustrated in Fig. 10.1, and in flow regions that receive this message there are velocity gradients. These velocity gradients lead to shear stresses that, in turn, accomplish the change from free-stream conditions to those dictated by the no-slip condition at the wall.

11.2 STEADY FULLY DEVELOPED FLOWS

In this section we present several cases of practical interest in which the N-S equations reduce to ordinary differential equations.

Case 1. Flow between Infinite Flat Plates

The physical implication of the problem statement is that any motion normal to the x-y plane (see Fig. 11.1) is negligible. The view should be taken that there is only one main direction to the flow and that flow out the gap in a direction normal to this main direction is so far away ("infinite" flat plates) from the area under consideration that it has no observable effect—end effects are negligible. In addition, the statement of fully developed flow means that flow has proceeded in the gap for a distance that is sufficient to allow the assumption that further changes in the main flow direction (x direction) are negligible—there are no further developments in this direction. In mathematical terms, then,

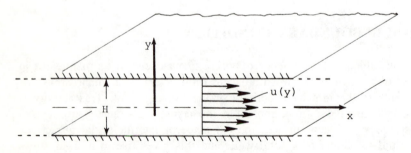

Figure 11.1 Geometry for Case 1.

$$w = 0 \qquad \frac{\partial}{\partial z} = 0 \qquad \text{and} \qquad u = u(y)$$

The latter expression should be fully interpreted: Not only is u a function of y, but it is not a function of anything else. Thus, $\partial u/\partial z = \partial u/\partial x = 0$. Dependency on time is ruled out by the steady hypothesis in the problem statement.

From the appropriate form of the continuity equation, $\partial v/\partial y$ must also be zero. But the velocity v is zero at the wall ($y = \pm H/2$) for all values of x, so we can only conclude that v is zero everywhere. The problem statement rules out one velocity component (w), and for fully developed flow the continuity equation rules out another component (v)—often the case in developing exact solutions to the N-S equations. Let us now look at the N-S equations in the appropriate form, Eq. (10.19), while incorporating the above observations: v, w, and $\partial/\partial t$ all zero, and $u = u(y)$.

For the x direction,

$$0 = -\frac{\partial}{\partial x}(P + \gamma h) + \mu \frac{d^2 u}{dy^2} \tag{11.1}$$

for the y direction,

$$0 = -\frac{\partial}{\partial y}(P + \gamma h) \tag{11.2}$$

and, finally, in the z direction, $0 = 0$.

Equation (11.2) states that the pressure sum, $P + \gamma h$, cannot vary with y. But $\partial/\partial z = 0$ also, so $P + \gamma h$ can depend only on x, so the partial derivative notation in Eq. (11.1) is redundant. Denoting $P + \gamma h$ by P', we have

$$\frac{dP'}{dx} = \mu \frac{d^2 u}{dy^2} \tag{11.3}$$

But *this* is telling us that a function of x only is equal to a function of y only: $f(x) = g(y)$, say. This is possible *only* if both "functions" are constants, and our final reduction of the N-S equations becomes

$$\frac{d^2 u}{dy^2} = \frac{1}{\mu} \frac{dP'}{dx} = \text{constant} \tag{11.4}$$

In this extremely simplified case the shear stress acting on a fluid element is exactly balanced by the pressure difference across it, and both act only in the x direction. The appropriate boundary conditions are obtained from application of the no-slip condition at the upper and lower surfaces:

$$u = 0 \qquad \text{at} \qquad y = \pm \frac{H}{2} \tag{11.4bc}$$

[The reader may wish to verify that dimensional consistency is satisfied in Eq. (11.4).]

Integrating Eq. (11.4) twice gives

$$u = \frac{1}{2\mu} \frac{dP'}{dx} y^2 + C_1 y + C_2$$

and application of the boundary conditions fixes the constants as

$$C_1 = 0 \quad \text{and} \quad C_2 = -\frac{1}{2\mu} \frac{dP'}{dx} \left(\frac{H}{2}\right)^2$$

The final result for velocity distribution across the gap is

$$\frac{u}{U_{max}} = 1 - \left(\frac{y}{H/2}\right)^2 \tag{11.5}$$

where U_{max} is the maximum velocity occurring at $y = 0$,

$$U_{max} = -\frac{1}{2\mu} \frac{dP'}{dx} \left(\frac{H}{2}\right)^2 \tag{11.6}$$

(The minus sign is physically significant. Since U_{max} must be positive, the pressure gradient must be negative. In order to overcome the shearing action of viscosity, the pressure must decrease in the direction of flow.)

Note that if the flow is horizontal, as inferred in Fig. 11.1, then $h = \text{constant} + y$ and gravity plays no role in driving the flow: $dP'/dx = dP/dx$. By calculating the derivative du/dy and evaluating it at the wall, it is found that the shear stress exerted *on the wall by the fluid* is

$$\tau_w = -\frac{H}{2} \frac{dP'}{dx}$$

The vorticity in the flow is $-du/dy$, or

$$\zeta_z = -\frac{1}{\mu} \frac{dP'}{dx} y$$

The vorticity is zero at the center of the gap and increases linearly in magnitude to maxima at the two walls.

Case 2. Flow in a Pipe: Hagen-Poiseuille Flow

Flow in a pipe is the cylindrical analog to Case 1. The mathematical implications of the problem statement are (see Fig. 11.2)

$$v_\theta = 0 \qquad \frac{\partial}{\partial \theta} = 0 \qquad v_z = v_z(r)$$

From continuity, Eq. (10.23), $\partial(rv_r)/\partial r = 0$ and, since v_r is zero at the wall, it is zero everywhere. From the N-S equation for the r direction, Eq. (10.20), $0 =$

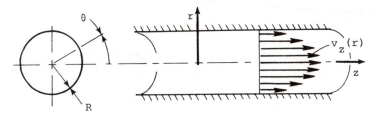

Figure 11.2 Geometry for Case 2.

$-\partial P'/\partial r$, so P' depends only on z. With these results incorporated into the z component of the N-S equations, Eq. (10.22), we have

$$\frac{1}{r}\frac{d}{dr}\left(r\frac{dv_z}{dr}\right) = \frac{1}{\mu}\frac{dP'}{dz} = \text{constant} \qquad (11.7)$$

with boundary conditions

$$v_z = 0 \qquad \text{at } r = R$$

$$\frac{dv_z}{dr} = 0 \qquad \text{at } r = 0 \qquad (11.7\text{bc})$$

The first of these is the no-slip condition. There is only one value of r in the pipe at which this condition is required, so the other boundary condition is obtained by imposing symmetry on the velocity distribution. (This would have worked in Case 1 as well, but it was not necessary there.) After integration and application of the boundary conditions (the reader should complete the details), we have

$$\frac{v_z}{W_{\text{max}}} = 1 - \left(\frac{r}{R}\right)^2 \qquad (11.8)$$

where

$$W_{\text{max}} = -\frac{R^2}{4\mu}\frac{dP'}{dz}$$

The velocity is seen to vary quadradically, as in Case 1.

Considerable additional information is available from this solution of the N-S equations. Evaluation of the velocity gradient at the wall gives $\tau_w = -(R/2)(dP'/dz)$. The volumetric flow rate, Q, is defined in terms of the mean flow velocity, $\bar{W}$, as follows:

$$Q = \int_A v_z \, dA = \pi R^2 \bar{W}$$

where A is the cross-sectional area of the pipe. From this, $\bar{W} = W_{\text{max}}/2$ and, with Eq. (11.8),

$$Q = -\frac{\pi R^4}{8\mu} \frac{dP'}{dz}$$

(Again, it is advisable at a point like this to check for dimensional consistency.)

If the Reynolds number for the flow is defined as $Re = \rho \bar{W} D / \mu$, then the pressure gradient is be given by

$$-\frac{dP'}{dz} = \frac{64}{Re} \frac{1}{D} \left(\frac{1}{2} \rho \bar{W}^2 \right)$$

and, with the d'Arcy friction factor,

$$f = \frac{4\tau_w}{(\frac{1}{2}\rho\bar{W}^2)} = \frac{64}{Re}$$

This classical result for laminar flow may be obtained by consideration of a force balance on a fluid element within the pipe, as is often done in a first course in fluid mechanics. Here it is important to realize that we have arrived at the same point by careful application of the boundary conditions to a very general expression, the Navier-Stokes equations.

Case 3. Couette Flow

In Cases 1 and 2 the flow is driven against viscous shear by either an applied pressure or the force of gravity, or both. Another mechanism for driving the flow stems from the boundary conditions themselves. Such flows are related to another classical solution of the N-S equations, Couette flow. Consider the geometry of Case 1, but in this case allow the two surfaces to be in motion relative to each other as illustrated in Fig. 11.3 (note that the origin of the coordinate system is now on the fixed surface). The *form* of the reduced version of the N-S equations will be exactly the same as Eq. (11.4):

$$\frac{d^2u}{dy^2} = \frac{1}{\mu} \frac{dP'}{dx} = \text{constant}$$

The difference from Case 1 lies in the boundary conditions, which for Couette

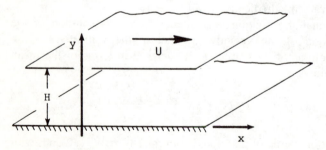

Figure 11.3 Geometry for Case 3.

flow are $u(0) = 0$ and $u(H) = U$, the speed of the upper surface relative to the lower. Integration of Eq. (11.4), followed by application of these new boundary conditions, gives

$$\frac{u}{U} = \frac{y}{H} - \frac{H^2}{2\mu U}\frac{dP'}{dx}\frac{y}{H}\left(1 - \frac{y}{H}\right) \tag{11.9}$$

If there is no pressure gradient, then this solution is $u/U = y/H$, the linear velocity profile associated with a constant shear stress. Flow occurs in this case for two reasons, either one of which may be set equal to zero. With no relative motion ($U = 0$), the solution reverts to that of Case 1. A wide variety of velocity profiles is possible, depending on the magnitude and sign of the dimensionless quantity $(H^2/2\mu U)(dP'/dx)$.

Case 4. Annular Flow along a Rod

We use the case of annular flow along a rod to illustrate a flow that is driven solely by gravity. After full development, a fluid film falls along the surface of a rod, as sketched in Fig. 11.4. There is no imposed pressure gradient. By analogy with Case 1, the governing equation has the same form as Eq. (11.7), but here we may write $h = h_{ref} - z$, where h_{ref} is the value of h when $z = 0$. Thus,

$$\frac{dP'}{dz} = \frac{d}{dz}(P + \gamma h) = \frac{dP}{dz} - \gamma$$

and, with the pressure in the surroundings taken as constant, $dP'/dz = -\gamma$, the hydrostatic pressure gradient. The flow is driven only by the force of gravity, and

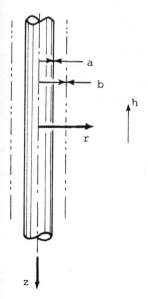

Figure 11.4 Geometry for Case 4.

$$\frac{1}{r}\frac{d}{dr}\left(r\frac{dv_z}{dr}\right) = -\frac{\gamma}{\mu} = -\frac{g}{\upsilon} \tag{11.10}$$

In addition to the no-slip condition at the rod surface, another boundary condition must be specified. This is accomplished by assuming that the fluid surrounding the falling film exerts no shear—an approximate but useful hypothesis. Thus,

$$\frac{dv_z}{dr} = 0 \text{ at } r = b \quad \text{and} \quad v_z = 0 \text{ at } r = a \tag{11.10bc}$$

Integration of Eq. (11.10), with application of these boundary conditions, leads to

$$v_z = \frac{ga^2}{2\upsilon}\left[\frac{b^2}{a^2}\ln\left(\frac{r}{a}\right) + \frac{1}{2}\left(1 - \frac{r^2}{a^2}\right)\right] \tag{11.11}$$

Evaluation of the shear stress exerted by the falling film on the surface of the rod will show that the force exerted is equal to the weight of the falling fluid.

11.3 SUDDENLY ACCELERATED FLAT PLATE: STOKES'S FIRST PROBLEM

Stokes's first problem is an example of an *unsteady* problem. It is like the others in that variations in only one *direction* are considered, but since time is also a factor, we arrive at a governing relationship that involves two independent variables—it will be a *partial* differential equation. We speculate about what might happen if we pulled an infinite (and weightless) flat plate, initially at rest, beneath a viscous fluid that is also initially at rest, as in Fig. 11.5. It is further specified that the fluid velocity parallel to the plate depends only on time and the coordinate y, $u = u(t, y)$, and the pressure is taken to be invariant with x and t.

From continuity it is concluded that $v = 0$ and, from the y component of the N-S equations, $\partial P'/\partial y = 0$ so that $P' = f(x, t)$, where, as before, $P' = P + \gamma h$. When this result is used in the x component of the N-S equations, we find that

$$\rho\frac{\partial u}{\partial t} = -\frac{\partial P'}{\partial x} + \mu\frac{\partial^2 u}{\partial y^2}$$

But $\partial P'/\partial x = \partial(P + \gamma h)/\partial x$ and, if h is parallel to y, as we have postulated here,

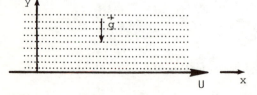

Figure 11.5 Suddenly accelerated flat plate.

then $\partial h/\partial x = 0$. Since it has already been specified that P is independent of x and t, it is concluded that $\partial P'/\partial x$ is zero. (The above reasoning may seem unnecessarily involved, but it helps to ensure that the obvious is supported by the mathematics.) The governing form of the N-S equations is therefore

$$\frac{\partial u}{\partial t} = v\,\frac{\partial^2 u}{\partial y^2} \tag{11.12}$$

The reader may wish to ponder the similarity between this expression and the heat conduction equation. There are many solutions to Eq. (11.12), depending on the boundary conditions. Those applicable here are

$$u(t, y) = 0 \qquad t \le 0$$

$$u(t, 0) = U_w$$

$$u(t, \infty) = 0 \tag{11.12bc}$$

Transformation of Eq. (11.12) into the Laplace domain leads to an ordinary differential equation in terms of the transformed velocity $U(s, y)$:

$$\frac{d^2 U}{dy^2} - \frac{s}{v} = 0 \qquad \text{with } U(s, \infty) = 0 \text{ and } U(s, 0) = \frac{U_w}{s}$$

The solution is forthcoming from this expression, with the level of difficulty depending on one's familiarity with such methods. In order to use this case to its best advantage in illustrating the fluid mechanical properties involved, we use a method that is popular in this field of study. We "seek" a *similarity variable* that depends on both y and t in such a way that its application in Eq. (11.12) will lead to an ordinary differential equation. The quotes are used here because we will not "seek" very long—a discussion of such methods is presented in the next chapter. For now, let us make a very good guess for such a variable,

$$\eta = \frac{y}{2\sqrt{vt}} \tag{11.13}$$

and operate under the assumption that $u(t, y) \rightarrow u(\eta)$ under the influence of this transformation. In particular, we write

$$u(t, y) = U_w f(\eta) \tag{11.14}$$

With Eq. (11.14), the various derivatives in Eq. (11.12) become

$$\frac{\partial u}{\partial t} = U_w\,\frac{df}{d\eta}\,\frac{\partial \eta}{\partial t} = U_w f'\,\frac{\partial \eta}{\partial t}$$

$$\frac{\partial u}{\partial y} = U_w f'\,\frac{\partial \eta}{\partial y}$$

$$\frac{\partial^2 u}{\partial y^2} = U_w f'\,\frac{\partial^2 \eta}{\partial y^2} + U_w f''\left(\frac{\partial \eta}{\partial y}\right)^2$$

From the definition of η, Eq. (11.13),

$$\frac{\partial \eta}{\partial t} = -\frac{\eta}{2t} \qquad \frac{\partial \eta}{\partial y} = \frac{1}{2\sqrt{vt}} \qquad \frac{\partial^2 \eta}{\partial y^2} = 0$$

and the transformed derivatives are found to be

$$\frac{\partial u}{\partial t} = -\frac{U_w \eta}{2t} f' \qquad \frac{\partial u}{\partial y} = \frac{U_w}{2\sqrt{vt}} f' \qquad \frac{\partial^2 u}{\partial y^2} = \frac{U_w}{4vt}$$

When these are substituted into the governing differential equation, Eq. (11.12), we obtain

$$f'' + 2\eta f' = 0 \tag{11.15}$$

Equation (11.13) was indeed a good guess! The transformed boundary conditions are

$$f(0) = 1$$
$$f(\infty) = 0 \tag{11.15bc}$$

The similarity transformation has really replaced one problem with another: the linear partial differential equation has been replaced with a nonlinear ordinary one (the same thing happened in the Laplace transform approach). In this case, however, the complexity of the nonlinear expression is not severe, and we can solve it directly. The first integral of Eq. (11.15) is easily obtained with the substitution $g = f'$, from which $g'/g = -2\eta$, and

$$g = f' = C_1 e^{-\eta^2}$$

where C_1 is an integration constant to be evaluated. A second integration gives

$$f = C_1 \int_0^\eta e^{-\eta^2} \, d\eta + C_2$$

and this may be expressed in terms of the tabulated error function, erf(η), which is defined as

$$\text{erf}(\eta) = \frac{2}{\sqrt{\pi}} \int_0^\eta e^{-\eta^2} \, d\eta$$

so that

$$f = \frac{\sqrt{\pi}}{2} C_1 \, \text{erf}(\eta) + C_2$$

The error function has the limiting values erf(0) = 0 and erf(∞) = 1. When these are applied to the above result, the constants of integration are obtained from the boundary conditions, Eq. (11.15bc), as $C_1 = -2/\sqrt{\pi}$ and $C_2 = 1$. Thus, $f = 1 - \text{erf}(\eta)$, but this is also tabulated and is known as the complimentary error function, erfc(η). Finally, combining this result with the definition of $f = u/U_w$, we have

$$\frac{u}{U_w} = \text{erfc}(\eta) \tag{11.16}$$

This solution is shown graphically in Fig. 11.6. The derivative f' is also useful, and is given by

$$f' = -\frac{2}{\sqrt{\pi}} e^{-\eta^2} \tag{11.17}$$

Figure 11.6 provides a single curve of u as a function of the single independent variable $\eta(t, y)$. This illustrates graphically the unifying nature of the variable η. For any selected value of t, a separate curve of u vs. y may be plotted. But each of these separate curves will be identical in *shape* to that of Fig. 11.6. Hence the identification η as a similarity variable.

Region of Viscous Action

We have identified the vorticity as an indication of the twisting and turning action that is induced by the presence of viscous action. Let us examine the results of the previous analysis from the point of view of the vorticity that is generated by virtue of the shear created at the plate and transmitted into the fluid by means of viscosity. For this simple motion the only component of vorticity is

$$\zeta_z = -\frac{\partial u}{\partial y} = -U_w f' \frac{\partial \eta}{\partial y} = -\frac{U_w}{2\sqrt{vt}} f'$$

and, from Eq. (11.16),

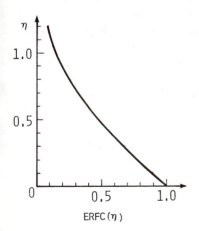

ERFC (η)

Figure 11.6 Velocity variation above a moving plate.

$$\zeta_z = \frac{U_w}{\sqrt{\pi v t}} \exp\left[-\left(\frac{y}{2\sqrt{vt}}\right)^2\right]$$

This may be simplified and made more meaningful if it is noted that the maximum value of the vorticity occurs at $y = 0$—the surface of the plate. Beginning with this maximum value, the vorticity decreases precipitously with y, for any selected value of t. Thus,

$$\frac{\zeta}{\zeta_w} = \exp\left[\left(\frac{y}{2\sqrt{vt}}\right)^2\right] \tag{11.18}$$

where the maximum value, at the wall, is

$$\zeta_w = \frac{U_w}{\sqrt{\pi v t}}$$

Since vorticity is evidence of the influence of viscosity, it is useful to study how this evidence is distributed throughout the fluid above the moving plate. This may be done by considering some representative value of the parameters in Eq. (11.18). Let us specify a typical value of kinematic viscosity of $v = 10^{-5}$ ft^2/s— roughly that for water. We consider regions and/or times in the flow where the vorticity is one-half its maximum value, that is, $\zeta = \frac{1}{2}\zeta_w$. With these hypotheses, Eq. (11.18) gives

$$\left(\frac{y}{\sqrt{t}}\right)_{50\%} = 2\sqrt{10^{-5}}\sqrt{\ln 2} \approx 0.005 \text{ ft/s}^{1/2}$$

in round numbers. Thus, if, for instance, 1 hour has elapsed from the initiation of the motion, then $y_{50\%}$ will be about 0.3 ft or about 4 in. The point at which the vorticity is 50% of the maximum value at the wall will only be 4 in. from the wall after the motion has proceeded for 1 hour! Clearly, there is a great proclivity for the vorticity to remain concentrated near the wall. From another point of view, consider *how long* it takes for the $\frac{1}{2}$-value point to reach 6 in. from the wall—about $2\frac{1}{2}$ hours! The point of all this is that when a plate is pulled through a viscous fluid that is otherwise at rest, the vorticity thus created stays very near its birthplace, the plate surface. There is an imaginary distance from the plate, in this unsteady problem, beyond which there is very little fluid rotation.

We now proceed to yet another point of view. The distance x is defined as the measure of the location of some reference point on the surface of the plate, which is being pulled along at U_w. That is, $x = U_w t$ and is initially zero. Using this reference dimension, a dimensionless group may be defined as

$$\text{Re} = \frac{U_w x}{v}$$

and the vorticity relationship becomes

$$\frac{\zeta}{\zeta_w} = \exp\left[-\left(\frac{y\sqrt{U_w}}{2\sqrt{vx}}\right)^2\right]$$

which may be rearranged to give

$$\frac{y}{x} = \frac{2}{\sqrt{Re}}\left(\ln\frac{\zeta_w}{\zeta}\right)^{1/2}$$

Now let us look for a particular value of y—that distance from the plate that marks the points where the vorticity is only 1% of its value at the surface. That is, we define a value $y = \delta$, beyond which the vorticity is less than 1% of the wall value. From the above result,

$$\frac{\delta}{x} = \frac{2}{\sqrt{Re}}(\ln 100)^{1/2} \approx \frac{4.3}{\sqrt{Re}} \tag{11.19}$$

For example, if the plate moves at 10 ft/s for 1 hour, in our typical fluid, the

Table 11.1 Some exact solutions to the Navier-Stokes equations (laminar flow of a Newtonian fluid)

Brief description	Steady	Incompressible	Constant viscosity	Dimension	Linear
1. Flow between parallel walls					
(a) Couette flow	x	x	x	1	x
(b) Poiseuille flow	x	x	x	1	x
(c) Combinations	x	x	x	1	x
2. Flow between rotating cylinders	x	x	x	1	x
3. Duct flows					
(a) Circular	x	x	x	1	x
(b) Annular	x	x	x	1	x
(c) Equilateral triangle	x	x	x	2	x
(d) Rectangular	x	x	x	2	x
4. Accelerating flows					
(a) Flat plate from rest		x	x	2	x
(b) Oscillating flat plate		x	x	2	x
(c) Starting flows		x	x	2	x
(d) Oscillating body of revolution		x	x	3	x
5. Vorticity diffusion around a vortex core		x	x	2	x
6. Stagnation flow (Hiemenz)	x	x	x	2	
7. Rotating disc	x	x	x	2	
8. Flow in converging/diverging channels	x	x	x	2	
9. Jet in an infinite fluid	x	x	x	2	
10. One-dimensional compressible flow	x		x	1	

reference point will have traveled 36,000 ft and the value of δ will be about 0.8 ft. Beyond that distance from the plate, the vorticity will be less than 1% of its value at the surface, at that point and time. We will use these insights to gain confidence that if a viscous fluid flows over a surface, the region within which most of the viscous action takes place, or *boundary layer,* may be very, very close to that surface; and the Navier-Stokes equations should contain this information.

11.4 SUMMARY

There are a great number of exact solutions to the N-S equations, many of which are minor extensions of one another, as we have seen in this chapter. The simplicity of the conditions imposed in order to obtain these solutions often makes them of rather academic interest. Perhaps the most important benefits that accrue from these results are the insights gained rather than the numerical values obtained. To give the reader a feeling that we have only scratched the surface here, Table 11.1 shows some of the other solutions that have been obtained. Note again, however, that we are still talking about the laminar flow of a Newtonian fluid.

PROBLEMS

11.1 Verify the results for wall shear stress and vorticity in Case 1.

11.2 Verify the results for wall shear stress, mean velocity, and flow rate in Case 2.

11.3 Express the solution for Couette flow in terms of the parameter $p = (H^2/2\mu U)$ $(-dP'/dx)$. Plot u/U vs. y/H for values of p ranging from -3 to $+3$ in unit increments.

11.4 Evaluate the velocity gradient at the surface of the rod in Case 4, and show that the shear stress there is $\tau_a = (\gamma/2a)(b^2 - a^2)$. Show that this stress exerts a force over a length of the rod that is equal in magnitude to the weight of the fluid film of that length.

11.5 In interpreting the results of the suddenly accelerated flat plate, the distance δ was evaluated on the basis of $\zeta/\zeta_w = 0.01$. Investigate the sensitivity of the result for δ/x to this choice of values.

11.6 Consider the steady laminar flow of a fluid between two concentric cylinders, with the inner cylinder rotating at an angular velocity ω, as shown in Fig. 11.7. There is no flow in the axial direction, and the tangential velocity depends only on r; that is, $v_z = 0$ and $v_\theta = v_\theta(r)$. Develop the appropriate form of the N-S equations for these conditions. Assume constant pressure and neglect hydrostatic effects. Include the necessary boundary conditions.

11.7 Show that the equation of motion developed in problem 11.6 integrates to give

$$\frac{v_\theta}{r} = \omega \frac{R_i^2}{r^2} \left(\frac{R_o^2 - r^2}{R_o^2 - R_i^2} \right)$$

11.8 If the cylinders in problem 11.6 are of length L and the fluid has a viscosity μ, find an expression for the torque necessary to turn the inner cylinder at angular velocity ω. Show that the units of your solution are compatible.

11.9 Consider flow vertically downward in the annular region between two concentric pipes (Fig. 11.8). The flow is fully developed and the pressure is constant. Give the reduced form of the N-S equations for this problem and state the boundary conditions.

11.10 Obtain the expression for shear stress in problem 11.9 as a function of r.

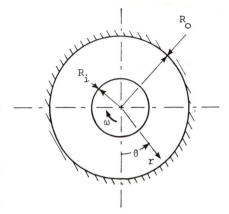

Figure 11.7 Problem 11.6.

11.11 Find the force due to shear acting on the rod of problem 11.9 over a length of 1 ft if the following values are given: $\rho = 2$ slug/ft^3, $a = 6$ in., $b = 12$ in.

11.12 Consider steady incompressible laminar flow between two plates separated by a distance H. The flow is fully developed. For a point in the flow at $y = H/4$ (measured from the centerline), find expressions for the following: (a) the normal and shear stresses; (b) the linear and angular rates of strain; (c) the rate of rotation.

11.13 Consider the flow induced when a thin wire, radius R_i, is drawn down the centerline of a pipe, radius R_o (Fig. 11.9). The flow is steady, incompressible, and laminar. Assume fully developed flow (end effects are negligible). Develop the appropriate form of the N-S equations governing the flow in the absence of an applied pressure gradient.

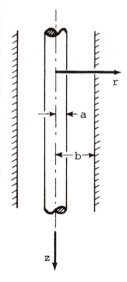

Figure 11.8 Problem 11.9.

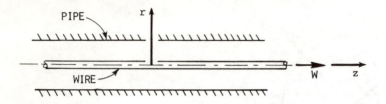

Figure 11.9 Problem 11.13.

11.14 Find an expression for the velocity distribution, $v_z(r)$, in the annular region of problem 11.13.

11.15 Find an expression for the force necessary to draw a wire of length L in problem 11.13.

11.16 A thin film falls downward along a flat plate under the influence of gravity (see Fig. 11.10). The pressure is constant, and there is no appreciable motion in the transverse (y) direction. If the flow is also laminar, incompressible, and steady, give the reduced form of the N-S equations for this problem.

11.17 Assume that the velocity in the x direction is negligible in problem 11.16. Find the appropriate form for the N-S equations for this case, and specify the boundary conditions.

11.18 Integrate the results of problem 11.17 to find the downward velocity, w, as a function of x. Evaluate the shear stress (magnitude and direction) acting on the wall.

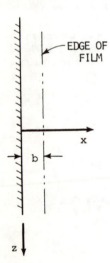

Figure 11.10 Problem 11.16.

TWELVE

THE BOUNDARY LAYER

Investigation of the flow associated with the suddenly accelerated flat plate has led to the insight that under certain circumstances the influence of viscosity—transport of momentum across fluid layers—is extremely confined in space. The presence of a bounding surface is "made known" to the viscous fluid through the no-slip condition, but the region that is influenced by this presence can be quite small.

These ideas were developed by means of an "exact" solution to the Navier-Stokes equations. We may expect, therefore, that a more detailed examination of these equations will lead to a more general view regarding the region of influence of viscosity—a region that is called the *boundary layer*. The concept of the boundary layer was first presented by Ludwig Prandtl (1875–1953) in 1904 to the Mathematical Congress in Heidelberg, in a paper titled (somewhat misleadingly), "Fluid Motion with Very Small Friction." The impact of these ideas on the field of fluid mechanics has been truly profound in allowing the solution of otherwise intractible viscous flow problems. Perhaps of equal significance is the fact that existence of the boundary layer region leads to the definition of a "nonboundary layer" region in which the flow may be analyzed by the methods of Part One: ideal fluid flow. Thus the boundary layer theory has brought to fluid mechanics an organized approach to a field of engineering that might otherwise have remained the highly theoretical province of the mathematician.

(Parenthetically, the reader might wish to contemplate what form the field of fluid mechanics would have taken if Prandtl had had access to the use of a high-speed digital computer. Today these machines allow the integration of the N-S equations in their most general form. If this had been made "easy" for Prandtl, the boundary layer concept, together with its many far-reaching insights, might have remained lost in a fog of digital number crunching.)

12.1 REDUCTION OF NAVIER–STOKES EQUATIONS

The boundary layer idea involves judgments as to which terms of the N-S equations are "large" and which are "small." Accordingly, it is necessary to identify some reference quantities so that the various terms can be compared on a common basis. The terms of the N-S equations are then divided by these reference quantities to create nondimensional forms that are independent of the particular system of dimensions being used. Thus, for example, to say that a length of "1" is large has no physical meaning if that length has a dimension—a centimeter may be large to a flea, but it is peanuts to an elephant.

Although three-dimensional boundary layers do exist, together with a considerable theoretical basis, only two-dimensional boundary layer flows will be considered in this text. Figure 12.1 illustrates the reference quantities that are relevant to the boundary layer. These are the free stream velocity $U(x)$ and the length L over which the terms of the N-S equations are evaluated. The coordinate x is measured along the surface of the boundary, and the coordinate y is directed normal to x from the surface outward, into the crossing flow. The boundary layer hypothesis embodies the notion that viscous effects are important only within the thin layer of thickness δ. For values of y greater than this, the flow is not influenced to any great extent by viscosity and, therefore, $U = U(x)$ and the methods of ideal flow analysis apply. Within the boundary layer, however, $u = u(x, y)$, and the viscous effects must be taken into account.

Reduction of the N-S equations is begun by scaling of the important variables such that

$$u_i^* = \frac{u_i}{U} \qquad x_i^* = \frac{x_i}{L} \qquad t^* = \frac{t}{L/U} \qquad P^* = \frac{P}{\rho U^2}$$

With the reference quantities so defined, nondimensional terms that are much less than unity are taken to be small relative to terms that are greater than unity. This

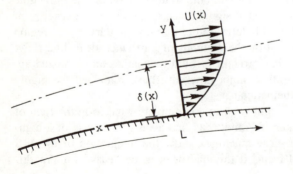

Figure 12.1 Region of boundary layer flow.

is because the boundary layer thickness, δ, is, by definition, very thin, so that δ/L is much less than unity. In nondimensional terms, $\delta^* = \delta/L$, the epitomy of small:

$$\delta^* \ll 1$$

On the other hand, the coordinate x is required to have values on the order of the reference length L. This is expressed by the notation

$$x^* = O(1)$$

which is read, "x-star has the order of magnitude unity." Within the boundary layer, y is always less than δ, so

$$y^* = O(\delta^*)$$

Also, within the boundary layer, the x-wise velocity varies from $u = 0$ at the wall to $u = U$ at the inner edge of the free stream flow. Therefore,

$$u^* = O(1)$$

It now becomes necessary to estimate the magnitudes of the derivatives that appear in the various terms of Eq. (10.19). (The reader should appreciate that these order-of-magnitude arguments are somewhat heuristic and that the results that ensue are dependent on their validity. Such validity has been verified by an extensive amount of experimental evidence.) The nondimensional rate of change in the x direction of the dominant velocity u, $\partial u^*/\partial x^* = (L/U)(\partial u/\partial x)$, is constrained to be no greater than unity by selection of the characteristic length L. Thus, $\partial u^*/\partial x^* = O(1)$. From continuity, $\partial v^*/\partial y^* = O(1)$ and, since $y^* = O(\delta^*)$, this can hold only if $v^* = O(\delta^*)$. The basic premise is that u^* is large relative to v^*, and that changes with x^* are small relative to those that occur across the narrow band encompassed by y^*. Keeping these in mind, it may be argued that $\partial u^*/\partial y^* = O(1/\delta^*)$ and that $\partial v^*/\partial x^* = O(\delta^*)$. Further, the magnitudes of the nondimensional second derivatives may be deduced from the first derivatives to be

$$\frac{\partial^2 u^*}{\partial x^{*2}} = O(1) \qquad \frac{\partial^2 u^*}{\partial y^{*2}} = O\left(\frac{1}{\delta^{*2}}\right)$$

and

$$\frac{\partial^2 v^*}{\partial x^{*2}} = O(\delta^*) \qquad \frac{\partial^2 v^*}{\partial y^{*2}} = O\left(\frac{1}{\delta^*}\right)$$

The derivative with respect to time, $\partial u^*/\partial t^*$, will be retained in the N-S equations if it is assigned a magnitude of unity. Accordingly, $\partial v^*/\partial t^* = O(\delta^*)$.

The preceding arguments should not be viewed as attempts to quantify the various velocities and their derivatives mathematically. The effort has been to identify *qualitatively* those quantities that are small relative to those that are not small. The order-of-magnitude notation is simply a device to systematically separate these quantities. In plain language, it would be just as well to say that u is large relative to v, and that variations across the boundary layer, if they exist, are much stronger than variations in the main flow direction. We will soon see that the order-of-magnitude analysis does allow some qualitative judgments to be made about the requirements necessary to validate the boundary layer hypothesis.

The Navier-Stokes equations for two-dimensional incompressible flow are brought forward below, from Eq. (10.19):

$$\frac{\partial u}{\partial t} + u\frac{\partial u}{\partial x} + v\frac{\partial u}{\partial y} = -\frac{1}{\rho}\frac{\partial P}{\partial x} + \upsilon\left(\frac{\partial^2 u}{\partial x^2} + \frac{\partial^2 u}{\partial y^2}\right)$$

$$\frac{\partial v}{\partial t} + u\frac{\partial v}{\partial x} + v\frac{\partial v}{\partial y} = -\frac{1}{\rho}\frac{\partial P}{\partial y} + \upsilon\left(\frac{\partial^2 v}{\partial x^2} + \frac{\partial^2 v}{\partial y^2}\right)$$

The hydrostatic terms, should they be important, may be thought of as being included in the pressure terms. These equations are now repeated below, in nondimensional terms, with each term assigned its appropriate relative magnitude:

$$\underset{1}{\frac{\partial u^*}{\partial t^*}} + \underset{1}{u^*\frac{\partial u^*}{\partial x^*}} + \underset{1}{v^*\frac{\partial u^*}{\partial y^*}} = -\frac{\partial P^*}{\partial x^*} + \frac{1}{\text{Re}}\left(\underset{1}{\frac{\partial^2 u^*}{\partial x^{*2}}} + \underset{1/\delta^{*2}}{\frac{\partial^2 u^*}{\partial y^{*2}}}\right) \quad (12.1)$$

with relative magnitudes 1, 1, $\delta^* \cdot 1/\delta^*$ under the left-hand terms.

$$\underset{\delta^*}{\frac{\partial v^*}{\partial t^*}} + \underset{1}{u^*\frac{\partial v^*}{\partial x^*}} + \underset{\delta^*}{v^*\frac{\partial v^*}{\partial y^*}} = -\frac{\partial P^*}{\partial y^*} + \frac{1}{\text{Re}}\left(\underset{\delta^*}{\frac{\partial^2 v^*}{\partial x^{*2}}} + \underset{1/\delta^*}{\frac{\partial^2 v^*}{\partial y^{*2}}}\right) \quad (12.2)$$

The Reynolds number, which appears upon nondimensionalization, is

$$\text{Re} = \frac{\rho UL}{\mu} = \frac{UL}{\upsilon}$$

From Eq. (12.1) it may be seen that the acceleration terms in the main flow direction (the x direction) are all of order of magnitude 1 in the scaling. Further, the $\partial^2 u^*/\partial y^{*2}$ $[O(1/\delta^{*2})]$ term is by far the largest of the terms in the viscous part of the right-hand side. Therefore the $\partial^2 u^*/\partial x^{*2}$ term may be neglected in the x-wise viscous effect. If we restrict applications to regions of flow where $\text{Re} = O(1/\delta^{*2})$, then the viscous terms will be of the same order of magnitude as that of the inertia terms on the left-hand side of Eq. (12.1)—there will be a *balance* between the inertia effects and viscous effects. *This, by definition, is the region of boundary layer flow.*

The major conclusions derived from the scaled version of the x-wise component of the N-S equations are

$$\frac{\partial^2 u^*}{\partial x^{*2}} \text{ is negligible}$$

and

$$\mathrm{Re} = \mathrm{O}\left(\frac{1}{\delta^{*2}}\right)$$

The latter result may be manipulated to give $\delta \approx \sqrt{vT}$, where $T = L/U$. This is directly analogous to the result obtained from the study of the suddenly accelerated flat plate, Eq. 11.19.

From the scaled version of the N-S equation for the y direction, Eq. (12.2):

$$\mathrm{O}(\delta^*) = -\frac{\partial P^*}{\partial y^*} + \mathrm{O}\left[\delta^{*2}\left(\delta^* + \frac{1}{\delta^*}\right)\right] = -\frac{\partial P^*}{\partial y^*} + \mathrm{O}(\delta^*)$$

From this it is deduced that

$$\frac{\partial P^*}{\partial y^*} = \mathrm{O}(\delta^*)$$

This is the third major result of the analysis: The pressure variation across the boundary layer (in the y direction) is negligibly small with respect to that in the x direction. The pressure in the free stream is "impressed" on the boundary layer so that very nearly the same pressure is felt at the wall. In other words, the only significant pressure variation is in the flow direction, and $\partial P^*/\partial x^* \approx dP^*/dx^*$.

Boundary Layer Equation

The previous analysis has led to the conclusion that, in boundary layer flow, the forces and accelerations in the direction normal to the bounding surface are negligible, and Eq. (12.2) need receive no further attention. The *boundary layer equation* is Eq. (12.1), which, after redimensionalization, appears as

$$\frac{\partial u}{\partial t} + u\frac{\partial u}{\partial x} + v\frac{\partial u}{\partial y} = -\frac{1}{\rho}\frac{dP}{dx} + v\frac{\partial^2 u}{\partial y^2} \tag{12.3}$$

with boundary conditions:

$$u(x, 0, t) = 0 \qquad v(x, 0, t) = 0 \qquad u(x, \infty, t) = U(x, t)$$

In unsteady flow, initial conditions would also be required. To these, lest it be forgotten, the equation of mass continuity should also be added:

$$\frac{\partial u}{\partial x} + \frac{\partial v}{\partial y} = 0 \tag{12.4}$$

Is the pressure in Eq. (12.3) an unknown? The answer is an emphatic "no," if the flow outside the boundary layer can be calculated by means of ideal flow analysis. We have said that the pressure is constant across the boundary layer (approximately), and that at the edge of the boundary layer the velocity must match that of the ideal flow there. The pressure must also match that in the flow at the boundary layer edge, so the pressure everywhere is prescribed by the outer boundary condition, $u(x, \infty, t) = U(x, t)$. In other words, as y approaches the edge of the boundary layer region, u ceases to depend on y and there is no further shear and fluid rotation. In such circumstances, outside the boundary layer, the boundary layer equation takes the form:

$$\frac{\partial U}{\partial t} + U\frac{\partial U}{\partial x} = -\frac{1}{\rho}\frac{dP}{dx} \tag{12.5}$$

It is pleasing to see the return of the Euler equation in the free stream outside the boundary layer—analyses such as those in Part One are now justified. When the ideal flow analysis is accomplished, the pressure distribution is available for use in Eq. (12.3).

The boundary equation may be viewed from a more general perspective if the viscous term is recast in its original form as a shear stress. If τ_{yx} is evaluated from Eq. (10.13),

$$\tau_{yx} = \tau_{21} = 2\mu e_{21} = \mu\left(\frac{\partial u_1}{\partial x_2} + \frac{\partial u_2}{\partial x_1}\right) = \mu\left(\frac{\partial u}{\partial y} + \frac{\partial v}{\partial x}\right)$$

$$\frac{\partial}{\partial y}(\tau_{yx}) = \mu\left(\frac{\partial^2 u}{\partial y^2}\right) + \mu\frac{\partial}{\partial x}\left(\frac{\partial v}{\partial y}\right)$$

But from the continuity equation for two-dimensional incompressible flow, $\partial v/\partial y = -(\partial u/\partial x)$ and the last term above becomes $-\mu(\partial^2 u/\partial x^2)$. This is the term that is neglected in the boundary layer approximation of the x component of the N-S equations. Accordingly, only one component of the shear stress is deemed to be important in boundary layer flow, and Eq. (12.3) may be written, for steady flow,

$$u\frac{\partial u}{\partial x} + v\frac{\partial u}{\partial y} = U\frac{dU}{dx} + \frac{1}{\rho}\frac{\partial}{\partial y}(\tau_{yx}) \tag{12.6}$$

Here we have used the free stream velocity $U(x)$ to demonstrate its use in replacing the pressure term through Eq. (12.5). Equation (12.6) is especially important in that it removes the constraint of laminar flow from the boundary layer equation. Of course, to proceed further with this result, constitutive relationships are needed to express the shear stress in terms of velocities.

Equations (12.3) and (12.6) are the forms of the boundary layer equations that will be addressed in this text. Other forms have been developed, and the boundary layer hypothesis has led to useful expressions in curvilinear coordinates as well as for compressible flows [Schlichting (1968), Thompson (1972)].

Remarks on the Mathematical Nature of the Boundary Layer Equations

The physical interpretation of the boundary layer hypothesis involves the rationale that things change much more dramatically across the boundary layer than along it; thus, the conclusion that $|\partial^2 u/\partial y^2| \gg |\partial^2 u/\partial x^2|$. In regions near the point of inception of a boundary layer, such as at the leading edge of a plate immersed in a flowing viscous fluid, gradients in the x direction are not negligible, and the boundary layer theory breaks down. In such cases the previous scaling of the N-S equations is no longer appropriate, and the full equations must be considered. The boundary layer equation is an *approximation* to the full N-S equations, and may be thought of as a term in a series expansion that becomes dominant at large values of the Reynolds number.

For another point of view, the boundary layer equation may be expressed in terms of the stream function, Eq. (3.11), to give

$$\frac{\partial}{\partial t}\left(\frac{\partial \psi}{\partial y}\right) + \frac{\partial}{\partial y}\frac{\partial^2 \psi}{\partial x \partial y} - \frac{\partial \psi}{\partial x}\frac{\partial^2 \psi}{\partial y^2} = -\frac{1}{\rho}\frac{dP}{dx} + v\frac{\partial^3 \psi}{\partial y^3}$$

This equation is third order with respect to the space coordinates, in contrast to the corresponding fourth-order expression of the N-S equations developed in Chapter 10. The reduction of the order is an important simplification, but the boundary layer equation cannot satisfy as many boundary conditions as can the full N-S equations.

From a computational point of view, the boundary layer equation may be integrated by a "marching" technique in which the unknown velocities may be explicitly determined at each point (or node point, in a finite difference grid) as the march proceeds in the predominant flow direction. Using the full N-S equations, on the other hand, solution for conditions at a point in the flow requires information at all surrounding points, upstream as well as downstream. The resulting difference equations are implicit, requiring iterative methods for their solution. In the general theory of partial differential equations, these characteristics are related to the fact that the boundary layer equations are of the parabolic type, whereas the N-S equations are elliptic. For further discussion, see Schlichting (1968, chap. 8), and Cebici and Bradshaw (1977).

Boundary Layer Separation in Steady Flow

When the steady flow version of Eq. (12.3) is evaluated at the wall, where both u and v are zero, the gradient of the shear stress there is found to be balanced by the pressure gradient (which, under the boundary layer assumption, is the same at the wall as it is in the free stream). Defining $\tau = \mu(\partial u/\partial y)$:

$$0 = -\frac{dP}{dx} + \left(\frac{\partial \tau}{\partial y}\right)_{y=0}$$

or

$$\left(\frac{\partial \tau}{\partial y}\right)_{y=0} = \frac{dP}{dx} \tag{12.7}$$

Equation (12.7) shows that the trend in shear stress near the wall depends on whether the pressure in the free stream is increasing or decreasing. Beginning with this observation, the nature of the variation of velocity in the boundary layer—the *velocity profile*—can be deduced. When, in the boundary layer, the velocity develops a tendency to move in a direction *opposite* to the main flow, the boundary layer is said to approach *separation* from the bounding surface.

Let us first consider a case in which the pressure is decreasing in the mainstream direction; that is, $dP/dx < 0$. In such cases Eq. (12.7) indicates that the shear stress tends to decrease from its value at the wall to lesser values in the nearby flow. It is known, by the boundary layer hypothesis, that the shear stress approaches zero at the outer edge of the boundary layer, and therefore it may be deduced that the shear stress in this case decreases monotonically with distance from the wall. This behavior of the shear stress is illustrated in Fig. 12.2a. Because the shear stress is steadily decreasing in the boundary layer, for $dP/dx < 0$, the velocity *gradient* is also maximum at the wall and steadily decreases with y. The velocity therefore begins at zero at the wall (the no-slip condition) and increases in the boundary layer until, at the outer edge, it matches the local free stream value, U. This increase is at a steadily decreasing rate until, eventually, $\partial u/\partial y$ reaches zero at the boundary layer edge, as shown in Fig. 12.2b. Such velocity profiles are said to be "full," and are indications of a negative or "favorable" pressure gradient.

An "adverse" pressure gradient is one that is positive and favors the onset of boundary layer separation. With $dP/dx > 0$, the shear stress *increases* near the wall, according to Eq. (12.7). In the free stream, it must still approach zero, so it is a necessary condition that if $dP/dx > 0$, the shear stress must reach a maximum at some point within the boundary layer. Such a trend is sketched in Fig. 12.3a. In regions of the boundary layer where $\tau = \mu(\partial u/\partial y)$ is increasing, the velocity increases from zero at the wall at an increasing rate. At the point of maximum shear stress, however, the rate of velocity increase in the boundary layer begins to decrease until, at the boundary layer edge, it again approaches zero. At the point of maximum shear stress, the shape of the velocity profile changes from concave downward to concave upward, as shown in Fig. 12.3b—it is a point of inflection of the velocity profile.

The point of separation of a boundary layer is defined as that point along the boundary at which the adjacent fluid particles cease to flow in the mainstream direction, that is, where $(\partial u/\partial y)$ ceases to be positive and becomes zero near the

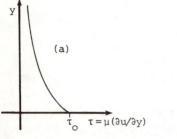

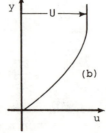

Figure 12.2 Profiles in a favorable pressure gradient: (a) shear stress; (b) velocity.

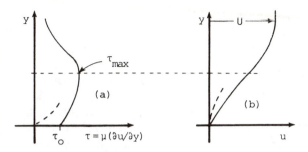

Figure 12.3 Profiles in an adverse pressure gradient: (a) shear stress; (b) velocity.

wall. This is also a point of zero wall shear stress. As the boundary layer thickens along a surface, in the presence of an adverse pressure gradient, the velocity profile becomes more nearly tangent to the y axis at the wall and the shear stress approaches zero there. When $\partial u/\partial y$ passes through zero at the wall, reverse flow occurs there and *boundary layer separation* occurs.

In the study of ideal fluid flows—those that obtain in regions outside the boundary layer—we have seen that increasing pressures are associated with decelerating flows such as those that occur as a flow curves around the downstream side of a blunt obstacle. In viscous flow the shear stresses acting within the boundary layer also exert a retarding effect on the fluid particles there. Thus the adverse pressure gradient and the viscous forces "team up" in the boundary layer until a point is reached when the fluid particles can no longer proceed downstream near the wall. A favorable pressure gradient, on the other hand, assists the boundary layer flow in overcoming the effects of viscous shear. The separation of the boundary layer marks a limit of validity of the boundary layer hypothesis. Upon separation, the region of viscous action grows abruptly, and even if the flow in the attached part of the boundary layer is laminar, the ensuing chaotic motion is highly turbulent. The expanded region of separated flow, or *wake,* behind a body displaces the ideal free stream flow. As a result, the entire flow field may no longer be thought of as a region of ideal fluid flow with a thin layer encompassing the viscous effects.

The separation of the boundary layer gives rise to what is termed the *form drag* of a blunt body in a viscous flow. We discuss this effect in more detail later. At this point it is important to note that viscosity, which gives rise to the boundary layer, also leads to its demise in the presence of a decelerating free stream flow. If it were not for viscosity, the fluid would be indifferent to the presence of the wall and no shear stresses or form drag effects would be present—here we are resolving d'Alembert's paradox.

12.2 LAMINAR BOUNDARY LAYER ON A FLAT PLATE

In Part One the tenets of ideal fluid flow allowed the solution of the equations of motion for bodies with arbitrary shape. We did not treat the case of a flat plate there because, without the influence of viscosity, such an "obstacle" is invisible to the flow. In viscous flow, however, the flat plate is the limiting case, from the point of view of simplicity, of a "birthplace" for viscous effects. The simplicity stems from the assumption that such a plate is vanishingly thin and therefore represents no obstruction to the passing ideal flow. The manifestation of this is the neglect of any pressure gradient in the flow. This is an approximation, of course, because if viscosity is present, there will be shear stresses and some pressure gradient will be necessary to sustain the flow. Suffice it to say that the hydrodynamicist's definition of a flat plate is a surface that does nothing more than provide a region where the no-slip condition is applied. The disturbance that is presented to the oncoming ideal flow (also called "potential flow," for obvious reasons) is insufficient to cause significant departure from the undisturbed solution outside the boundary layer: $U = U_\infty$ = constant, $dP/dx = 0$. The flow configuration is illustrated in Fig. 12.4.

With the approximation that the pressure variation along the boundary layer is negligible, the boundary layer equation becomes

$$u \frac{\partial u}{\partial x} + v \frac{\partial u}{\partial y} = v \frac{\partial^2 u}{\partial y^2} \tag{12.8}$$

with the boundary conditions

$$u(x, 0) = 0 = v(x, 0)$$

$$u(x, \infty) = U_\infty \tag{12.8bc}$$

We first discuss the so-called exact solution to this equation.

Solution Using a Similarity Variable

Equation (12.8) is a partial differential equation relating the two dependent velocity components to the two spacial coordinates x and y. In the search for an appropriate similarity variable, the goal is to express these velocities in terms of a single independent variable so that the governing equation reduces to an ordinary differential equation. We have already seen this process in the study of the suddenly accelerated flat plate in Chapter 11. The dominant velocity, u, is the component of central interest, so the purpose of the similarity variable, $\eta(x, y)$ is to accomplish the following transformation:

$$u(x, y) \rightarrow u[\eta(x, y)]$$

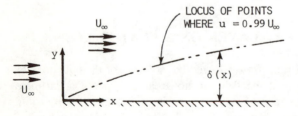

Figure 12.4 Flat plate boundary layer flow.

We have already gained some insight into this matter, because in the development of the boundary layer equation it was found that the boundary layer thickness behaves such that $\delta\alpha(vL/U_\infty)^{1/2}$. Since the plate is infinitely long, in theory, we are without a characteristic length L, and this expression must be replaced by

$$\delta\alpha\left(\frac{vx}{U_\infty}\right)^{1/2} \tag{12.9}$$

The idea of similarity is that at any point along the flat plate the *shape* of the velocity profile will be the same. This is represented schematically in Fig. 12.5, and may be expressed mathematically as

$$\frac{u}{U_\infty} = g\left(\frac{y}{\delta}\right)$$

For instance, if at some point on the plate the velocity is 50% of the free stream velocity at 30% of the boundary layer depth, then at some other point on the plate, if the boundary layers are similar, the same functional dependency applies and at $y/\delta = 0.3$ the value of u/U_∞ will also be 0.5. In yet other words,

$$\frac{u_1}{U_\infty} = g\left(\frac{y_1}{\delta_1}\right) \quad \text{and} \quad \frac{u_2}{U_\infty} = g\left(\frac{y_2}{\delta_2}\right)$$

At each point along the plate, the velocity varies from zero at the surface to U_∞ at the boundary layer outer edge, so there must be values of y that can be found such that $u_2 = u_1$. In such a case y_2/δ_2 must equal y_1/δ_1, and for equal values of u,

$$y_2 = y_1\left(\frac{\delta_2}{\delta_1}\right)$$

The boundary layer thickness is a scaling parameter for the velocity distribution within the boundary layer, and with the dependency expressed in Eq. (12.9), this scaling is related to the x coordinate so that

$$\frac{u}{U_\infty} = g\left(\frac{y}{\delta}\right) = g\left[\frac{y}{(vx/U_\infty)^{1/2}}\right]$$

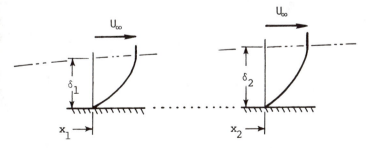

Figure 12.5

The "logical" choice for the similarity variable is, in this case,

$$\eta = y\left(\frac{U_\infty}{vx}\right)^{1/2} \tag{12.10}$$

considerably more logic than this is available in the literature on similarity variables, but here we rely on these arguments—we will not address the general questions of the existence of similarity variables, and methods of finding them. The interested reader might consult Schlichting (1968, chap. 8), or Hansen (1964). Substituting x/U_∞ for the time, t, in Eq. (11.13), the reader will note the analogy between this selection of the similarity variable and that used for the analysis of Stokes's first problem.

With the substitution of the stream function, $u = \partial\psi/\partial y$,

$$u = \frac{\partial\psi}{\partial\eta}\frac{\partial\eta}{\partial y} = \frac{\partial\psi}{\partial\eta}\left(\frac{U_\infty}{vx}\right)^{1/2}$$

and, since it is the goal that u depend only on η, the x dependency is absorbed into the stream function by defining

$$\frac{\partial\psi}{\partial\eta} = f'(\eta)(U_\infty vx)^{1/2}$$

such that

$$\frac{u}{U_\infty} = f'(\eta) \tag{12.11}$$

and

$$\psi = (U_\infty vx)^{1/2}f(\eta) \tag{12.12}$$

The problem is now posed in terms of the selected similarity variable. Through repeated applications of the chain rule of differentiation, the various terms in the boundary layer equation may be derived in terms of this new variable:

$$\frac{\partial u}{\partial x} = U_\infty f''\frac{\partial\eta}{\partial x} = -\frac{1}{2}U_\infty f''\left(\frac{\eta}{x}\right) \tag{12.13}$$

$$\frac{\partial u}{\partial y} = U_\infty f''\frac{\partial\eta}{\partial y} = U_\infty f''\left(\frac{U_\infty}{vx}\right)^{1/2} \tag{12.14}$$

$$\frac{\partial^2 u}{\partial y^2} = \frac{U_\infty^2}{vx}f''' \tag{12.15}$$

and, for the transverse velocity,

$$v = -\frac{\partial\psi}{\partial x} = -\frac{\partial}{\partial x}[(U_\infty vx)^{1/2}f(\eta)]$$

from which

$$v = -\frac{1}{2}\left(\frac{\upsilon U_\infty}{x}\right)^{1/2}(f - \eta f') \qquad (12.16)$$

Equations (12.13)–(12.16) are now substituted into Eq. (12.8), with the result

$$ff'' + 2f''' = 0 \qquad (12.17)$$

The corresponding boundary conditions are

$$f(0) = f'(0) = 0$$

and at the boundary layer edge,

$$f'(\infty) = 1 \qquad (12.17bc)$$

The search for a suitable similarity variable is not always successful, but in this case of a flat plate, we see that it has indeed been possible to transform the partial differential boundary layer equation to the ordinary form of Eq. (12.17). This latter expression is still nonlinear, however, and no closed-form solution is known to exist. Equation (12.17) can be integrated formally, accounting for the boundary conditions, by a technique that involves an incremental calculation beginning at the wall. The following set of difference equations illustrate the procedure:

$$f(\eta + \Delta\eta) = f(\eta) + f'(\eta)\,\Delta\eta$$

$$f'(\eta + \Delta\eta) = f'(\eta) + f''(\eta)\,\Delta\eta$$

$$f''(\eta + \Delta\eta) = f''(\eta) + f'''(\eta)\,\Delta\eta$$

At the wall, it is known from the boundary conditions that both $f(0)$ and $f'(0)$ are zero. From these and Eq. (12.17), it is found that $f'''(0)$ is also zero. The quantity $f''(0)$ is not known, however, and in order to begin the procedure, a value must be guessed. By repeated applications of the difference equations, values away from the wall are calculated until, at a suitably large value of η, the value of $f'(\infty)$ is found. To the extent that this value does not match the boundary condition of unity, the guessed value of $f''(0)$ is in error and the procedure must be repeated until suitable accuracy is obtained.

Other, more classical methods have been used to solve Eq. (12.17), and one of these has caused the expression to become known as the *Blasius equation*. Blasius (1908) used the surface boundary conditions to define a series solution of Eq. (12.17), valid for small values of η. An outer asymptotic expansion was also postulated, and the matching of the two expressions at intermediate values of η led to determination of the various coefficients of the series. The solution of Eq. (12.17) is presented in Figs. 12.6 and 12.7; these results have been amply verified by a number of experimenters [see Schlichting (1968, pp. 131–133)].

Skin friction and drag. The *total* resistance to flow *in the case of a flat plate* is that due to frictional effects. This arises because of the adherence of the fluid

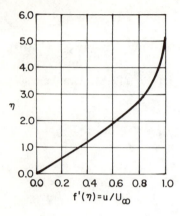

Figure 12.6 Nondimensional velocity variation across a laminar flat plate boundary layer.

particles to the surface of the plate and can be determined by evaluating the shear stress there. Thus, using Eq. (12.14),

$$\tau_0 = \mu\left(\frac{\partial u}{\partial y}\right)_{y=0} = \mu U_\infty\left(\frac{U_\infty}{\nu x}\right)^{1/2} f''(0)$$

This is sometimes termed the *local* wall shear stress, because it depends on the particular value of x along the plate. Note that the local shear stress *decreases* with x, because the growth of the boundary layer leads to a spreading of the region of momentum transport and a decrease in the velocity gradient at the wall. The nondimensional version of the wall shear stress leads to the local *skin friction coefficient*, c_f, defined as

$$c_f = \frac{\tau_0}{\frac{1}{2}\rho U_\infty^2} \tag{12.18}$$

The result for the laminar flat plate boundary layer is

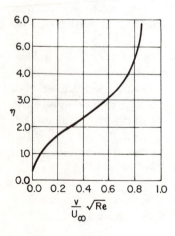

Figure 12.7 Transverse velocity component in a laminar flat plate boundary layer.

$$c_f = \frac{2f''(0)}{\text{Re}^{1/2}}$$

where the Reynolds number is defined as $\text{Re} = U_\infty x/v$. The quantity $f''(0)$ may be estimated from Fig. 12.6 to be about 0.33 [the exact solution to Eq. (12.17) gives $f''(0) = 0.33206$], and we have

$$c_f = \frac{0.664}{\text{Re}^{1/2}} \tag{12.19}$$

The frictional *drag, D_f,* exerted on the surface of a plate of length L and width b is found from

$$D_f = \int_0^L \tau_0 b \, dx = \tfrac{1}{2}\rho U_\infty^2 b \int_0^L c_f \, dx$$

and the friction drag coefficient is defined as this quantity divided by the free stream dynamic pressure and the reference area, which in this case is bL.

$$C_f = \frac{D_f}{\frac{1}{2}\rho U_\infty^2 bL} = \frac{1}{L} \int_0^L c_f \, dx$$

After substitution of Eq. (12.19) and integration,

$$C_f = \frac{1.328}{\text{Re}_L^{1/2}} \tag{12.20}$$

where $\text{Re}_L = U_\infty L/v$, based on the length L over which the effects of viscous shear are integrated.

Boundary layer thickness. The merging of the viscous region with the ideal free stream flow takes place, in theory, in an asymptotic way as η approaches infinity. In fact, however (and this is in the nature of the boundary layer concept), this adjustment is virtually complete at a much smaller value of η, as Fig. 12.6 shows. From this figure, a value of $\eta \approx 5$ is appropriate to designate the outer edge of the boundary layer—the exact solution gives $f' = u/U_\infty = 0.99155$ at this value of η. The value of y at this boundary is designated as the boundary layer thickness, δ, and, from Eq. (12.10),

$$5.0 \approx \delta \left(\frac{U_\infty}{vx}\right)^{1/2}$$

from which

$$\frac{\delta}{x} \approx \frac{5.0}{\text{Re}^{1/2}} \tag{12.21}$$

where, as before, $\text{Re} = U_\infty x/v$. Note that the boundary layer grows at a rate proportional to $x^{1/2}$, as has been hypothesized in earlier analyses. From Fig. 12.7, at this value of η, it may be estimated that the transverse velocity at the boundary layer edge is

$$\frac{v(\delta)}{U_\infty} = \frac{0.83}{Re^{1/2}}$$

Since the boundary layer existence requires that $Re \gg 1$, the transverse velocity is indeed small.

Exercise A thin flat plate is immersed in a flow with a velocity of 5 ft/s. The fluid has a kinematic viscosity of $v = 10^{-5}$ ft^2/s and a density of 2 slug/ft^3. (*a*) Estimate the boundary layer thickness and skin friction coefficient at a point 1 ft from the plate leading edge. (*b*) Estimate the distance from the plate surface at this point, where the velocity is one-half that of the free stream. (*c*) What will be the velocity at this distance from the plate but at a value of $x = 2$ ft? (*d*) What is the friction drag on a 2-ft length of plate?

(*a*) At a distance of 1 ft down the plate, the Reynolds number is

$$Re = \frac{U_\infty x}{v} = \frac{5(1)}{10^{-5}} = 5 \times 10^5$$

From Eq. (12.21), $\delta/x = 0.0071$, so $\delta = 0.0071$ ft or about 0.085 in. From Eq. (12.19), $c_f = 0.00094$.

(*b*) With $u/U_\infty = 0.5$, Fig. 12.6 indicates a value of about 1.6. From Eq. (12.10),

$$y = \eta\left(\frac{vx}{U_\infty}\right)^{1/2} \qquad \text{or} \qquad \frac{y}{x} = \frac{\eta}{Re^{1/2}}$$

For the given value of Re, $y/x = 1.6/(10^5)^{1/2} = 0.00226$. This corresponds to a value of y of about 0.027 in., or about one-third of the boundary layer thickness at that point.

(*c*) With the same value of y as in part (*b*), but at a distance of $x = 2$ ft down the plate, $y/x = 0.00113$ and $Re = 10^6$. (It should be noted that at such a large Reynolds number there is considerable doubt that the boundary layer would remain laminar, as will be further discussed.) From the relationship above, the corresponding value of η is about 1.13, and from Fig. (12.6), u/U_∞ may be estimated at 0.33, so that $u \approx 1.7$ ft/s.

(*d*) For a plate length of 2 ft, $Re_L = 10^6$, and from Eq. (12.20), $C_f = 0.001328$. The frictional drag is therefore

$$D_f = C_f(\tfrac{1}{2}\rho U_\infty^2 bL)$$

or

$$\frac{D_f}{b} = (0.001328)\left(\frac{1}{2}\right)(2)(25)(2)$$

which gives a value of about 0.066 lb/ft of plate width.

12.3 APPROXIMATE METHODS OF EVALUATING INTEGRALS OF BOUNDARY LAYER EQUATIONS

In the previous section the simplest form of the boundary layer equation was integrated by means of a similarity variable. In many cases, if the free stream flow is not well behaved, such variables are not available. Even in the case of the flat plate, where no changes occurred in the free stream flow, the mathematical complexity was considerable. With the availability of the high-speed digital computer, much of the mathematical difficulty may be overcome by numerical methods (which, it might be pointed out, are themselves approximate). The methods introduced here are still important, however, since they provide additional insights into these momentum transport processes. They also serve as benchmark results for determining the validity of computational schemes.

As we will see, it is possible to cast the boundary layer equation in terms of integral quantities that express average values over the boundary layer. In this process, the boundary conditions are satisfied, but the conditions of individual fluid particles within the boundary layer are not evaluated directly. Because the boundaries are "where it's at" in the boundary layer, these approximate methods lead to results that are quite accurate in determining the overall effects of viscous action, such as the frictional resistance. In addition, these "integral" methods are widely used when it is necessary to include the further complications that are associated with turbulent motion.

Formal Integration across the Boundary Layer

The problem is first approached in a purely mathematical way. In a subsequent section the same result will be obtained in a more physical way, through application of momentum principles. In either case, it is important to bear in mind the boundary layer concept, in which variations with respect to y (the coordinate normal to the surface, see Fig. 12.1) occur only within the boundary layer. Outside this layer, only variations with x are allowed, in accordance with the rules of ideal fluid flow. Although generalizations are possible, we consider cases of two-dimensional, steady, incompressible flow with constant viscosity.

The process begins with Eq. (12.6), and this is integrated, *at a fixed value of x,* over the boundary layer depth, δ—the integration is with respect to y *only.* In integral form, Eq. (12.6) is written

$$\int_0^\delta \left(u\frac{\partial u}{\partial x} + v\frac{\partial u}{\partial y} - U\frac{dU}{dx} \right) dy = \frac{1}{\rho}\tau_{yx}\big|_0^\delta = -\frac{\tau_0}{\rho} \tag{12.22}$$

The completion at the end of this expression takes into account the fact that the shear stress is zero at the boundary layer outer edge. The transverse velocity, v, is eliminated by means of the continuity equation, Eq. (12.4), in integral form,

$$v = -\int_0^y \frac{\partial u}{\partial x}\,dy$$

so that

$$\int_0^\delta v \frac{\partial u}{\partial y}\, dy = -\int_0^\delta \frac{\partial u}{\partial y}\left(\int_0^y \frac{\partial u}{\partial x}\, dy\right) dy$$

This may be reduced by means of integration by parts, using the identity

$$\int_0^\delta a\, db = ab\big|_0^\delta - \int_0^\delta b\, da$$

The appropriate substitutions are

$$a = \int_0^y \frac{\partial u}{\partial x}\, dy \qquad \text{and} \qquad b = u$$

so that

$$\int_0^\delta v \frac{\partial u}{\partial y}\, dy = -\left(u \int_0^y \frac{\partial u}{\partial x}\, dy\right)\bigg|_0^\delta + \int_0^\delta u \frac{\partial u}{\partial x}\, dy$$

$$= -U \int_0^\delta \frac{\partial u}{\partial x}\, dy + \int_0^\delta u \frac{\partial u}{\partial x}\, dy \qquad (12.23)$$

With Eq. (12.23), Eq. (12.22) becomes

$$\int_0^\delta \left(2u \frac{\partial u}{\partial x} - U \frac{\partial u}{\partial x} - U \frac{dU}{dx}\right) dy = -\frac{\tau_0}{\rho} \qquad (12.24)$$

But note that since U is independent of y, the second term in this expression may be written

$$\int_0^\delta U \frac{\partial u}{\partial x}\, dy = \frac{\partial}{\partial x}\int_0^\delta uU\, dy - \frac{dU}{dx}\int_0^\delta u\, dy$$

After this substitution in Eq. (12.24) and regrouping of terms [always bearing in mind that $U = U(x)$], the *momentum integral equation* is obtained:

$$\frac{d}{dx}\left[U^2 \int_0^\delta \frac{u}{U}\left(1 - \frac{u}{U}\right) dy\right] + U \frac{dU}{dx}\int_0^\delta \left(1 - \frac{u}{U}\right) dy = \frac{\tau_0}{\rho} \qquad (12.25)$$

Note that the integral expressions in Eq. (12.25) yield functions that depend *only* on the integration limit, δ, and they are therefore functions only of x. If the velocity distributions within the boundary layer are known, or estimated, then these integrals may be evaluated to give length quantities that are independent of y. These lengths have definite physical interpretations, which will become apparent in the following analysis. It may also be observed that, in Eq. (12.25), the ideal flow behavior is manifested in the free stream velocity, U, and its derivatives. These must be considered as available from the application of potential flow methods such as those introduced in Part One. In the case of the flat plate, the second term on the left-hand side is zero, since the free stream velocity is constant.

Momentum Considerations

To this point, the boundary layer equation and its integrals have been evaluated by methods that are largely mathematical. The underlying principles have included Newton's second law and mass continuity, both in differential form. The integral form of Newton's second law is the expression of conservation of momentum, and this principle can be used to arrive at the integral form of the boundary layer equation directly. To apply the momentum principle it is necessary to construct a system, or control volume, around and within which an accounting for momentum is made. Naturally enough, the control volume used here contains a segment of the boundary layer, as shown in Fig. 12.8. The control volume is a differential element of the boundary layer, and the relevant changes across it are therefore expressible as differential quantities.

The forces acting on the surfaces of the control volume are those due to pressure and viscous shear. This latter quantity is exerted on the fluid at the control volume surface represented by the flow boundary. At the outer edge of the boundary layer, which is also the outer edge of the control volume, shear stresses are absent and only pressure forces are felt by the fluid. This pressure is expressed as the average of the upstream and downstream pressure, an approximation that becomes unnecessary as the element becomes sufficiently small. In this analysis, only the forces acting in the x direction are of particular importance. On a basis of unit depth normal to the plane of flow, a summation of these forces gives

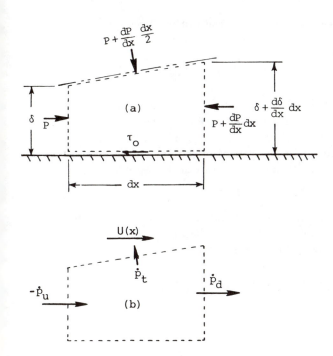

Figure 12.8 Element of a boundary layer: (a) forces; (b) flows.

$$dF_x = P\delta - \left(P + \frac{dP}{dx}\,dx\right)\left(\delta + \frac{d\delta}{dx}\,dx\right) + \left(P + \frac{dP}{dx}\frac{dx}{2}\right)\frac{d\delta}{dx}\,dx - \tau_0\,dx$$

When this is expanded and then reduced by eliminating second-order differential terms,

$$dF_x = -\delta\,\frac{dP}{dx}\,dx - \tau_0\,dx \tag{12.26}$$

According to the principle of conservation of momentum, this force increment exerted on the fluid in the control volume must be balanced by the *net* efflux of momentum across the control volume boundaries. (In steady flow, as is considered here, there can be no accumulation of momentum with time.) The x component of the momentum efflux at the upstream boundary may be evaluated as

$$\dot{p}_u = -\int_0^\delta \rho u^2\,dy$$

where the minus sign is required because, at the upstream boundary, the momentum flux is into the control volume (a negative efflux). At the downstream side of the control volume, the momentum efflux is

$$\dot{p}_d = \int_0^\delta \rho u^2\,dy + \frac{\partial}{\partial x}\left(\int_0^\delta \rho u^2\,dy\right)dx$$

and the net efflux between the upstream and downstream faces of the boundary layer element is

$$\dot{p}_u + \dot{p}_d = \frac{\partial}{\partial x}\left(\int_0^\delta \rho u^2\,dy\right)dx \tag{12.27}$$

This is not the total efflux from the control volume, however, since the flow has a y component of velocity at the upper surface of the control volume. The associated mass flow rate carries with it an x component of momentum. At the top of the control volume, the x component of the velocity is the free stream velocity, U. The x component of momentum flux at the top is therefore

$$\dot{p}_t = \rho v U\,dx$$

The quantity $\rho v\,dx$ is the mass flow rate per unit depth out of the top of the control volume, and this must be equal to the difference between the mass flow rates into and out of the control volume through the upstream and downstream sides. Thus,

$$\rho v\,dx = \int_0^\delta \rho u\,dy - \left[\int_0^\delta \rho u\,dy + \frac{\partial}{\partial x}\left(\int_0^\delta \rho u\,dy\right)dx\right]$$

or

$$\rho v\,dx = -\frac{\partial}{\partial x}\left(\int_0^\delta \rho u\,dy\right)dx$$

and the x component of momentum flux out of the top of the control volume may be written

$$\dot{p}_t = -U \frac{\partial}{\partial x} \left(\int_0^\delta \rho u \, dy \right) dx \tag{12.28}$$

The sum of Eqs. (12.27) and (12.28) gives the net x component of momentum flux for the control volume, and according to the principle of conservation of momentum, this must be equal to the sum of the forces acting on the fluid in that direction. With Eq. (12.26), therefore,

$$\frac{\partial}{\partial x} \left(\int_0^\delta \rho u^2 \, dy \right) - U \frac{\partial}{\partial x} \left(\int_0^\delta \rho u \, dy \right) = -\delta \frac{dP}{dx} - \tau_0 \tag{12.29}$$

From the Euler equation, $dP/dx = -\rho U \, dU/dx$, and the pressure term in Eq. (12.29) may be written

$$-\delta \frac{dP}{dx} = \frac{dU}{dx} \int_0^\delta \rho U \, dy$$

The second term on the left-hand side of Eq. (12.29) may be expanded to give

$$-U \frac{\partial}{\partial x} \left(\int_0^\delta \rho u \, dy \right) = -\frac{\partial}{\partial x} \left(U \int_0^\delta \rho u \, dy \right) + \frac{dU}{dx} \int_0^\delta \rho u \, dy$$

These two substitutions allow Eq. (12.29) to be written as follows:

$$\frac{\partial}{\partial x} \left[\int_0^\delta \rho(uU - u^2) \, dy \right] + \frac{dU}{dx} \int_0^\delta \rho(U - u) \, dy = \tau_0 \tag{12.30}$$

After invoking constant density, and some rearrangement,

$$\frac{d}{dx} \left[U^2 \int_0^\delta \frac{u}{U} \left(1 - \frac{u}{U} \right) dy \right] + U \frac{dU}{dx} \int_0^\delta \left(1 - \frac{u}{U} \right) dy = \frac{\tau_0}{\rho} \tag{12.25}$$

This result is seen to be identical with that obtained from the reduced version of the Navier-Stokes equations, Eq. (12.6).

Integral boundary layer quantities. As we have seen, analytical methods employing momentum considerations are, in general, concerned with integral quantities evaluated across the boundary layer thickness. There are two such quantities that are of special importance here: the *displacement thickness*, δ_1, and the *momentum thickness*, δ_2. Let us consider the x component of velocity within the boundary layer, where it varies from zero at the wall to $U(x)$ at some distance δ, the boundary layer thickness. (The actual distance δ is defined as the distance beyond which the difference between the velocity u and the velocity U in the unaffected flow is negligible. What is "negligible" is a matter for philosophers, but most engineering investigators have agreed that $u/U = 0.99$ is a reasonable limit.) The mass flow within the boundary layer is, on a per-unit-depth basis,

$$\int_0^\delta \rho u \, dy = \rho \int_0^\delta u \, dy$$

for an incompressible flow. If the flow were inviscid at velocity U, the mass flow through the same height would be

$$\rho U \delta = \rho \int_0^\delta U \, dy$$

The difference between these two quantities is

$$\Delta \dot{m} = \rho \int_0^\delta (U - u) \, dy$$

This is a reduction in mass flow due to the retarding effect of the no-slip condition, which is diffused across the boundary layer by viscosity, and if it is present, turbulent mixing. The displacement thickness is defined as that distance, δ_1, that would have the same amount of mass flow as $\Delta \dot{m}$ but at a uniform velocity U. Thus,

$$\rho U \delta_1 = \Delta \dot{m} = \rho \int_0^\delta (U - u) \, dy$$

and

$$\delta_1 = \int_0^\delta \left(1 - \frac{u}{U}\right) dy \tag{12.31}$$

Physically, the displacement thickness is the amount that a boundary in inviscid flow would have to be displaced to cause the same mass flow reduction as that due to the influence of viscosity and turbulent mixing. Put another way, the effectiveness of a duct, in terms of its ability to pass a mass rate at a uniform velocity, is reduced as if its boundaries are shrunk by an amount equal to the displacement thickness.

The momentum thickness, δ_2, is derived from similar logic but deals with the momentum deficiency due to shear in the boundary layer. The momentum flux in the x direction across the boundary layer is, on a per-unit-depth basis,

$$\int_0^\delta u(\rho u) \, dy = \rho \int_0^\delta u^2 \, dy$$

The momentum flux through a strip of inviscid flow at velocity U and *having the same mass flow rate as that through the boundary layer* is

$$\int_{\delta_1}^\delta U(\rho U) \, dy = \rho \int_{\delta_1}^\delta U^2 \, dy$$

Note the use of the displacement thickness so that the momentum integrals for uniform flow carry the same mass flow as that of the boundary layer. The momentum deficit is therefore exclusive of that caused by the deficit in mass flow, and is given by

$$\Delta \dot{p} = \rho \left(\int_{\delta_1}^{\delta} U^2 \, dy - \int_0^{\delta} u^2 \, dy \right)$$

The first integral may be manipulated as follows:

$$\int_{\delta_1}^{\delta} U^2 \, dy = U^2(\delta - \delta_1) = U^2 \left[\int_0^{\delta} dy - \int_0^{\delta} \left(1 - \frac{u}{U} \right) dy \right] = \int_0^{\delta} uU \, dy$$

so that

$$\Delta \dot{p} = \rho \int_0^{\delta} (uU - u^2) \, dy$$

The momentum thickness is defined as that thickness of uniform flow that would have the same momentum deficit; that is,

$$\rho U^2 \delta_2 = \Delta \dot{p} = \rho \int_0^{\delta} (uU - u^2) \, dy$$

or

$$\delta_2 = \int_0^{\delta} \frac{u}{U} \left(1 - \frac{u}{U} \right) dy \tag{12.32}$$

A particularly useful form of this result may be obtained by dividing both sides by the boundary layer thickness, $\delta(x)$. In doing so, the variable of integration is y/δ, which ranges from 0 to 1. Thus,

$$\frac{\delta_2}{\delta} = \int_0^1 \frac{u}{U} \left(1 - \frac{u}{U} \right) d\left(\frac{y}{\delta} \right) \tag{12.32a}$$

If the velocity profile can be expressed as a function of y/δ, as with similar velocity profiles, this integral yields nothing more than a number expressing the constant ratio of the two thicknesses.

The integral boundary layer equation may now be written in terms of these thicknesses. With Eqs. (12.31) and (12.32), Eq. (12.25) becomes

$$\frac{d}{dx}(U^2 \delta_2) + U \frac{dU}{dx} \delta_1 = \frac{\tau_0}{\rho} \tag{12.33}$$

In the event that the free stream flow is relatively constant with x, as in the flat plate, the second term above is negligible and Eq. (12.33) becomes

$$\tau_0 = \rho U^2 \frac{d\delta_2}{dx} \tag{12.34}$$

The shear stress at the wall is seen to be related directly to the change in the momentum thickness along the flow direction. This change in turn reflects the influence of viscosity as it leads to a momentum deficit within the boundary layer region.

Semiempirical Models for Flat Plate Skin Friction

The local skin friction coefficient, defined in Eq. (12.18), is obtained from Eq. (12.34) as

$$c_f = 2 \frac{d\delta_2}{dx} \tag{12.35}$$

and the friction drag coefficient for a plate of length L is

$$C_f = 2 \frac{\delta_2}{L} \tag{12.36}$$

In order to proceed further, it must first be realized that the number of unknowns exceeds the number of available relationships. In Eq. (12.35) the unknown quantities are c_f and δ_2. The displacement thickness, as defined by Eq. (12.32), also contains the as-yet-unknown velocity distribution, as well as the boundary layer thickness, δ. Thus the two equations cited contain the four unknown quantities, c_f, δ, δ_2, and $u(y)$. Clearly, two additional pieces of information are required to complete a solution for either friction coefficients. The standard procedure is to provide a constitutive relationship and an approximate distribution for the velocity within the boundary layer. We discuss each of these below.

Turbulent flows are treated in the following chapter. For laminar flow of a Newtonian fluid, the constitutive relationship is at hand:

$$\tau_0 = \mu \left(\frac{\partial u}{\partial y} \right)_{y=0}$$

In the particular case of the flat plate, it is possible to take advantage of the fact that the velocity profiles are similar, so that

$$\frac{u}{U_\infty} = f \left[\frac{y}{\delta(x)} \right] \tag{12.37}$$

and the wall shear stress may be written

$$\tau_0 = \frac{\mu U_\infty}{\delta} \left[\frac{\partial(u/U_\infty)}{\partial(y/\delta)} \right]_{y=0} = \frac{\mu U_\infty}{\delta} f'(0) \tag{12.38}$$

Equation (12.34) may be recast in the form

$$\tau_0 = \rho U_\infty^2 \frac{d}{dx} \left[\delta \left(\frac{\delta_2}{\delta} \right) \right] = \rho U_\infty^2 \frac{\delta_2}{\delta} \frac{d\delta}{dx} \tag{12.39}$$

because δ_2/δ is a constant. When these two equations are combined to eliminate the wall shear stress, a simple ordinary differential equation in the boundary layer thickness is obtained:

$$\delta \frac{d\delta}{dx} = \frac{\upsilon}{U_\infty(\delta_2/\delta)} f'(0) \tag{12.40}$$

The term on the right-hand side is a constant for a specified function f, and when Eq. (12.40) is integrated with $\delta(0) = 0$, the result is

$$\frac{\delta}{x} = \frac{C_1}{\mathrm{Re}^{1/2}} \quad \text{where} \quad C_1 = \left[\frac{2f'(0)}{\delta_2/\delta}\right]^{1/2} \tag{12.41}$$

The inverse-square-root dependency on the Reynolds number is in accordance with several previous observations for laminar flow. As will be shown in the exercise below, specification of the velocity profile determines the value of the constant C_1. With δ/x thus determined, $\delta_2/x = (\delta/x)(\delta_2/\delta)$, and the friction drag coefficient is found from Eq. (12.36). For the local skin friction coefficient, Eq. (12.41) may be combined with Eq. (12.35) to give

$$c_f = \frac{\delta_2}{\delta} \frac{C_1}{\mathrm{Re}^{1/2}} = \frac{\delta_2}{x} = \frac{C_f}{2} \tag{12.42}$$

Specification of velocity profile. The remaining task is to provide the functional relationship denoted in Eq. (12.37). Because of the specific conditions required at the edges of the boundary layer, this is not an arbitrary process. In fact, quite good results can be obtained by specifying a velocity profile that satisfies only the velocity conditions at these points: $u(0) = 0$ and $u(\delta) = U_\infty$. These two conditions embody the essence of the boundary layer. The given profile may also be made to satisfy the asymptotic nature of the approach to the free stream velocity, that is, $d^n u/dy^n = 0$ at $y = \delta$. It turns out that polynomials of third degree or less are quite satisfactory in predicting the wall shear coefficients. As an exercise, we evaluate a polynomial of zeroth order, i.e., a straight line.

Exercise Assume a linear velocity profile for the flat plate boundary layer. (*a*) Define the functional relationship, Eq. (12.37), necessary to satisfy the boundary conditions. (*b*) Find the appropriate expression for $\delta(x)$. (*c*) Find the friction drag coefficient and the local skin friction coefficient.

(*a*) A general linear velocity profile is given by

$$u = a + by$$

At the wall, $u(0) = 0$, so $a = 0$. At the outer edge of the boundary layer, $u(\delta) = U_\infty$, with the result that $b = U_\infty/\delta$. The necessary functional relationship is therefore

$$\frac{u}{U_\infty} = f\left(\frac{y}{\delta}\right) = \frac{y}{\delta}$$

(b) With the linear profile, $f'(0) = 1$. (In fact, with this profile, f' has this value everywhere.) Letting $y/\delta = \eta$, Eq. (12.32a) becomes

$$\frac{\delta_2}{\delta} = \int_0^1 \eta(1 - \eta)\, d\eta = \frac{1}{6}$$

[Evaluation of Eq. (12.31) will show that the displacement thickness is given by $\delta_1/\delta = \frac{1}{2}$.] For the linear profile, therefore, the constant in Eq. (12.41) is $C_1 = (12)^{1/2}$, and the boundary layer thickness is given by

$$\frac{\delta}{x} = \frac{3.46}{Re^{1/2}}$$

This result may be compared with Eq. (12.21).
(c) The momentum thickness is found to be

$$\frac{\delta_2}{x} = \left(\frac{\delta}{x}\right)\left(\frac{\delta_2}{\delta}\right) = \frac{0.577}{Re^{1/2}}$$

From Eq. (12.36), therefore, $C_f = 1.154/Re^{1/2}$, and from Eq. (12.42), $c_f = 0.577/Re^{1/2}$. Considering the crudeness of the velocity profile (especially the constant slope), these two values are remarkably close to those given by the exact solution, Eqs. (12.19) and (12.20).

12.4 SUMMARY

At the outset of this chapter the Navier-Stokes equations were approximated for conditions in which the region of viscous action is confined to a narrow strip of flow near the boundary. The approximation procedure yielded several important results: the dominance of the lateral shear stress, the necessity for a relatively large Reynolds number, and the insight that the pressure variation across the boundary layer may be neglected. This latter observation coupled the flow in the boundary layer to the external ideal flow. The boundary layer equation, together with its boundary conditions, was thus developed as the governing differential equation for such flows.

It has been demonstrated that in the presence of an adverse pressure gradient—that is, when the external ideal flow is decelerating—the boundary layer cannot be expected to persist along a body. This is the notion of boundary layer separation, and it will play an important role in the discussion of forms of resistance other than those due to surface friction.

The behavior of the laminar boundary layer on a flat plate has been analyzed with both exact and approximate approaches. The latter method can lead to results that are quite useful, especially in light of its simplicity relative to the exact approach. The integration of the boundary layer equation for flow in which the external pressure gradient is not constant has only been touched upon. Detailed treatment of these problems is beyond the scope of this text, but the approach has been indicated—it is a matter of matching the viscous flow to the ideal flow at the boundary layer edge. The most important aspect of these flows, with pressure gradient, is that of boundary layer separation. The operative result of this is the existence of form drag, and this will be discussed later.

Although the analysis of laminar flow is of significant theoretical interest, its main values lie in the development of insights as to the nature of viscous action and in the establishment of methods of solution of the governing equations. All laminar flows, if they persist at all, eventually become turbulent. This is a fact of life, albeit an unfortunate one, and the solution of engineering problems in fluid mechanics cannot proceed very far without taking turbulence into account. We provide an introduction to turbulent viscous flows in the next chapter.

REFERENCES

Blasius, H.: Grenzschichten in Flüssigkeiten mit leiner Reibung, Z. *Math. u. Phys.*, 56, 1, 1908. Translated into English in NACA TM 1256.

Cebeci, T., and P. Bradshaw: *Momentum Transfer in Boundary Layers*, Hemisphere, Washington, D.C., 1977.

Hansen, A. G.: *Similarity Analysis of Boundary Value Problems in Engineering*, Prentice-Hall, Englewood Cliffs, N.J., 1964.

Schlichting, H.: *Boundary Layer Theory*, 6th ed., McGraw-Hill, New York, 1968.

Thompson, P. A.: *Compressible-Fluid Dynamics*, McGraw-Hill, New York, 1972.

PROBLEMS

12.1 Develop a numerical algorithm to solve Eq. (12.17). Construct a computer program using this algorithm to verify Figs. 12.6 and 12.7.

(Problems 12.2 through 12.6 relate to the exact solution to the boundary layer equation.)

12.2 Air flows past a smooth flat plate at a speed of 32 ft/s. Assuming laminar flow, compute the components of the velocity along and normal to the plate at point A in Fig. 12.9.

12.3 Calculate the inclination angle of the streamline passing through the point A in problem 12.2.

12.4 At the edge of the boundary layer and in line with the downstream edge of the plate of problem 12.2, compute the inclination angle of the locus of points given by $y = \delta(x)$.

12.5 The exact solution of the boundary layer equation yields $f(5) = 3.28329$. Considering $\eta = 5$ to represent the outer edge of the boundary layer, show that the transverse velocity there is given by

$$\frac{v_\infty}{U_\infty} = \frac{0.86}{\mathrm{Re}^{1/2}}$$

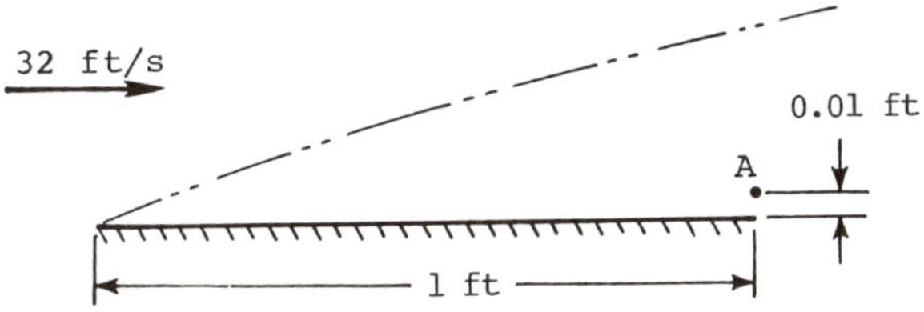

Figure 12.9 Problem 12.2.

12.6 Use the result of problem 12.5 to find the inclination angle of the flow at the point described in problem 12.4. Compare your result with that of problem 12.4.

12.7 In order to solve the boundary layer problem for laminar flow, the following velocity profile is specified:

$$\frac{u}{U_\infty} = a + b\eta - \eta^2 \quad \text{where} \quad \eta = \frac{y}{\delta}$$

What are the necessary values of the constants a and b?

12.8 For the velocity distribution found in problem 12.7, show that $\delta_2/\delta = 2/15$ and evaluate the boundary layer thickness as a function of x.

12.9 For the conditions of problem 12.8, find the friction drag coefficient, C_f, as a function of Reynolds number.

12.10 For laminar flow, the velocity profile in the boundary layer over flat plate is closely approximated by the Prandtl profile,

$$\frac{u}{U_\infty} = \frac{3}{2}\eta - \frac{1}{2}\eta^3 \quad \text{where} \quad \eta = \frac{y}{\delta}$$

State the boundary conditions satisfied by this equation at the plate surface and at the boundary layer edge.

12.11 Find the friction drag coefficient for the velocity profile given in problem 12.10.

12.12 Consider the laminar flow past a flat plate for which the local skin friction coefficient is given by

$$c_f = \frac{C_2}{Re_\delta^{(1/m)}}$$

where $Re_\delta = U_\infty\delta/\nu$. Using the laminar shear stress law and assuming similar velocity profiles, deduce the necessary values for the constants C_2 and m.

12.13 A thin plate is held parallel to the flow of water with a free stream velocity of 20 ft/s. The plate dimensions are 2 ft by 4 ft. Compute the drag on the plate for both orientations (long and short sides parallel to the flow) assuming that the boundary layer is entirely laminar. (Use the results of problem 12.11.)

THIRTEEN

TURBULENT FLOWS

In the study of viscous flows thus far, the transport of momentum across and between fluid layers has been taken to be due solely to activity at the molecular level. In this chapter we introduce the concept of turbulence, which is a much more effective means of momentum transport. By its very nature, turbulence is an extremely complex motion that is very difficult to characterize. In fact, the subject of turbulent flows really constitutes another major branch of the field of fluid mechanics and could easily constitute a fourth part for this text. Instead, we will make an effort to develop an appreciation of the difficulty of the problem while still arriving at some useful engineering applications.

Following a discussion of the occurrence of turbulence as it evolves from a laminar flow, we present the usual method of modeling turbulent flows as consisting of fluctuating quantities superimposed on an otherwise well-behaved motion. This point of view allows identification of the turbulent components that are most influential in the momentum transport process. From these insights it is possible to formulate expressions for shear stresses in terms of the turbulent motions. The motions themselves, however, are of a stochastic nature, and for practical purposes it is necessary to rely on constitutive relationships based on logic as much as fact. These phenomenological models will be discussed, and their use will be demonstrated for the solution of practical engineering problems.

Note on Transition and Stability

Turbulence in fluid flow is the rule more than the exception. In much the same sense as that dictated by the second law of thermodynamics, laminar flows may be expected to increase in randomness, disorder, and uncertainty unless steps are taken to maintain their relative serenity. This evolution from order to disorder may occur as the result of natural or intentional disturbances, and the study of *transition to turbulence* is yet another area of intensive investigation. The issue of transition to turbulence is intimately associated with determination of the factors that affect the *stability* of laminar flows. This problem has been approached from both the theoretical and experimental points of view; in this section we review briefly the main results of such studies. [For further information the reader is again referred to what is considered by many to be the "bible" for advanced studies in viscous flows: Schlichting (1968, chaps. 16, 17).]

Theoretical investigations of the stability of laminar flows are most often based on the Navier-Stokes equations for unsteady flow. The unsteady terms arise from a mathematical perturbation, in the form of velocity fluctuations, which is introduced into the equations for otherwise steady flow. When the perturbation components are separated from the mean flow terms, the result is a fourth-order, ordinary, linear differential equation for the amplitude of the perturbation stream function. This equation is the famous *Orr-Sommerfeld equation,* and its solution constitutes an eigenvalue problem involving four parameters: the wavelength and propagation velocity of the disturbance, a damping factor, and the Reynolds number of the undisturbed flow. When the Reynolds number and disturbance wavelength are specified, the Orr-Sommerfeld equation yields a complex eigenvalue whose imaginary part is the damping factor. The sign of the damping factor determines whether the disturbance exponentially grows or diminishes with time, and the issue of stability is resolved by determining those values of amplitude and Reynolds number corresponding to zero damping (corresponding to neutral stability). The results of the analysis are neutral stability curves such as the one illustrated in Fig. 13.1. In the figure, δ_1 is a characteristic viscous length for the flow, in this case the displacement thickness (in flow between two parallel plates, this length would be the separation distance).

The shape of these neutral stability curves is characteristic. For instance, there is a Reynolds number below which the flow is stable regardless of the wavelength of the disturbance. In the case of Fig. 13.1, this Reynolds number is about 420 based on the displacement thickness, or about 60,000 based on distance along the plate. It should be noted that this point of neutral stability predicts a Reynolds number (or point on the plate, for a given velocity and kinematic viscosity) below which transition to turbulence will not occur. The point of transition to turbulence will be somewhat downstream of this theoretically calculated limit of stability, because an unstable disturbance will have traveled some distance downstream before transition to turbulence is fully completed. The wavelength corresponding to the critical Reynolds number of Fig. 13.1 is about $17\delta_1$ or 6δ. For disturbances of longer wavelengths, there exists a Reynolds number above which transition may be expected. For disturbances of slightly shorter wavelengths, the laminar flow may be stable (within the limits imposed by the theory) for all Reynolds numbers. At intermediate wavelengths, there are corresponding intermediate ranges of the Reynolds number where disturbances will be amplified. (In a way, these results with respect to wavelength are of rather academic interest, because in ac-

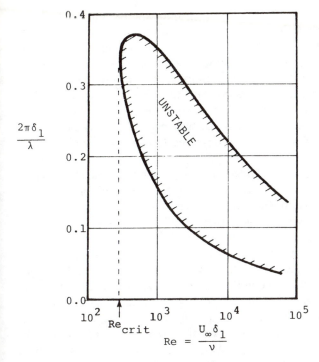

Figure 13.1 Neutral stability curve for a flat plate at zero incidence (λ = disturbance wavelength).

tual practice the wavelength content of most natural disturbances spans a spectrum of values.)

Actually, the analysis outlined above leads to two aspects of stability. One of these is termed *frictionless stability,* because it leads to a neutral stability curve that does not account for the damping effect of viscous action. This frictionless approximation to the Orr-Sommerfeld equation is quite important, however, because the associated analysis shows that points of inflection in laminar velocity profiles are necessary and sufficient for the occurrence of instability (in the absence of friction). In the previous chapter we noted that the existence of such inflection points is related to the direction of change of the pressure gradient in the flow. According to those results, in regions where the pressure increases in the direction of the flow, velocity profiles possess inflection points. Thus a positive pressure gradient, which was termed adverse from the point of view of boundary layer separation, is also adverse from the point of view of stability: Adverse pressure gradients promote instability, whereas favorable pressure gradients stabilize the flow. In spite of the general nature of the effect of pressure gradient on both stability and separation, the reader should take care not to confuse the two issues. Separation of the boundary layer may either precede or follow transition to turbulence—a boundary layer may separate (and therefore cease to exist) whether it is laminar or turbulent.

In the theory of stability the effect of friction is often estimated as a correction to the results for frictionless stability. In general, it has been found that the amplification of small disturbances is much less in viscous instability than in the hypothetical frictionless case. For the purposes of this discussion it is sufficient to note that there is considerable theoretical information available for the prediction of the onset of turbulent flows. These results have been largely verified by experiment, and on the presumption that the presence of amplification of small disturbances leads to the existence of larger disturbances, the study of flow instability provides the theoretical link to the observed fact of transition from laminar to turbulent flow.

The landmark experimental investigation of transition to turbulence was reported by Osborne Reynolds (1883). These experiments consisted of the visualization of the behavior of a dye filament introduced into the flow within a carefully constructed smooth tube. Reynolds found that at sufficiently low velocities the streak of dye within the tube remained laminar, with no "appearance of sinuosity." At increased velocities he observed that ". . . at some point in the tube, always at a considerable distance from the trumpet of intake, the colour band would all at once mix with the surrounding water." Reynolds deduced from his experiments that the breakdown of laminar flow occurred at a value of $\bar{u}d/v \approx$ 13,000, where $\bar{u}$ is the mean flow velocity (Q/A) and d is the tube diameter. These and other similar experiments have been often repeated, and values of the transition Reynolds number for pipe flows have been found that are both larger and smaller than that obtained by Reynolds. Very careful control of disturbances has led to laminar flows at Reynolds numbers up to about 50,000, while it has been verified that below Re = 2000 the flow remains laminar even in the presence of very strong disturbances. Because, as mentioned above, a wide spectrum of disturbances is usually present in engineering applications of pipe flows, it has become accepted practice to consider a low value of Reynolds number as critical:

$$\mathrm{Re_{crit}} = \frac{\bar{u}d}{v} = 2300 \tag{13.1}$$

At Reynolds numbers below this value, pipe flows are presumed to be laminar; while flows at higher Reynolds numbers may be expected to be turbulent. The value given in Eq. 13.1 is for round pipes. Noncircular cross sections, open channel flows, and other situations with differing characteristic lengths will have different critical Reynolds numbers.

In flat plate boundary layer flows, a range of critical Reynolds numbers has been found, depending on the intensity of turbulence in the free stream flow. This range is, roughly,

$$3 \times 10^5 < \mathrm{Re_{crit}} < 3 \times 10^6 \tag{13.2}$$

where $\mathrm{Re} = U_\infty x/v$. On the flat plate, as in a pipe, the critical Reynolds number can be increased by controlling the turbulence level of the external flow, and the larger value given in Eq. (13.2) is for a free stream flow that is virtually free of turbulence. In boundary layer flows over a bluff body, such as a cylinder, the stability of the laminar boundary layer near the upstream portions of the body is assured because of the favorable pressure gradients there. The point of instability is strongly dependent on the point of minimum pressure in the external flow, and has been calculated for bodies of various shapes. As a general rule, the point of transition to turbulence lies at or behind the point of minimum pressure at all except very large Reynolds numbers [Schlichting (1968, p. 477)].

We will see that turbulent boundary layers are associated with relatively large values of skin friction so that it is often beneficial to design body shapes (such as airfoils) such that the laminar boundary layer is prolonged along the body. These designs are based on control of the body thickness so that the point of minimum pressure lies well downstream on the body. There are a number of ways of controlling such boundary layer behaviors as transition and separation. Treatment of these methods, which is beyond the scope of this text, is well documented in the fluid mechanics literature. We will see, however, that turbulent boundary layers are less likely to separate than laminar ones. In cases where the main source of total body drag is due to separation (and not due to skin friction), it can be profitable to intentionally promote transition to turbulence.

This brief review has indicated that transition of laminar flows to turbulent ones is common in a wide variety of applications. Some guidelines have been suggested in the form of critical Reynolds numbers for common flows. In addition, it is well to remember that transition depends on the nature of the disturbance, the level of free stream turbulence, and the pressure gradient in the external flow. In subsequent sections we introduce some of the methods that have been developed for the analysis and prediction of the effects of turbulence.

13.1 AVERAGE VALUES AND FLUCTUATING QUANTITIES

Perhaps the shortest definition of turbulence is that it is a random motion with discernible mean values. This is a convenient definition in that it allows the treatment of turbulent motion to be partitioned into the analysis of a mean motion on which a random fluctuation is superimposed. (Although some motions are even more complicated than this, the problem is tough enough as it is, and the above definition has led to results that are applicable to many practical problems.) Turbulent flows have been classified further, according to the nature of the boundary conditions. Such classifications include wall turbulence associated with boundary layer flows, and free turbulence such as that observed in jet and wake flows. In addition, the basic nature of the turbulent motion may be *homogeneous,* in which the average properties of the random motion are independent of position in the flow. Homogeneous turbulence may also be isotropic if the fluctuations exhibit no preferred orientation. In the discussions that follow, many of these distinctions will not be necessary. We will only touch upon the statistical nature of turbulence, which must be treated in careful detail for an extensive (but still incomplete) understanding of such flows. Complete texts on the subject are available, such as Hinze's (1959).

The maximum size of turbulent eddies is basically limited by the size of the flow apparatus. The lower limit on eddy size is dictated by viscous effects that increase with the large velocity gradients present in the small-scale motions. Sizes of discernable turbulent eddies are typically on the order of a few millimeters or more, and the associated fluctuating velocity components may be on the order of 10% of the mean flow velocity.

Even though the presence of turbulence may be invisible from a overall gross view of the flow, these motions have profound influence on the internal workings of the viscous dissipative processes within the fluid layers. In fact, "layers" of fluid can no longer be expected to exist, except in regions influenced by the no-slip condition at a boundary. The bottom line, from an engineering point of view, is that the stresses acting on a fluid element are no longer proportional to the velocity gradients. Because of this, the Navier-Stokes equations no longer hold when they are applied only to the mean flow. It therefore becomes necessary to treat both the mean flow and the fluctuating quantities together, as is done in the following sections.

Some Rules Governing Averaging

In treating turbulence it is useful to consider the *instantaneous* value of a quantity to consist of the sum of the *mean* or average value of that quantity and a *fluctuating* value that causes the instantaneous value to differ from its mean. Here we consider only time averages, at fixed locations in space, although the complete theory of turbulence involves several other such averaging processes. The concept of such a partitioning of the instantaneous value is illustrated in Fig. 13.2.

In general, the mean value of a quantity will be designated by a overscore, while fluctuating components are denoted by a prime. Thus the instantaneous value of a quantity a is given by

$$a = \bar{a} + a' \tag{13.3}$$

where the mean value is defined *at a specified location in space* as

$$\bar{a} = \frac{1}{T} \int_{t_0}^{t_0+T} a \, dt \tag{13.4}$$

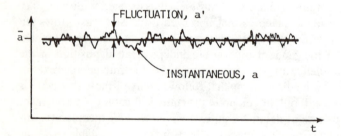

Figure 13.2 Instantaneous, mean, and fluctuating quantities.

There is some arbitrariness associated with the selection of the averaging period T, but the main requirement is that this interval be sufficiently long that the value of the mean is independent of time. Thus, if the mean of the same instantaneous quantity is integrated over another interval,

$$\frac{1}{T} \int_{t_0+T}^{t_0+2T} \bar{a} \, dt = \frac{1}{T} \bar{a}[(t_0 + 2T) - (t_0 + T)] = \bar{a}$$

The time mean of the fluctuation must be zero, as can be shown as follows:

$$\bar{a} = \frac{1}{T} \int_{t_0}^{t_0+T} (\bar{a} + a') \, dt = \bar{a} + \frac{1}{T} \int_{t_0}^{t_0+T} a' \, dt = \bar{a} + \bar{a}'$$

and therefore

$$\bar{a}' = 0 \tag{13.5}$$

With the definition of the time mean, the following useful rules may be proved:

$$\overline{a + b} = \bar{a} + \bar{b}$$

$$\overline{\bar{a}b} = \bar{a}\bar{b}$$

Further, if s is one of the space dimensions and $f = f(s)$,

$$\overline{\frac{\partial f}{\partial s}} = \frac{\partial \bar{f}}{\partial s} \qquad \text{and} \qquad \overline{\int f \, ds} = \int \bar{f} \, ds$$

If the variation of the mean value with time is slow relative to the frequency of variation of the fluctuating quantity (which is often the case), then the derivative and integral formulas given above are also valid for the time dimension.

Consider the mean of the product of two instantaneous values:

$$\overline{ab} = \frac{1}{T} \int_{t_0}^{t_0+T} (\bar{a} + a')(\bar{b} + b') \, dt = \frac{1}{T} \int_{t_0}^{t_0+T} (\bar{a}\bar{b} + a'\bar{b} + b'\bar{a} + a'b') \, dt$$

or

$$\overline{ab} = \bar{a}\bar{b} + \overline{a'\bar{b}} + \overline{b'\bar{a}} + \overline{a'b'}$$

Since the mean of the fluctuating quantities is zero, the second and third terms on the right-hand side of this expression are zero, and

$$\overline{ab} = \bar{a}\bar{b} + \overline{a'b'} \tag{13.6}$$

Note that, in general, the mean of the product of two instantaneous values is not equal to the product of their means. This is an important result from the point of view of turbulence, since otherwise the fluctuating quantities would not appear in the above formula. In such a case, when $\overline{a'b'} = 0$, the two fluctuations are said to be *uncorrelated*. We will see that if these are turbulent velocity fluctuations, the mean value of their product is generally not zero, and some *correlation* between them may be expected.

Example Consider two periodically fluctuating components, a' and b', given by

$$a' = a_0 \sin \omega t \quad \text{and} \quad b' = b_0 \sin (\omega t + \phi)$$

The product of these is

$$a'b' = a_0 b_0 (\sin^2 \omega t \cos \phi + \sin \omega t \cos \omega t \sin \phi)$$

and the mean of this product will be found, after integration, to be

$$\overline{a'b'} = \frac{a_0 b_0}{T} \left[\cos \phi \left(\frac{t}{2} - \frac{\sin 2\omega t}{4\omega} \right) + \sin \phi \left(\frac{\sin^2 \omega t}{2\omega} \right) \right]_{t_0}^{t_0 + T}$$

As the sampling period T is allowed to become large, only the term involving $t/2$ will survive the division by T, so that

$$\overline{a'b'} = \frac{a_0 b_0}{2} \cos \phi$$

This function is plotted in Fig. 13.3, and it is seen that the functions are uncorrelated if the phase angle is $\pm\pi/2$. The maximum correlation occurs at phase angle of $-\pi$, 0, and π. In this latter case the fluctuations are said to be completely correlated (a sketch will show that their time shapes are identical in this case).

The reader may verify that the mean square fluctuations are given by

$$\overline{a'^2} = \frac{a_0^2}{2} \quad \text{and} \quad \overline{b'^2} = \frac{b_0^2}{2}$$

These allow evaluation of the *correlation coefficient*, defined as

$$\frac{\overline{a'b'}}{(\overline{a'^2})^{1/2} (\overline{b'^2})^{1/2}} = \cos \phi$$

which is independent of the amplitudes of the fluctuations.

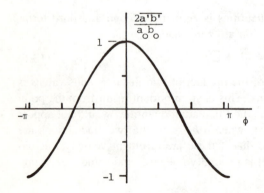

Figure 13.3

13.2 MOMENTUM TRANSPORT VIA TURBULENCE

In this section we examine the effects of turbulence by means of the definition given above—a random fluctuation accompanying a discernable mean flow. The fluctuations of primary interest are, of course, those that occur in the instantaneous velocity components. When the Navier-Stokes equations are written to account for both the mean and fluctuating aspects of turbulent flows, certain mathematical combinations of terms occur that have great physical significance. With this expectation, we first examine the nature of the effect of velocity fluctuations on the transport of momentum.

Physical Influence of Velocity Fluctuations

Let us now consider the effect of fluctuations of the velocity components in a turbulent flow. The view taken is much the same as that discussed in the beginning of Chapter 10, where the nature of molecular viscosity was considered. As shown in Fig. 13.4, we consider the transport of x-wise momentum across a reference plane of area A. In laminar flow this transport was accomplished by the random motion of molecules in the presence of a velocity gradient. Here, however, the transport is accomplished in a similar way, but on a much grander scale. With the velocities disposed as in Fig. 13.4, the instantaneous momentum flux across the reference plane is

$$\frac{\dot{p}}{A} = \rho vu = \rho(\bar{v} + v')(\bar{u} + u') = \rho(\bar{u}\bar{v} + v'\bar{u} + u'\bar{v} + u'v')$$

and the mean value of this flux is

$$\frac{\bar{\dot{p}}}{A} = \rho\bar{u}\bar{v} + \rho\overline{u'v'}$$

Even if there is no transverse mean velocity component—that is, if $\bar{v} = 0$—the y component of the turbulent fluctuation serves as a carrier of momentum across fluid layers. The quantity $\rho\overline{u'v'}$ gives rise to a shear stress, as does momentum flux on a molecular level. But this turbulent quantity is a property of the *flow* (i.e., the turbulent content) and not of the *fluid*. In turbulent flows it is much more influential than molecular transport in spreading the effects of viscosity. Because of this effectiveness of turbulent mixing, external flow velocities are brought much closer to bounding surfaces—turbulent boundary layers have "full" velocity profiles. At a boundary, however, the no-slip condition must still be satisfied.

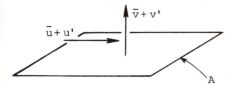

Figure 13.4 Momentum transport via turbulence.

Accordingly, there are much larger velocity gradients near the wall in a turbulent flow, and the accompanying shear stresses are much larger than those associated with laminar flows. We will have more to say on this matter.

Consider a shear flow in which the velocity is increasing in the y direction, as sketched in Fig. 13.5. A positive transverse velocity fluctuation, v', will tend to cause slower-moving fluid particles to interact with those that are moving faster in the layer above, and at this upper layer this is felt as a fluctuation in the negative x direction. Thus, a positive v' is associated with a negative u', and the turbulent shear stress, $\rho\overline{u'v'}$, is generally not zero and is negative in sign.

Equations of Motion for Turbulent Flow

We are now in a position to apply the rules of averaging to Eq. (10.19), which is the form of the Navier-Stokes equations selected for this purpose: incompressible flow with constant *molecular* viscosity. Before doing this, however, it is useful to recast the substantive derivative in a slightly different form. This is accomplished by noting that

$$u_j \frac{\partial u_i}{\partial x_j} = \frac{\partial}{\partial x_j}(u_i u_j) - u_i \frac{\partial u_j}{\partial x_j}$$

The last term is zero, from the equation of continuity for an incompressible flow, and the acceleration part of the N-S equations may now be written as

$$\frac{Du_i}{Dt} = \frac{\partial u_i}{\partial t} + u_j \frac{\partial u_i}{\partial x_j} = \frac{\partial u_i}{\partial t} + \frac{\partial}{\partial x_j}(u_i u_j)$$

The N-S equations are now written for the instantaneous quantities, and these are given as sums of the appropriate means and fluctuations. (As has been done before, for compactness we group the hydrostatic term with the static pressure.) From Eq. (10.19),

$$\rho \left\{ \frac{\partial}{\partial t}(\bar{u}_i + u_i') + \frac{\partial}{\partial x_j}[(\bar{u}_i + u_i')(\bar{u}_j + u_j')] \right\} = -\frac{\partial}{\partial x_i}(\bar{p} + p') + \mu \nabla^2(\bar{u}_i + u_i')$$

This may be expanded and rearranged to give

$$\left.\begin{array}{c} \rho\left[\dfrac{\partial \bar{u}_i}{\partial t} + \dfrac{\partial}{\partial x_j}(\bar{u}_i\bar{u}_j)\right] \\[2ex] +\rho\left[\dfrac{\partial u_i'}{\partial t} + \dfrac{\partial}{\partial x_j}(\bar{u}_j u_i' + \bar{u}_i u_j' + u_i' u_j')\right] \end{array}\right\} = \left\{\begin{array}{c} -\dfrac{\partial \bar{p}}{\partial x_i} + \mu\nabla^2\bar{u}_i \\[2ex] -\dfrac{\partial p'}{\partial x_i} + \mu\nabla^2 u_i' \end{array}\right.$$

Figure 13.5 Transport across a velocity gradient.

This expression has been arranged in this fashion to illustrate that the N-S equation for the mean motion is included in that for the overall instantaneous motion. If the upper level of the above expression is time averaged, it will not change because it is already expressed in mean quantities. If the second set of terms, on the lower level, is also averaged, the only surviving turbulence quantity is $\overline{u_i' u_j'}$. On the premise that this equation governs the motion on an average basis as well as instantaneously, the following result is obtained:

$$\rho \left[\frac{\partial \bar{u}_i}{\partial t} + \frac{\partial}{\partial x_j} \left(\bar{u}_i \bar{u}_j + \overline{u_i' u_j'} \right) \right] = - \frac{\partial \bar{p}}{\partial x_i} + \mu \, \nabla^2 \bar{u}_i$$

Notice that the turbulent fluctuation quantities appear on the left-hand side as part of the acceleration term in this expression. They are, in fact, kinematic effects due to the turbulent motion. As far as the acceleration of the mean flow is concerned, however, they appear as additional shear stress terms. This is illustrated by the more popular form of the result:

$$\rho \frac{D\bar{u}_i}{Dt} = - \frac{\partial \bar{p}}{\partial x_i} + \mu \, \nabla^2 \bar{u}_i - \rho \frac{\partial}{\partial x_j} (\overline{u_i' u_j'}) \tag{13.7}$$

Equations (13.7), consisting of three coordinate components, form a basis for the analysis of turbulent flows. These equations, which stem from the N-S equations and the definition of turbulence, are sometimes called the *Reynolds equations*. The mean fluctuation terms, now located on the right-hand side, are called the *Reynolds stresses*. The Reynolds stresses make up the components of the *apparent shear stress* tensor. (The latter term, being more descriptive, will be used in future reference to these stresses.) In interpreting these "stresses" it is well to bear in mind that, in fact, they represent momentum transport due to the turbulent fluctuations.

Equations (13.7), together with the equation of mass continuity, are a set of four differential relationships for the following dependent variables:

1. The three components of the mean velocity,

$$\bar{u}_i \qquad (i = 1, 2, 3)$$

2. The mean pressure, $\bar{p}$
3. The apparent stresses, $\rho \overline{u_i' u_j'}$, consisting of

$$\rho \overline{u_1'^2}, \ \rho \overline{u_2'^2}, \ \rho \overline{u_3'^2}, \ \rho \overline{u_1' u_2'}, \ \rho \overline{u_1' u_3'}, \ \rho \overline{u_2' u_3'}$$

An accounting reveals that there are four equations available to describe the behavior of *ten unknowns*. Here we have arrived at the basic hitch in the turbulence "git-along." In more scientific terms, this is called the *turbulence closure problem*, and it has led to a continuing search, on a worldwide basis, for *turbulence closure models*. These efforts require an in-depth understanding of the details of turbulent motion, and in many cases the proposed models are of a stochastic nature. Although much progress has been achieved, there are not generally applicable turbulence closure models, and many of them are impractical from an engineering or design point of view. An alternative to detailed theoretical modeling is based on efforts to interpret experimental results in such a way that the basic analytical concepts, such as those developed above, are preserved. These approaches, which have led to results of much practical value, are discussed in subsequent sections.

In order to clarify the nature of the apparent stresses, the component of Eq. (13.7) for motion in the x direction is examined further:

$$\rho \frac{D\bar{u}}{Dt} = -\frac{\partial \bar{p}}{\partial x} + \mu \, \nabla^2 \bar{u} - \rho \left[\frac{\partial}{\partial x}\left(\overline{u'^2}\right) + \frac{\partial}{\partial y}\left(\overline{u'v'}\right) + \frac{\partial}{\partial z}\left(\overline{u'w'}\right) \right]$$

Expanding the Laplacian and rearranging terms, this may be written

$$\rho \frac{D\bar{u}}{Dt} = -\frac{\partial \bar{p}}{\partial x} + \frac{\partial}{\partial x}\left(\mu \frac{\partial \bar{u}}{\partial x} - \rho\overline{u'^2} \right)$$

$$+ \frac{\partial}{\partial y}\left(\mu \frac{\partial \bar{u}}{\partial y} - \rho\overline{u'v'} \right) + \frac{\partial}{\partial z}\left(\mu \frac{\partial \bar{u}}{\partial z} - \rho\overline{u'w'} \right) \tag{13.8}$$

The shear stress terms in this expression illustrate how the apparent stresses may be grouped with the stresses that arise at the molecular level. Over large portions of a turbulent flow, the apparent stresses are orders of magnitude greater than their molecular counterparts, and in such regions the latter effects may be safely neglected. The apparent stresses also differ conceptually from the laminar shear terms because, by themselves, they are not dissipation terms—they are diffusers, producers, and purveyors of turbulent motion that spread dissipation, through viscosity, over the flow region. In addition, the so-called eddy viscosity, arising from the apparent stresses, cannot be a scalar (like μ) because the turbulent fluctuations are directional and likely to vary throughout the flow. This aspect of the apparent stresses is clearly indicated in experiments. Figure 13.6 is an illustration of such results, adapted from the work reported by Reichardt (1938) and cited in Schlichting (1968, p. 542).

There are a number of aspects of the results shown in Fig. 13.6 that are fundamental to an understanding of turbulent flows. The mean fluctuation quantities are seen to vary considerably throughout the flow. In particular, they die out near the wall so that the eddy viscosity is zero there. But this "viscous sublayer" is where viscosity acts to give wall shear stress. The turbulent velocity profile is much fuller than it would have been had the flow been laminar, and this is evidence of the effectiveness of turbulent motions in transporting momentum from the tunnel centerline to regions near the wall. Thus, even though the fluctuations die out at the wall, they have "done their thing" in causing a large mean velocity near the wall. Accordingly, the large velocity gradient near the wall gives rise to a shear stress there that is due to viscosity, but greatly increased over that which would be found in purely laminar flow. These results are the sort that have been used in developing the so-called phenomenological models for the apparent shear stresses. It should also be observed that the shear stress in Fig. 13.6 varies linearly across the channel.

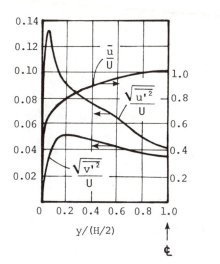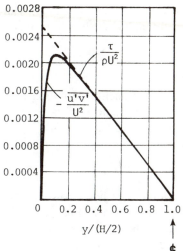

Figure 13.6 Variation of mean velocity and turbulence components in a wind tunnel ($U = 100$ cm/s). Measurements of Reichardt (1938).

In these experiments the shear stress at the wall, divided by the density, was about 24.4 cm²/s². The flow geometry was identical to that studied in Case 1 of Chapter 11, where it was found that for *laminar* flow, the wall shear stress could be predicted by

$$\tau_w = -\left(\frac{H}{2}\right)\left(\frac{dP}{dx}\right) = \frac{4\mu U_{max}}{H}$$

or

$$\frac{\tau_w}{\rho} = \frac{4\upsilon U_{max}}{H}$$

In the experiment, the channel height was $H = 24.4$ cm, and the centerline velocity was $U_{max} = 100$ cm/s. With these values and a standard value for the kinematic viscosity of air ($\upsilon \approx 0.15$ cm²/s), the expected laminar value for wall shear stress is found to be

$$\frac{\tau_w}{\rho} = 2.4 \text{ cm}^2/\text{s}^2 \qquad \text{(laminar)}$$

Thus, a tenfold increase in wall shear stress may be attributed to the turbulence in the flow.

13.3 MODELS FOR APPARENT STRESSES

Early attempts to deduce useful relationships for the apparent shear stresses succumbed to the temptation to model these effects in ways that were mathematically analogous to the molecular viscosity. Thus an eddy viscosity (also called "apparent," or "virtual" viscosity), mentioned above, was defined such that

$$\rho\overline{u'v'} = -A_\tau \frac{\partial \bar{u}}{\partial y}$$

The convenience of such a definition lies in the close coupling of the apparent shear stress with the molecular value. In Eq. (13.8), for instance, the dominant shear stress term could be written

$$\frac{\partial}{\partial y}(\tau_{yx}) = \frac{\partial}{\partial y}\left(\mu \frac{\partial \bar{u}}{\partial y} - \rho\overline{u'v'}\right) = \frac{\partial}{\partial y}\left[(\mu + A_\tau)\frac{\partial \bar{u}}{\partial y}\right]$$

The simplification stops here, however, because the eddy viscosity, A_τ, is clearly not a constant, as shown in Fig. 13.6. The variation of the eddy viscosity with velocity therefore becomes a central issue, and this is the thrust of the mixing length theory developed by Prandtl.

Prandtl's Mixing Length Theory

Prandtl resorted to the analogy between molecular and turbulent momentum transport. As discussed in connection with Fig. 13.5, turbulent transport of "lumps" of fluid across a gradient in mean velocity is somehow related to a fluctuation in the x-wise component of the instantaneous velocity. Thus, it is surmised that

$$u'\alpha \frac{\partial \bar{u}}{\partial y}\Delta y$$

where Δy refers to an incremental step across the gradient of mean velocity. The "mixing length," ℓ, is defined in such a way as to make this relationship an equality:

$$u' = -\ell \frac{\partial \bar{u}}{\partial y}$$

The imagination is further extended by the reasoning that the u' fluctuations, when convected in a transverse direction across the velocity gradient, give rise to proportional v' fluctuations. This is necessary from the point of view of mass conservation at the fluctuating level so that the apparent shear stress is modeled as

$$\tau = -\rho\overline{u'v'} = \rho\ell^2 \left|\frac{d\bar{u}}{dy}\right| \frac{d\bar{u}}{dy} \tag{13.9}$$

The absolute value of one of the gradients must be taken in order to preserve the sign dependency of the shear stress on velocity gradient. (The partial derivative notation and the subscript on τ are dropped because the previous logic applies only to variations in the transverse y direction.)

In these arguments a number of proportionality constants accrue, and these may be thought of as having been absorbed in the characteristic mixing length—the mixing length is the quantity that makes Eq. (13.9) work. That Eq. (13.9) does work in many instances is quite remarkable. Basically, it says that turbulent fluctuations give rise to the velocity gradient *and vice versa,* and that the two components of fluctuation are proportional because of mass continuity (the third component, normal to the x and y directions, is hopefully overlooked). If the shear stress is expressed in terms of a turbulent viscosity, μ_t, that is, $\tau = \mu_t(d\bar{u}/dy)$, then another form is obtained for the eddy kinematic viscosity, namely,

$$\epsilon = \frac{\mu_t}{\rho} = \ell^2 \left| \frac{d\bar{u}}{dy} \right| \tag{13.10}$$

Neither ϵ nor ℓ is amenable to analytical description, but if they can be related to velocity profiles and then compared with experimental observations, it is possible to make a pretty good guess at their values.

Universal Velocity Distribution Laws

The von Karman similarity hypothesis provided a further contribution to attempts to relate the mixing length to flow quantities. This hypothesis is based on the premise that for a flow that is steady in the mean, the fluctuations must be similar at different points in the flow field—that is, they can be scaled between various points by means of a characteristic length and a characteristic intensity (velocity). The upshot of this analysis is the deduction that the mixing length should obey the following relationship:

$$\ell = A \left| \frac{d\bar{u}/dy}{d^2\bar{u}/dy^2} \right| \tag{13.11}$$

where A is a constant thought to be universally valid in regions of the flow that are fully turbulent and thus outside of the thin laminar sublayer near the wall.

When Eqs. (13.9) and (13.11) are combined, the mixing length may be eliminated to obtain

$$v_*^2 = -A \frac{(d\bar{u}/dy)^2}{d^2\bar{u}/dy^2} \tag{13.12}$$

where $v_* = (\tau/\rho)^{1/2}$ is termed the *friction velocity.* (The minus sign in this expression is necessary to account for the negative value of $d^2\bar{u}/dy^2$ assuming that y is measured from the wall.) In order to obtain expressions describing the velocity variation within turbulent boundary flows, Eq. (13.12) may be integrated if simple assumptions may be made for the variation of v_* (i.e., τ) with y. Using different assumptions, both von Karman and Prandtl pursued this approach. Von Karman concentrated on flow between parallel plates (Case 1, Chapter 11), and made the logical premise, supported by data such as that in Fig. 13.6, that $\tau = \tau_0(y/H)$. In this case y is measured from the duct centerline, and τ_0 is the shear stress at the wall. Prandtl, on the other hand, assumed that near the wall the mixing length should vary in accordance with $\ell \alpha y$; a logical assumption since, at the wall, the mixing length must go to zero. Prandtl, whose analysis was based on boundary layer flow, made use of the experimental observation that the shear stress could be taken as constant and equal to the wall shear stress in a region near but outside the edge of the laminar sublayer.

In both the procedures outlined above, a number of constants accumulate, such as in attempting to match the turbulent velocity profiles with the rapid decrease in velocity beginning at the edge of the laminar sublayer. Resolution of these constants relies heavily on experimental evidence, and as a result, these analyses are of value in rationalizing and extending such evidence. Both analyses are only concerned with regions of the flow where the shear stress is due, in the main, to the turbulence. They do not apply in the thin laminar sublayer near the wall, and in fact they lead to a logarithmic singularity at the wall.

From the Prandtl approach, for instance, the following form is found for the velocity distribution:

$$\frac{\bar{u}}{\bar{v}_{*0}} = \frac{1}{A} \ln\left(\frac{yv_{*0}}{v}\right) + B \tag{13.13}$$

where the constants A and B are determined from comparisons with measurements. The constants will be specified later, but it is useful to note at this point that Eq. (13.13) may be written

$$\frac{\bar{u}}{Bv_{*0}} = \frac{1}{AB} \ln\left(\frac{yv_{*0}}{v}\right) + 1 \approx \left(\frac{yv_{*0}}{v}\right)^{1/AB}$$

where the approximation is valid for $1/AB \ll 1$. It has been found that the latter form of the velocity distribution does indeed hold for turbulent boundary flows. Most significantly, the quantity AB is found to increase with Reynolds number so that the logarithmic form of Eq. (13.13) may be regarded as an asymptotic law valid, or "universal," for high Reynolds numbers. Observations such as these also lend credence to the basic logic embraced by Prandtl's mixing length theory and von Karman's similarity hypothesis. For further discussion of the universal velocity distribution laws and related matters, the reader may wish to consult Schlichting (1968, chap. 19).

13.4 TURBULENT FLOW IN PIPES

In the preceding discussions we outlined some of the fundamental aspects of turbulent flows, including the governing differential equations and the turbulence closure problem. It is hoped that at this point the reader has some appreciation of the roles played by the fluctuating parts of such flows and the difficulty of formulating a complete analytical solution to the problem. In this brief introduction to turbulent flows, we now proceed to descriptions of some of the methods that have been developed to allow engineering calculations to be made. In particular, we discuss internal flows (pipe flows) and external flows. In all cases we treat flows that are steady in the mean ($\partial/\partial t = 0$), and the dependent variables are time-averaged values—the overbar to denote this will be omitted.

As in Case 2 of Chapter 11, for laminar flow, we first examine fully developed flow in a pipe for which dP/dx is a constant and the averaged x component of the flow varies only with distance normal to the pipe centerline. The configuration is that of Fig. 13.7. As before, the postulation that u is a function only of r leads to the conclusion (from mass continuity) that the transverse velocity must be zero. We could proceed with a reduction of the Reynolds equations, Eq. (13.7), but this would lead to a need to treat the fluctuating quantities in some detail. Instead we treat both the laminar and turbulent contributions of the shear stress lumped together in a single stress term that varies only with r. (This is a direct outgrowth of the fully developed hypothesis and is supported by measurements such as those presented in Fig. 13.6.) Because $u = u(r)$ and $v = 0$, the substantive derivative of u is zero—there is no acceleration of the flow, local or convective, and the pressure forces form a balance with the viscous forces.

If a force balance is written for the annular strip shown in Fig. 13.7, the result is as follows:

$$[P - (P + \Delta P](2\pi r \, \Delta r) + \tau(2\pi r) \, \Delta x - (\tau + \Delta\tau)[2\pi(r + \Delta r)] \, \Delta x = 0$$

and, when this is divided by $2\pi \, \Delta r \, \Delta x$,

$$r \frac{\Delta P}{\Delta x} + r \frac{\Delta\tau}{\Delta r} + \tau + \Delta\tau = 0$$

Proceeding to differentially small quantities, the last term drops out and

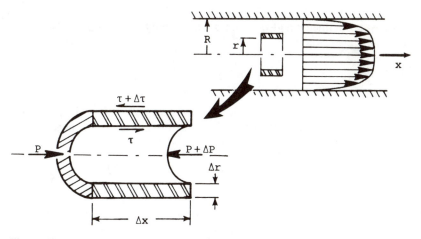

Figure 13.7 Fully developed turbulent flow in a pipe.

$$r\frac{dP}{dx} + r\frac{d\tau}{dr} + \tau = 0$$

or

$$-\frac{1}{r}\frac{d}{dr}(r\tau) = \frac{dP}{dx} \tag{13.14}$$

Equation (13.14) is to be compared with the laminar version, Eq. (11.7), with the substitution of the laminar shear stress law, $\tau = -\mu\, du/dr$ (here we have chosen the x coordinate as the streamwise direction, instead of z). With dP/dx a constant, integration leads to

$$-r\tau = \frac{r^2}{2}\frac{dP}{dx} + C_1$$

but setting r to zero (the pipe centerline) the integration constant, C_1, is found to be zero, and

$$\tau = -\frac{r}{2}\frac{dP}{dx} \tag{13.15}$$

The shear stress is seen to vary linearly across the pipe and, in particular, at the pipe wall, where $r = R$,

$$\tau_0 = -\frac{R}{2}\frac{dP}{dx} \tag{13.16}$$

It is now convenient to introduce a dimensionless wall shear stress, or *coefficient of resistance*, f, such that

$$f = -\frac{dP/dx}{q/D} \tag{13.17}$$

where D is the pipe diameter and q is the dynamic pressure based on the mean velocity. That is,

$$q = \frac{1}{2}\rho\bar{u}^2$$

and the mean velocity is defined as the flow rate per unit area; $\bar{u} = Q/\pi R^2$. (The overbar must not be confused with the designation of time average, used previously.) With Eq. (13.16), Eq. (13.17) becomes

$$f = 4\frac{\tau_0}{q}$$

which is in accordance with the definition used in Case 2 of Chapter 11. (In Chapter 9 we used this quantity in the analysis of Fanno flow. There it was called the d'Arcy friction factor.)

In the discussion of laminar flow, the laminar shear stress law allowed expression of the velocity as a function of r and subsequent calculation of $\bar{u}$ and f. The result was $f = 64/\text{Re}$. In this case, however, the turbulent velocity profile is not yet specified, and the coefficient of resistance cannot be evaluated directly. Note, however, that in fully developed pipe flow the wall shear stress may be deduced directly from the pressure drop. From an experimental point of view this is an enormous simplification, and many experiments have been conducted with this conclusion as a basis. As a result, the *laws of resistance*—that is, correlations of experimental data to obtain formulas for f—have preceded the deduction of turbulent velocity profiles.

One such correlation—a classical one—dates back to 1811 and is called the *Blasius correlation:*

$$f = \frac{0.3164}{\text{Re}^{1/4}} \tag{13.18}$$

where $\text{Re} = \rho\bar{u}D/\mu = \bar{u}D/\upsilon$. *Correlations of this form are restricted in their application to certain values of the Reynolds number.* The Blasius correlation has been found to be valid for turbulent flow with Reynolds numbers of *up to* 10^5, but at higher Reynolds numbers it differs considerably from experimental observations. Note that in Eq. (13.18) the shear stress (and therefore the pressure gradient) varies with the 7/4th power of the velocity. In laminar flow, on the other hand, the dependency was found to be linear with velocity. Accordingly, the wall shear stress and rate of pressure decrease in a pipe are considerably greater in turbulent flows.

Let us now examine the connection between such laws of resistance and the velocity gradient across the pipe. From Eq. (13.18) we may write

$$\tau_0 = \rho v_{*0}^2 = 0.03325\rho\bar{u}^{7/4}\upsilon^{1/4}R^{-1/4}$$

In the preceding section we found that it is logical to seek a velocity distribution such that

$$\frac{u}{v_{*0}} = f(\eta) \qquad \text{where} \qquad \eta = \frac{yv_{*0}}{\upsilon}$$

and from the previous expression,

$$\frac{v_{*0}}{\bar{u}} = (0.03325)^{4/7}\left(\frac{\upsilon}{Rv_{*0}}\right)^{1/7}$$

or

$$\frac{\bar{u}}{v_{*0}} = 6.99\left(\frac{Rv_{*0}}{\upsilon}\right)^{1/7} \tag{13.19}$$

This result, together with the reasoning associated with the universal velocity distributions, suggests a *power law* velocity profile of the form

$$\frac{u}{U} = \left(\frac{y}{R}\right)^{1/n} \tag{13.20}$$

where y is measured from the pipe wall and U is the maximum velocity at the pipe centerline. Note that this assumption, which is based on considerable experimental evidence, satisfies the essential boundary conditions at the pipe wall $[u(0) = 0]$ and at the centerline $[u(R) = U]$. [Early measurements of the velocity distribution in turbulent pipe flow were conducted by J. Nikuradse (1932), cited in Schlichting (1968, p. 563).] The mean velocity may be found from its definition and Eq. (13.20) as

$$\bar{u} = \frac{U}{\pi R^2} \int_0^R \left(\frac{R - r}{R}\right)^{1/n} (2\pi r \, dr)$$

and, upon integration,

$$\frac{\bar{u}}{U} = \frac{2n^2}{(n + 1)(2n + 1)} \tag{13.21}$$

With $n = 7$, as expected from the Blasius correlation, $\bar{u}/U \approx 0.8$, and with Eq. (13.19),

$$\frac{U}{v_{*0}} = \frac{\bar{u}/v_{*0}}{\bar{u}/U} = 8.74\left(\frac{Rv_{*0}}{v}\right)^{1/7}$$

The centerline velocity U corresponds to $y = R$; to extend this result to intermediate values of y, we need only write

$$\frac{u}{v_{*0}} = 8.74\left(\frac{yv_{*0}}{v}\right)^{1/7} \tag{13.22}$$

Thus the link is achieved between the power law velocity distribution and resistance formulas of the Blasius type.

Just as the coefficient and exponent of Eq. (13.18) depend on Reynolds number, so do those of Eq. (13.22). This is a central difficulty with such formulas, but the logic associated with the universal velocity distributions has led to the expectation that these polynomial expressions lead to an asymptotic relationship valid for higher Reynolds number. That is, as n increases with Reynolds number, as has been experimentally verified, the logarithmic form of Eq. (13.13) should be valid. From measurements, the best values of the free constants in Eq. (13.13) have been found to be $A = 0.4$ and $B = 5.5$, so that

$$\frac{u}{v_{*0}} = 2.5 \ln \eta + 5.5 \tag{13.23}$$

where η is as defined above. Prandtl further extended this result to determine a friction factor given by

$$\frac{1}{f^{1/2}} = 0.87 \ln(\mathrm{Re}\, f^{1/9}) \quad 0.8 \tag{13.24}$$

The relationship is known as *Prandtl's universal law of friction,* and it has been found to be valid at Reynolds numbers up to 3.4×10^6. Most significant, however, is that the universality of the velocity distribution on which it is based allows use of this expression at even higher Reynolds numbers. Equation (13.24) does have the drawback that it is implicit in the friction factor. An iterative procedure is therefore necessary when the Reynolds number is the given quantity. Figure 13.8 illustrates these results for both turbulent and laminar flows.

These discussions have all had to do with turbulent flows in smooth, straight, round pipes. Extensions to rough pipes, curved pipes, and pipes of noncircular cross section may be found in the literature. For practical calculations involving piping systems, the overall pressure drop must include that due to wall shear, discussed here, in addition to pressure losses through elbows, valves, and other plumbing discontinuities. These so-called minor losses (in some piping systems they are in fact a major portion of the total) are largely empirical in nature, and recommended values may be found in engineering handbooks. We proceed now to a brief discussion of external flows.

13.5 FLAT PLATE BOUNDARY LAYER FLOWS

From an applications point of view, the primary goal of the analysis of pipe flows is to derive reliable methods for predicting pumping requirements, that is, pressure drops due to friction. We have seen how the pressure drop and shear stress are intimately related in such flows, and the analogy carries over to external flow in that here the main goal is to develop means to predict the shear stress acting on bounding surfaces. These predictions lead, in turn, to the frictional resistance, as we have seen in the study of laminar flows.

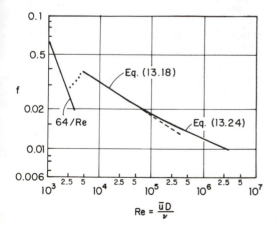

Figure 13.8 Friction factor for incompressible steady flow in smooth pipes.

Just as the flow velocity increases from zero at the wall to a maximum at the centerline of a pipe flow, the velocity in a turbulent boundary layer increases from zero at the boundary to a maximum at the boundary layer edge. Early models for turbulent boundary layer flows were therefore strongly linked to the results obtained for pipe flows. We outline these procedures here. Recall that from momentum considerations the local skin friction coefficient could be expressed in terms of the rate of change of the momentum thickness. This result, previously designated as Eq. (12.35), was independent of whether or not the flow was laminar, and is repeated here:

$$c_f = \frac{\tau_0}{\rho U_\infty^2 / 2} = 2 \frac{d\delta_2}{dx} \tag{13.25}$$

With the momentum thickness defined as in Eq. (12.36), the problem had been reduced to two equations in the four unknowns: wall shear stress (or c_f), δ, δ_2, and the velocity profile $u(y)$. Recall that in Chapter 12 it was stated that the general approach was to specify a "reasonable" velocity profile together with a shear stress law. The latter relationship was available for laminar flow, but in turbulent flow the constitutive relationship is yet to be defined.

For boundary layer flow the velocity profile is obtained by analogy with Eq. (13.20) for pipe flow. Thus,

$$\frac{u}{U_\infty} = \left(\frac{y}{\delta}\right)^{1/n} \tag{13.26}$$

Here we have drawn the parallels that $U \approx U_\infty$ and $R \approx \delta$. Using Eqs. (12.31) and (12.32), this velocity profile leads to the following results for the displacement and momentum thicknesses:

$$\frac{\delta_1}{\delta} = \frac{1}{1 + n} \tag{13.27}$$

$$\frac{\delta_2}{\delta} = \frac{n}{(1 + n)(2 + n)} \tag{13.28}$$

and we are again brought to the realization that such ratios are merely numbers. Accordingly, we may rewrite Eq. (13.25) as

$$c_f = 2 \frac{\delta_2}{\delta} \frac{d\delta}{dx}$$

or

$$\tau_0 = \rho U_\infty^2 \frac{\delta_2}{\delta} \frac{d\delta}{dx} \tag{13.29}$$

The procedure follows that used in analyzing the laminar boundary layer, the only difference thus far being the choice of velocity profiles.

The second major difference is the specification of the resistance law, and this is again achieved by analogy with turbulent pipe flows. Selecting the Blasius correlation, the boundary layer form of Eq. (13.22) is

$$\frac{u}{v_{*0}} = 8.74\left(\frac{yv_{*0}}{v}\right)^{1/7}$$

and at the boundary layer edge,

$$\frac{U_\infty}{v_{*0}} = 8.74\left(\frac{\delta v_{*0}}{v}\right)^{1/7} \tag{13.30}$$

This may be sorted out for the friction velocity,

$$v_{*0} = \left(\frac{\tau_0}{\rho}\right)^{1/2} = 0.15U_\infty\left(\frac{v}{U_\infty\delta}\right)^{1/8}$$

and it is seen, not surprisingly, that the Reynolds number based on the boundary layer thickness emerges as the characteristic nondimensional ratio for boundary layer flow. Further, the skin friction coefficient for this correlation may now be written as

$$c_f = \frac{0.045}{(\mathrm{Re}_\delta)^{1/4}} \tag{13.31}$$

where $\mathrm{Re}_\delta = U_\infty\delta/v$. Realizing that the constants in this relationship are dependent on Reynolds number, just as in its pipe flow analogy, a general form for the wall shear stress may be written as

$$\frac{\tau_0}{\rho U_\infty^2} = a\left(\frac{v}{U_\infty\delta}\right)^{1/m} \tag{13.32}$$

If $a = 0.0225$ and $m = 4$ in this expression, then we are describing the Blasius correlation for turbulent boundary layer flow on a flat plate.

Exercise Find the appropriate values for a and m in Eq. (13.32) if the flow is laminar.

From Eq. (12.38) we had

$$\tau_0 = \frac{\mu U_\infty}{\delta}f'(0)$$

or

$$\frac{\tau_0}{\rho U_\infty^2} = f'(0)\left(\frac{v}{U_\infty\delta}\right)$$

The form of the Reynolds is therefore appropriate in laminar boundary layer flow as well, and $a = f'(0)$ and $m = 1$. [Recall that $f'(0)$ was found from the selected velocity profile for the laminar case.]

Following the procedures used in Chapter 12, we now have, from Eqs. (13.29) and (13.32),

$$a\left(\frac{v}{U_\infty\delta}\right)^{1/m} = \frac{\delta_2}{\delta}\frac{d\delta}{dx}$$

or

$$a\left(\frac{v}{U_\infty}\right)^{1/m}\left(\frac{\delta}{\delta_2}\right) = \delta^{1/m}\frac{d\delta}{dx}$$

The left-hand side of this relation is just a conglomeration of constants, and so the expression may be integrated, with $\delta(0) = 0$, to get

$$\frac{\delta}{x} = C_t\frac{1}{(Re_x)^{1/(m+1)}} \tag{13.33}$$

where

$$C_t = \left\{\frac{a[(m+1)/m]}{\delta_2/\delta}\right\}^{m/(m+1)}$$

and

$$Re_x = \frac{U_\infty x}{v}$$

Using the values of a and m derived in the previous exercise, the reader may compare this result with the laminar flow relation given in Eq. (12.41). With δ/x thus determined, other boundary layer quantities may be found using Eqs. (13.27), (13.28), and (12.36). In addition, the skin friction coefficient is given by

$$c_f = \frac{2m}{m+1}\left(\frac{\delta_2}{\delta}\right) \tag{13.34}$$

With the values of a, m, and n given, all quantities are thus determined. In the case of the Blasius correlation, $a = 0.0225$, $m = 4$, and $n = 7$, and the reader may verify that

$$C_t = 0.37 \tag{13.35a}$$

$$\frac{\delta}{x} = \frac{0.37}{Re_x^{1/5}} \tag{13.35b}$$

$$\frac{\delta_1}{x} = \frac{0.046}{Re_x^{1/5}} \tag{13.35c}$$

$$\frac{\delta_2}{x} = \frac{0.036}{Re_x^{1/5}} \tag{13.35d}$$

$$c_f - \frac{0.058}{\text{Re}_x^{1/5}} \qquad (13.35e)$$

$$C_f = \frac{0.072}{\text{Re}_x^{1/5}} \qquad (13.35f)$$

When used within the range of Reynolds numbers in which they are valid, these expressions give remarkably good agreement with theory. For instance, experimental evidence recommends a coefficient of 0.74 instead of 0.72 in the result for C_f, the friction drag coefficient. This range of validity may be estimated using the previous analogies (see problems). It is generally recommended that use of the Blasius correlation—that is, the results above (stemming from the corresponding values of a, m, and n)—should be restricted to the range $5 \times 10^5 < \text{Re}_x < 10^7$. The lower limit corresponds to a nominal critical Reynolds number below which the flow may be expected to be laminar. Other correlations have been proposed for use at higher Reynolds numbers, but these are preempted by universal skin friction laws in much the same way as was found for pipe flows.

The logarithmic velocity profiles have been used by a number of investigators to extend the flat plate friction results over large Reynolds number ranges (see Schlichting, 1968, pp. 601–603). A particular application requiring such large Reynolds numbers is that of determining the frictional resistance of ship hulls. The so-called Shoenherr formula has been the basis for such calculations by naval architects, and is given by

$$\frac{0.557}{C_f^{1/2}} = \ln(\text{Re}_x \, C_f) \qquad (13.36)$$

The similarity with the expression for pipe flow friction factor, Eq. (13.24), will be noted, and again, such expressions are implicit for the resistance coefficient. Relationships such as Eq. (13.36) have been found to be valid at Reynolds numbers up to 10^9! Figure 13.9 illustrates these results for both laminar and turbulent flat plate flow.

Initial Laminar Boundary Layer

In many instances the turbulent boundary spans such a large portion of the length of a body that the laminar portion may be neglected. In water, for instance, with a kinematic viscosity on the order of 10^{-5} ft^2/s,

$$U_\infty x_{\text{crit}} = \text{Re}_{\text{crit}}(10^{-5})$$

Taking the range of critical Reynolds number to be that given by Eq. (13.2), this becomes

$$3 < U_\infty x_{\text{crit}} < 30 \text{ ft}^2/\text{s}$$

Thus it is seen that for a speed of more than $U_\infty = 10$ ft/s, say, the point of transition to turbulence will be no more than 3 ft from the leading edge of the

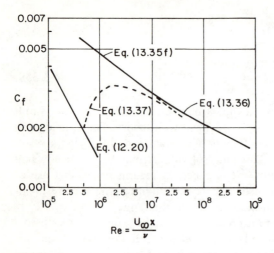

Figure 13.9 Coefficient of friction drag for flow over a smooth flat plate at zero incidence.

plate. On a ship that is 300 ft in length, this is certainly not appreciable, and it is acceptable to estimate the friction resistance as due to a turbulent boundary layer. On the other hand, if the body under consideration is a 10-ft rowboat at 3 ft/s, the laminar portion of the boundary layer may well span a major portion of the hull surface.

In order to account for a combined laminar-turbulent boundary layer, it is customary to correct the drag calculated for an entirely turbulent boundary layer by subtracting from it the turbulent drag for a length equal to the laminar length, and then adding back the laminar resistance for that length. Thus,

$$D_c = D_{tt} - D_{tx} + D_{lx}$$

Here we use the subscript notations c = combined, tt = total turbulent, tx = turbulent to x_{crit}, and lx = laminar to x_{crit}. In coefficient form, the left-hand side, and the first term on the right, are based on total length of the body, ℓ, while the last two terms on the right are based on the distance to transition, x_{crit}. Therefore the expression above may be written

$$C_{f_c} = C_{f_{tt}} - \frac{x_{crit}}{\ell}(C_{f_{tx}} - C_{f_{lx}})$$

The ratio of the lengths, x_{crit}/ℓ, may be expressed as the ratio of the Reynolds numbers based on these lengths, and when this is done,

$$C_{f_c} = C_{f_{tt}} - \frac{A}{Re_\ell} \tag{13.37}$$

where

$$A = Re_{crit}(C_{f_{tx}} - C_{f_{lx}})$$

Both the coefficients in A are based on the critical length, and if we use the Blasius relationship for laminar and turbulent flow, Eqs. (12.20) and (13.35*f*), respectively,

$$A = 1.328(\text{Re}_{\text{crit}})^{1/2}[0.056(\text{Re}_{\text{crit}})^{0.3} - 1.0] \qquad (13.38)$$

For a nominal value of the critical Reynolds number, say, 5×10^5, $A \approx 1750$. [For the 300-ft ship at 10 ft/s in water, $\text{Re} \approx 3 \times 10^8$ and the correction factor in Eq. (13.37) is indeed negligible.]

In our discussions of viscous flow past bodies, we have emphasized the results that have been obtained for the flat plate. It has been pointed out that the major hydrodynamical feature of a flat plate is that the pressure variation in the surrounding ideal fluid flow is negligible. This has allowed a major simplification to the equations of motion, and several useful—if approximate—solutions have been found. The resistance to such flows has been consistently termed *frictional* resistance because it arises directly from the shear stresses acting at the surface of the plate. In flows past nonslender, or "blunt," bodies, there are other forms of resistance that can be as important as frictional resistance. In some cases, in fact, these other forms are by far the major part of the total resistance of a blunt body.

One such "other" form of resistance is the *wave-making resistance* associated with the motion of bodies near a free surface. When a ship moves through the water, the pressure distribution on the submerged portion of the hull may be estimated by the methods introduced in Part One. Because the pressure above the free surface is a constant, this underwater pressure distribution must be offset by corresponding increases and decreases in the level of the free surface. We are aware, for instance, that the maximum underwater pressure occurs at the leading edge of a body where stagnation pressures are found. Thus the most prominent wave associated with ship motion is its bow wave. The production of such wave patterns requires an expenditure of energy that must be provided by the propulsion system—it is a form of resistance to the motion. At relatively low speeds, on the order of 70% of the total drag is likely to be made up of frictional resistance. At high speeds, however, the wave-making resistance dominates the total. This component of resistance increases dramatically with ship speed, and it is the main reason why conventional displacement hulls are limited to speeds of about 50 ft/s. Beyond this brief description, we will have no more to say about wave-making resistance in this text.

A more common form of resistance, to be added to frictional resistance, is that which arises from the separation of the boundary layer from bodies with adverse pressure gradients. In the next section we provide a few guidelines for the understanding and accounting of this type of resistance.

13.6 FLOWS WITH PRESSURE GRADIENTS

The existence of lateral momentum transport in a flowing viscous fluid, either through molecular motion or through turbulent mixing, has several profound effects on the flow. We have already discussed several of these effects, and the major retarding influence they have on the flow near boundaries. The flat plate results that have been presented here are often used to estimate the drag of relatively slender bodies. For instance, the formulas for turbulent flow past a flat plate are of primary importance in estimating the drag of wings, tails, rudders, ship hulls, and other such shapes that present relatively mild obstructions to the oncoming flow. In Part One, where viscous effects were neglected, it was demonstrated that a fluid flows around a body in a very balanced way—pressure exchanged for velocity in accelerating regions must be subsequently recovered in regions where the flow decelerates. The presence of the boundary layer gives rise to frictional resistance. There is also a small additional displacement effect of the body or, more to the point, a reduction of the size of the region that may be thought of as inviscid.

For bodies that extend much further in the flow direction than normal to it, these are the main effects of viscosity. The pressure variation around the body remains much the same as that predicted for ideal flows, as in Part One. This is illustrated further in Figs. 13.10 and 13.11, in which measured pressure distributions are compared with those given by ideal flow theory. Note that the pressure variations about both of these bodies are very much in agreement with the theoretical ideal flow predictions. This agreement is the hydrodynamic definition of "streamlined." Departures of theory from experiment, where they can be discerned, are in regions of rapidly increasing pressure (adverse pressure gradients).

Figure 13.12 illustrates the flow past an airfoil in a situation very similar to that of Fig. 13.11. The photograph is the result of the flow visualization work of Morrison (1984), in which streamlines are made visible by smoke particles. The airfoil in Fig. 13.12 is an NACA 65 series blade at a relatively small angle of attack of 4°. The Reynolds number is 150,000. Because of the slenderness of the foil and its low angle of attack, streamlines near the foil are of very much the same shape as the foil itself. The external flow, outside the influence of viscosity,

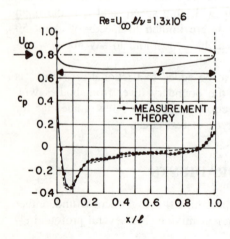

Figure 13.10 Pressure distribution about a streamlined body of revolution. Ideal flow prediction compared measurements due to Fuhrmann (1910), reported in Schlichting (1968).

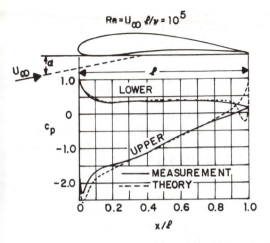

Figure 13.11 Pressure distribution about an airfoil at 6° angle of attack. Ideal flow prediction compared with measurements due to Betz (1915), as reported in Schlichting (1968).

is essentially the same as if there were no such influence, and the pressure agreement illustrated in Fig. 13.11 is to be expected.

If we now consider a blunt body, such as a cylinder, we can expect large pressure gradients in the flow outside the boundary layer. In Chapter 4 we analyzed the ideal flow past a cylinder, with the result that the velocity on the cylinder

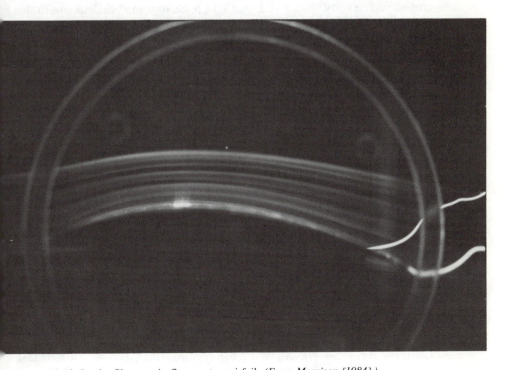

Figure 13.12 Smoke filaments in flow past an airfoil. *(From Morrison [1984].)*

surface is given by $q = -2U_\infty \sin \theta$. The ideal flow pressure distribution is found from the Bernoulli equation, Eq. (3.7), which may be written in the following form:

$$c_p \equiv \frac{p - p_\infty}{\rho U_\infty^2 / 2} = 1 - \frac{q^2}{U_\infty^2}$$

where P_∞ is the pressure in the undisturbed flow and, as before, we have lumped the hydrostatic pressure head, $\rho g h$, with the static pressure. The dimensionless term c_p is called the *pressure coefficient* and is a convenient form for presenting pressure distributions. For the case of the cylinder it is

$$c_p = 1 - 4 \sin^2 \theta \qquad (13.39)$$

This functional relationship is illustrated in Fig. 13.13, where $\phi = \pi - \theta$. At the upstream and downstream stagnation points of the cylinder, $\phi = 0, \pi$, the pressure coefficient is seen to have a maximum value of unity, indicating that at *both* these points the stagnation pressure is realized: $P = P_\infty + \rho U_\infty^2 / 2$. The stagnation pressure upstream is "recovered" at the symmetrical downstream point on the cylinder. This pressure recovery would be very beneficial if it actually occurred, for then the only component of resistance would be frictional resistance due to shear stresses acting at the cylinder surface.

Will experiment continue to validate the inviscid theory for the pressure variation around such a body? Study of Fig. 13.13 leads to a "no" in response to this question. In Chapter 12, in the discussion of boundary layer separation, we observed that viscosity, which gives birth to the boundary layer, is also responsible

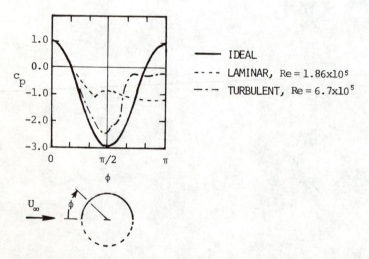

Figure 13.13 Pressure distribution about a cylinder. Ideal flow prediction compared with measurements with laminar and turbulent boundary layer. *(From Flachsbart [1927], as reported in Schlichting [1968].)*

for its demise. From the physical point of view, to repeat, viscosity acts within the boundary layer to decelerate the flow near the wall and dissipate energy in the frictional shearing processes there. These viscous retarding effects are opposed by a pressure that decreases in the direction of the flow and, depending on which effect wins, there may or may not be a continuous downstream motion near the wall. When the external pressure gradient is adverse—that is, increasing in the flow direction—both pressure and viscous effects tend to retard the flow. At some point along the wall, where the viscous effects are accumulating, the fluid eventually loses its ability to continue downstream along the wall. This can happen even in a region of negative pressure gradient, as the experimental results of Fig. 13.13 show.

The exact point of separation is difficult to predict. As shown in Chapter 12, it is related to a point of inflection in the boundary layer velocity profile. When this point eventually reaches the body surface, the shear stress there is zero and separation occurs. The inflection point was seen to be related to the sign and magnitude of the pressure gradient, so the shape of the body plays an important role. In addition, laminar boundary layers are more prone to separation than are turbulent ones. This is because the turbulent mixing produces a fuller velocity profile, as we have seen, with larger velocity gradients near the wall and associated larger shear stresses. It is harder to reverse the flow in the wall region of a turbulent boundary layer, because the velocities are larger there than in a corresponding laminar boundary layer.

Form Drag

When major regions of separated flow exist on a body, a large wake region develops and, as far as the external flow is concerned, there is no longer any need to converge around the base of the body. But this convergence is what gave rise to deceleration and pressure recovery in the inviscid case! The result is that the inviscid pressure balance is drastically modified, and the relatively high pressures upstream are no longer offset by high pressures downstream. The increase in unbalanced force in the downstream direction is what is termed *form drag*. It has also been called eddy resistance, in recognition of the nature of the chaotic separated-flow wake region, and a more descriptive but less common term is viscous pressure drag.

Figure 13.14 illustrates the fact that it is not just the shape of a body that determines its bluntness, from a hydrodynamic point of view. This photograph shows the same airfoil as Fig. 13.12, but in this case the angle of attack is 20°. The upper surface of the foil is seen to be fully separated, and as a result, the drag of the foil is made up largely of form drag. In addition, the lift production of the foil is greatly reduced because the low pressures normally found on this upper surface, at low angles of attack (see Fig. 13.11), are no longer present. The foil is said to be *stalled*—an occasion of some trauma for airplanes and their occupants.

In Figs. 13.15 and 13.16, some results are shown for the drag coefficient

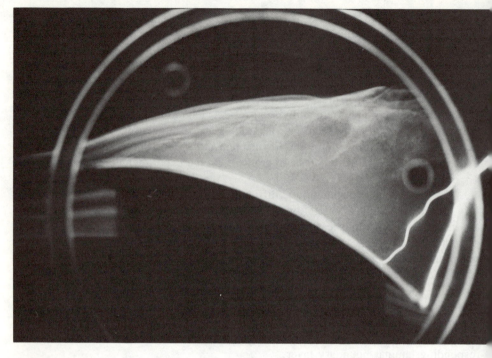

Figure 13.14 Smoke filaments in flow past a stalled airfoil. *(From Morrison [1984].)*

associated with flow past cylinders and spheres. The drag coefficient is defined in terms of the drag, D, as

$$C_D = \frac{D}{(\rho U_\infty^2/2)S} \tag{13.40}$$

where S is a reference area based on a measure of the frontal area of the body. In the case of the sphere, Fig. 13.15, this area is $S = \pi d^2/4$, where d is the sphere diameter. In the case of the cylinder this area is $d\ell$, where d is the cylinder diameter and ℓ is the length. [In drag measurements of cylinders, great care is taken to minimize end effects so that the results may be interpreted as being for a cylinder of infinite length. In this case the length dimension is attached to the drag term and Eq. (13.40) is interpreted on a drag-per-unit-length basis—D/ℓ.] In deducing the total drag of a body from coefficient data, great care must always be taken to apply the correct reference area. In the case of frictional resistance, for instance, this area is the wetted surface area, and not the frontal area.

Both the sphere and the cylinder are about as blunt as one can get from the point of view of obstruction of the flow. Their drag coefficient characteristics are quite similar, and are representative of other blunt shapes. For the cylinder, for example, Fig. 13.16 shows that there is an inverse logarithmic relationship between Re and C_D at very low values of the Reynolds number. (For the sphere the drag coefficient is given by 24/Re for Re < 1.) Such low Reynolds numbers are

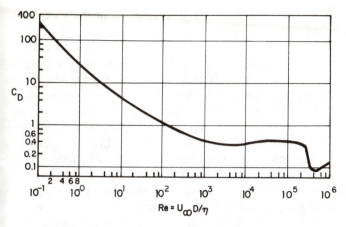

Figure 13.15 Drag coefficient of a sphere. *(From Schlichting [1968], by permission of McGraw-Hill Book Company.)*

associated with what we have called slow flow, and for bodies of any appreciable size, the practical applications are limited. At intermediate Reynolds numbers the drag coefficient continues to decrease as Reynolds number increases, but less dramatically. In this range the flow over much of the body is of a laminar boundary layer nature, and even though separation is present, a significant part of the total drag is due to skin friction, as indicated by the dependency on Reynolds number—qualitatively similar to the flat plate dependency shown in Fig. 13.9. As the Reynolds number approaches a value of about 1000, the laminar boundary layer becomes fully separated, and the skin friction contributes only a small percentage of the total drag. The measurements show that at Reynolds numbers beyond those associated with the onset of fully separated flow, there is a broad region in which the drag coefficient is approximately independent of Reynolds

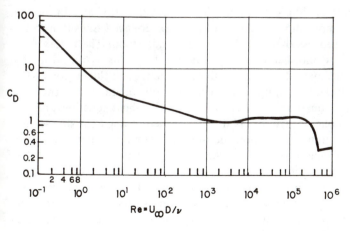

Figure 13.16 Drag coefficient of a cylinder. *(From Schlichting [1968], by permission of McGraw-Hill Book Company.)*

number. At these relatively high Reynolds numbers, which commonly occur in practical flow situations, the total drag of a blunt body is made up virtually entirely of form drag—the friction resistance is a negligible fraction of the total.

The existence of broad ranges of Reynolds number independence justifies the common assumption of a constant drag coefficient. It must be appreciated, however, that the validity of this assumption does depend on the Reynolds number. For the cylinder the drag coefficient is roughly constant at a value of 1.1 for Reynolds numbers in the range of 1000 to 300,000. At the upper limit of this range, a dramatic drop in C_D is observed. This marks the point of transition to a turbulent boundary layer flow and, as has been mentioned, such a boundary layer is less prone to separate. Accordingly, the separation point on the body shifts rearward, allowing more of a recovery of pressure on downstream portions of the body. Mental integration of the pressure distributions shown in Fig. 13.13 for the laminar and turbulent case will indicate the nature of this drag reduction. Note also the rearward shift of the region of separation on the cylinder for the turbulent case. For situations in which the drag is largely form drag, it may be useful to intentionally induce boundary layer turbulence. This has led to the "fur" on tennis balls and the dimples on golf balls. On aircraft, turbulence generators are sometimes placed on the suction side of their wings in order to maintain a larger region of attached flow at high angles of attack. It should be remarked, however, that turbulence is to be avoided on slender bodies, where the total drag is largely frictional.

Much experimentation has been undertaken to determine the drag of various blunt shapes. The results of these tests are reported in the literature and in handbooks, such as Hoerner's (1965). Some recommended values of drag coefficients, largely representative of form drag effects, are shown in Fig. 13.17.

13.7 SUMMARY

In this chapter an introduction to turbulent flows has been provided. In the majority of engineering applications, flows will have experienced transition to turbulence—a complex process that is only partially understood. The theory associated with turbulence has been greatly advanced by the assumption that instantaneous quantities may be replaced with mean values and fluctuations. Having done this, it is a relatively simple matter to develop equations governing the motion—the Reynolds equations. The principal feature of these equations, beyond those of the Navier-Stokes equations, is the appearance of apparent stresses that express the convective transport of momentum by the turbulent fluctuations. The apparent stresses constitute a set of six additional dependent variables, and their existence leads to the turbulence closure problem.

Despite the difficulties introduced by turbulence in fluid motion, much has been deduced about the physical nature of the apparent stresses. Near the boundaries of a turbulent flow, the fluctuations die out and viscosity accomplishes the deceleration to the no-slip condition at the wall. This deceleration is much more severe than in laminar flows, because of the relatively high velocities transported

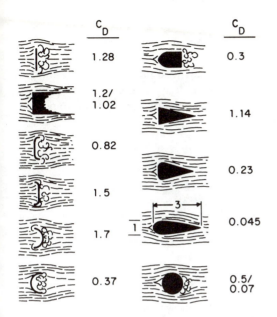

	c_D		c_D
	1.28		0.3
	1.2/ 1.02		1.14
	0.82		0.23
	1.5		0.045
	1.7		
	0.37		0.5/ 0.07

Figure 13.17 Approximate drag coefficients for various blunt shapes. *(Adapted from Dommasch et al. [1957].)*

to the near-wall regions by turbulence. Through the analysis of momentum transport, the concept of the mixing length has been developed as a measure of the characteristic length scale for the turbulent mixing. In the regions outside the laminar sublayer, the form of the velocity gradient, valid at large Reynolds numbers, may be deduced from the mixing length theory.

Fully developed turbulent pipe flow has provided a major vehicle for the development of formulas useful in practical applications. This is largely because of the direct relationship between the axial pressure gradient and the transverse shear-stress gradient, both of which are constant in such flows. From experimental measurements, coupled with analytical insights, it has been possible to develop reliable formulas for velocity gradients, shear stresses, and pressure drops in pipes.

By analogy with fully developed pipe flow, formulas have also been derived for the turbulent flat plate boundary layer. Again, velocity gradients and shear stresses may be accurately predicted, and the latter lead to the frictional resistance for flat plates. In external flow cases where the effects of pressure gradient cannot be ignored, boundary layer separation becomes a critical factor and frictional resistance can be only a small part of the total body drag. The added drag, called form drag, has been determined experimentally for many blunt-body shapes. At relatively high Reynolds numbers, where form drag dominates, the drag coefficient for blunt bodies is almost constant.

Some of the issues that are *not* treated here have already been mentioned. These include roughness effects and complex geometric configurations. The basic concepts retain their validity—mean and fluctuating quantities, momentum transport on molecular and bulk scales, the influence of pressure gradient, the no-slip condition, etc.. We have concentrated on coming to grips with the question of drag. In Parts One and Two, some treatment has been given to the problem of determining the *lift* on bodies immersed in a flow, and in Part One it was found that lift is linked to circulation. The link between circulation and viscous effects has been established, but we leave this subject to more advanced texts. In general, the lift is determined by the component of the integrated pressure distribution that is aligned with the free stream flow, normal to the drag component. In this sense, the reader will understand that boundary layer separation can have an important effect on body lift, as well as drag.

REFERENCES

Betz, A: Untersuchung einer Joukowskyschen Tragfläche, *ZFM, 6,* pp. 173–179, 1915.

Dommasch, D. O., S. S. Sherby, and T. F. Connolly: *Airplane Aerodynamics,* 2nd ed., Pitman Pub. Corp., New York, 1957.

Flachsbart, O.: Winddruck auf Gasbehälter, *Reports of the Aerody. Versuchsanstalt, IV series,* Göttingen, pp. 134–138, 1927.

Fuhrmann, G.: Theoretische und experimentelle Untersuchungen an Ballonmodellen, Diss., Göttingen, 1910.

Hinze, J. O.: *Turbulence,* McGraw-Hill, New York, 1959.

Hoerner, S. F.: *Fluid Dynamic Drag,* 2nd ed., published by the author, Midland Park, N.J., 1965.

Morrison, G. A.: On the Use of Liquid Crystal Thermography as a Technique of Flow Visualization, MS thesis, Naval Postgraduate School, Monterey, Calif., 1984.

Nikuradse, J.: Gesetzmässigkeit der Turbulenten Strömung in Glatten Rohren, *Forsch. Arb. Ing.-Wes.,* no. 356, 1932.

Reichardt, H.: Messungen turbulenter Schwankungen, *Naturwissenschaften* 404, 1938. (See also *ZAMM, 13,* pp. 177–180, 1933, and *ZAMM, 18,* pp. 358–361, 1938.)

Reynolds, O.: On the Experimental Investigation of Circumstances Which Determine Whether the Motion of Water Shall Be Direct or Sinuous, and the Law of Resistance in Parallel Channels, *Phil. Trans. Roy. Soc., 174,* pp. 935–982, 1883. (See also *Collected Papers,* vol. II, p. 51.)

Schlichting, H.: *Boundary Layer Theory,* 6th ed., McGraw-Hill, New York, 1968.

PROBLEMS

Some of the following problems require the use of fluid properties. Unless otherwise specified, use the following:

Air: $v = 1.6 \times 10^{-4}$ ft^2/s, $\rho = 0.00234$ slug/ft^3
Water: $v = 1.2 \times 10^{-5}$ ft^2/s, $\rho = 1.94$ slug/ft^3

13.1 Consider the fully developed turbulent flow of air at a Reynolds number of 10^5 in a 6-in.-diameter smooth pipe. Determine the wall shear stress and the pressure drop per unit length of pipe.

13.2 For the conditions of problem 13.1, plot the shear stress divided by density, τ/ρ, vs. the distance from the pipe wall divided by the radius, y/R. On the same scale, plot $v \, du/dy$ vs. y/R. (Assume a 1/7 power law velocity profile.)

13.3 Noting that $\tau/\rho = -\overline{u'v'} + v \, du/dy$, plot the ratio

$$-\frac{\rho \overline{u'v'}}{\mu(du/dy)}$$

vs. y/R for the conditions of problem 13.1.

13.4 In turbulent pipe flow the Blasius correlation, Eq. (13.18), is said to be valid for Reynolds numbers up to Re $= \bar{u}D/v = 100{,}000$. Using the analogies leading to Eq. (13.35) for the flat plate boundary layer, show that the corresponding upper limit of validity is about $Re_x = U_\infty x/v = 3.3 \times 10^6$. Compare this value with that given in the text.

13.5 Repeat problem 12.13 assuming an entirely turbulent boundary layer. Can you explain why the plate has less drag with its long side parallel to the flow?

13.6 Repeat problem 12.13 taking into account a combined laminar and turbulent boundary layer. Assume a critical Reynolds number of 8×10^5.

13.7 Repeat problem 13.6 using a critical Reynolds number of 3×10^5.

13.8 Turbulent flow over a flat plate is found to have a shear stress correlation such that $a = 0.025$ and $m = 4$ in Eq. (13.32). The power law velocity profile gives $n = 8$. Find the ratio of the momentum thickness to displacement thickness, δ_2/δ_1. Find δ/x as a function of the plate Reynolds number.

13.9 Consider a ship cruising at 5 knots. The ship is 150 ft long and has a wetted surface area of 10,000 ft^2. Using the values given in problem 13.8, estimate the frictional resistance acting on the ship. (1 knot $= 1.689$ ft/s.)

13.10 Consider a flat plate with length 1 ft parallel to the flow and a width of 5 ft. The flow velocity is 12.5 ft/s. Calculate the drag on the plate. The fluid is water.

13.11 For the plate of problem 13.10, estimate the thickness of the boundary layer at the trailing edge.

13.12 On the plate of problem 13.10, where will the shear stress be 0.7 lb/ft^2?

13.13 Consider the flow of a viscous fluid over a flat plate from x_1 to x_2. Show that the friction drag, D, acting on the surface over this distance is given by

$$D = \rho U_\infty^2 b[\delta_2(x_2) - \delta_2(x_1)]$$

where b is the plate width.

13.14 A ship hull is 200 ft long and has a wetted surface area of 5000 ft^2. Estimate the frictional resistance of the hull at a speed of 20 ft/s. Assuming a 1/7 power law velocity profile, estimate the thickness of the boundary layer at the stern of the ship.

13.15 Consider a ship with length 312 ft and wetted surface area of 15,600 ft^2. Using Eq. (13.35f), estimate the frictional resistance at 10 knots. Comment on the validity of your result and give an improved estimate of the frictional resistance. What is the percent error in the initial estimate?

13.16 Consider the flow of air past a smooth flat plate. At a Reynolds number of 500,000, flow in the boundary layer could be either laminar or turbulent. Assuming that the flow is (a) laminar and then (b) turbulent; determine and plot, on the same axes, u/U_∞ vs. y/δ.

13.17 For the conditions of problem 13.16, find the ratio $\delta_{turb}/\delta_{lam}$. Repeat for the ratios of displacement thickness and momentum thickness.

13.18 The accompanying table gives approximate values obtained from Fig. 13.6. At the locations given, determine the nondimensional mixing length, $\ell/(H/2)$ and the nondimensional eddy diffusivity, $\epsilon/(UH/2)$. Here the nondimensional mean velocity profile is $f'(\eta) = d(u/U)/d(2y/H)$. The experiments were conducted at a centerline speed of $U = 100$ cm/s.

$y/(H/2)$	$-\overline{u'v'}$ (cm^2/s^2)	$f'(\eta)$
0.05	19.0	1.93
0.10	21.5	0.91
0.25	18.5	0.53
0.50	12.8	0.28

MACH NUMBER FUNCTIONS FOR ISENTROPIC FLOW (PERFECT GAS, $K = 1.400$)

Mach	A/A^*	P/P_0	T/T_0	$(P/P_0) \times (A/A^*)$
0.00	INF	1.00000	1.00000	INF
0.01	57.87367	0.99993	0.99998	57.86980
0.02	28.94205	0.99972	0.99992	28.93402
0.03	19.30049	0.99937	0.99982	19.28838
0.04	14.48146	0.99888	0.99968	14.46528
0.05	11.59143	0.99825	0.99950	11.57118
0.06	9.66588	0.99749	0.99928	9.64159
0.07	8.29151	0.99658	0.99902	8.26315
0.08	7.26161	0.99553	0.99872	7.22917
0.09	6.46133	0.99435	0.99838	6.42484
0.10	5.82183	0.99303	0.99800	5.78126
0.11	5.29922	0.99158	0.99759	5.25459
0.12	4.86431	0.98999	0.99713	4.81560
0.13	4.49686	0.98826	0.99663	4.44406
0.14	4.18239	0.98640	0.99610	4.12552
0.15	3.91034	0.98441	0.99552	3.84937
0.16	3.67273	0.98229	0.99491	3.60768
0.17	3.46350	0.98003	0.99425	3.39435
0.18	3.27792	0.97765	0.99356	3.20466
0.19	3.11225	0.97514	0.99283	3.03487
0.20	2.96351	0.97250	0.99206	2.88201
0.21	2.82929	0.96973	0.99126	2.74366
0.22	2.70760	0.96685	0.99041	2.61783
0.23	2.59680	0.96384	0.98953	2.50290
0.24	2.49556	0.96071	0.98861	2.39750

Mach	A/A^*	P/P_0	T/T_0	$(P/P_0) \times (A/A^*)$
0.25	2.40271	0.95745	0.98765	2.30048
0.26	2.31728	0.95409	0.98666	2.21089
0.27	2.23847	0.95060	0.98563	2.12789
0.28	2.16555	0.94700	0.98456	2.05078
0.29	2.09793	0.94329	0.98346	1.97896
0.30	2.03506	0.93947	0.98232	1.91188
0.31	1.97650	0.93554	0.98114	1.84910
0.32	1.92185	0.93151	0.97993	1.79021
0.33	1.87074	0.92736	0.97869	1.73486
0.34	1.82288	0.92312	0.97740	1.68273
0.35	1.77797	0.91877	0.97609	1.63355
0.36	1.73578	0.91433	0.97473	1.58707
0.37	1.69608	0.90979	0.97335	1.54308
0.38	1.65869	0.90516	0.97193	1.50138
0.39	1.62343	0.90043	0.97048	1.46179
0.40	1.59014	0.89562	0.96899	1.42415
0.41	1.55867	0.89071	0.96747	1.38833
0.42	1.52890	0.88572	0.96592	1.35419
0.43	1.50072	0.88065	0.96434	1.32161
0.44	1.47400	0.87550	0.96272	1.29049
0.45	1.44867	0.87027	0.96108	1.26073
0.46	1.42463	0.86496	0.95940	1.23225
0.47	1.40179	0.85958	0.95769	1.20495
0.48	1.38010	0.85413	0.95595	1.17878
0.49	1.35947	0.84861	0.95418	1.15365
0.50	1.33984	0.84302	0.95238	1.12951
0.51	1.32116	0.83737	0.95055	1.10630
0.52	1.30338	0.83166	0.94870	1.08397
0.53	1.28645	0.82588	0.94681	1.06246
0.54	1.27032	0.82005	0.94489	1.04173
0.55	1.25494	0.81417	0.94295	1.02173
0.56	1.24029	0.80823	0.94098	1.00244
0.57	1.22632	0.80224	0.93899	0.98381
0.58	1.21301	0.79621	0.93696	0.96580
0.59	1.20030	0.79013	0.93491	0.94840
0.60	1.18820	0.78400	0.93284	0.93155
0.61	1.17665	0.77784	0.93074	0.91525
0.62	1.16564	0.77164	0.92861	0.89946
0.63	1.15515	0.76540	0.92646	0.88416
0.64	1.14515	0.75913	0.92428	0.86932

Mach	A/A^*	P/P_0	T/T_0	$(P/P_0) \times (A/A^*)$
0.65	1.13561	0.75283	0.92208	0.85492
0.66	1.12653	0.74650	0.91986	0.84096
0.67	1.11789	0.74014	0.91762	0.82739
0.68	1.10965	0.73376	0.91535	0.81422
0.69	1.10182	0.72735	0.91306	0.80141
0.70	1.09437	0.72093	0.91075	0.78896
0.71	1.08729	0.71449	0.90841	0.77685
0.72	1.08057	0.70803	0.90606	0.76507
0.73	1.07419	0.70155	0.90369	0.75360
0.74	1.06814	0.69507	0.90129	0.74243
0.75	1.06241	0.68857	0.89888	0.73155
0.76	1.05700	0.68207	0.89644	0.72095
0.77	1.05188	0.67556	0.89399	0.71061
0.78	1.04705	0.66905	0.89152	0.70053
0.79	1.04250	0.66254	0.88903	0.69070
0.80	1.03823	0.65602	0.88653	0.68110
0.81	1.03421	0.64951	0.88400	0.67173
0.82	1.03046	0.64300	0.88146	0.66259
0.83	1.02696	0.63650	0.87891	0.65365
0.84	1.02369	0.63000	0.87633	0.64493
0.85	1.02067	0.62351	0.87374	0.63640
0.86	1.01787	0.61703	0.87114	0.62806
0.87	1.01530	0.61057	0.86852	0.61991
0.88	1.01294	0.60412	0.86589	0.61193
0.89	1.01080	0.59768	0.86325	0.60413
0.90	1.00886	0.59126	0.86059	0.59650
0.91	1.00713	0.58486	0.85791	0.58903
0.92	1.00559	0.57848	0.85523	0.58171
0.93	1.00426	0.57212	0.85253	0.57455
0.94	1.00311	0.56578	0.84982	0.56753
0.95	1.00214	0.55946	0.84710	0.56066
0.96	1.00136	0.55317	0.84437	0.55392
0.97	1.00076	0.54691	0.84162	0.54732
0.98	1.00034	0.54067	0.83887	0.54085
0.99	1.00008	0.53446	0.83611	0.53450
1.00	1.00000	0.52828	0.83333	0.52828
1.01	1.00008	0.52214	0.83055	0.52218
1.02	1.00033	0.51602	0.82776	0.51619
1.03	1.00073	0.50994	0.82496	0.51031
1.04	1.00130	0.50389	0.82215	0.50454

Mach	A/A^*	P/P_0	T/T_0	$(P/P_0) \times (A/A^*)$
1.05	1.00203	0.49787	0.81934	0.49888
1.06	1.00290	0.49189	0.81651	0.49332
1.07	1.00393	0.48595	0.81368	0.48786
1.08	1.00511	0.48005	0.81085	0.48250
1.09	1.00644	0.47418	0.80800	0.47724
1.10	1.00792	0.46836	0.80515	0.47207
1.11	1.00954	0.46257	0.80230	0.46698
1.12	1.01131	0.45682	0.79944	0.46199
1.13	1.01321	0.45112	0.79657	0.45708
1.14	1.01527	0.44545	0.79370	0.45225
1.15	1.01745	0.43983	0.79083	0.44750
1.16	1.01978	0.43425	0.78795	0.44284
1.17	1.02224	0.42872	0.78507	0.43825
1.18	1.02484	0.42323	0.78218	0.43374
1.19	1.02757	0.41778	0.77929	0.42930
1.20	1.03044	0.41238	0.77640	0.42493
1.21	1.03344	0.40702	0.77350	0.42063
1.22	1.03657	0.40171	0.77061	0.41640
1.23	1.03983	0.39645	0.76771	0.41224
1.24	1.04322	0.39123	0.76481	0.40814
1.25	1.04675	0.38606	0.76191	0.40411
1.26	1.05040	0.38093	0.75900	0.40013
1.27	1.05419	0.37586	0.75610	0.39622
1.28	1.05809	0.37083	0.75319	0.39237
1.29	1.06214	0.36585	0.75029	0.38858
1.30	1.06630	0.36092	0.74739	0.38484
1.31	1.07059	0.35603	0.74448	0.38116
1.32	1.07502	0.35119	0.74158	0.37754
1.33	1.07956	0.34640	0.73867	0.37396
1.34	1.08423	0.34166	0.73577	0.37044
1.35	1.08903	0.33697	0.73287	0.36697
1.36	1.09396	0.33233	0.72997	0.36355
1.37	1.09901	0.32773	0.72707	0.36018
1.38	1.10419	0.32319	0.72418	0.35686
1.39	1.10949	0.31869	0.72128	0.35358
1.40	1.11492	0.31424	0.71839	0.35035
1.41	1.12048	0.30984	0.71550	0.34717
1.42	1.12616	0.30549	0.71262	0.34403
1.43	1.13197	0.30119	0.70973	0.34093
1.44	1.13790	0.29693	0.70685	0.33788

Mach	A/A^*	P/P_0	T/T_0	$(P/P_0) \times (A/A^*)$
1.45	1.14396	0.29272	0.70398	0.33486
1.46	1.15014	0.28856	0.70111	0.33189
1.47	1.15646	0.28445	0.69824	0.32896
1.48	1.16290	0.28039	0.69537	0.32606
1.49	1.16947	0.27637	0.69251	0.32321
1.50	1.17616	0.27240	0.68966	0.32039
1.51	1.18299	0.26848	0.68680	0.31761
1.52	1.18994	0.26461	0.68396	0.31487
1.53	1.19702	0.26078	0.68112	0.31216
1.54	1.20423	0.25700	0.67828	0.30948
1.55	1.21157	0.25326	0.67545	0.30685
1.56	1.21904	0.24957	0.67262	0.30424
1.57	1.22664	0.24593	0.66980	0.30167
1.58	1.23437	0.24233	0.66699	0.29913
1.59	1.24223	0.23878	0.66418	0.29662
1.60	1.25023	0.23527	0.66138	0.29414
1.61	1.25836	0.23181	0.65858	0.29170
1.62	1.26662	0.22839	0.65579	0.28928
1.63	1.27502	0.22501	0.65301	0.28690
1.64	1.28355	0.22168	0.65023	0.28454
1.65	1.29221	0.21840	0.64746	0.28221
1.66	1.30102	0.21515	0.64470	0.27991
1.67	1.30996	0.21195	0.64194	0.27764
1.68	1.31903	0.20879	0.63919	0.27540
1.69	1.32825	0.20567	0.63645	0.27318
1.70	1.33760	0.20259	0.63371	0.27099
1.71	1.34710	0.19956	0.63099	0.26883
1.72	1.35673	0.19656	0.62827	0.26669
1.73	1.36651	0.19361	0.62556	0.26457
1.74	1.37642	0.19070	0.62285	0.26248
1.75	1.38649	0.18782	0.62016	0.26042
1.76	1.39669	0.18499	0.61747	0.25837
1.77	1.40705	0.18220	0.61479	0.25636
1.78	1.41754	0.17944	0.61212	0.25436
1.79	1.42819	0.17672	0.60945	0.25239
1.80	1.43898	0.17404	0.60680	0.25044
1.81	1.44992	0.17140	0.60415	0.24851
1.82	1.46101	0.16879	0.60151	0.24661
1.83	1.47225	0.16622	0.59888	0.24472
1.84	1.48364	0.16369	0.59626	0.24286

Mach	A/A^*	P/P_0	T/T_0	$(P/P_0) \times (A/A^*)$
1.85	1.49519	0.16120	0.59365	0.24102
1.86	1.50689	0.15873	0.59105	0.23920
1.87	1.51875	0.15631	0.58845	0.23739
1.88	1.53075	0.15392	0.58587	0.23561
1.89	1.54292	0.15156	0.58329	0.23385
1.90	1.55525	0.14924	0.58072	0.23211
1.91	1.56774	0.14695	0.57816	0.23038
1.92	1.58038	0.14470	0.57561	0.22868
1.93	1.59319	0.14247	0.57307	0.22699
1.94	1.60617	0.14028	0.57054	0.22532
1.95	1.61930	0.13813	0.56802	0.22367
1.96	1.63261	0.13600	0.56551	0.22203
1.97	1.64607	0.13390	0.56301	0.22042
1.98	1.65971	0.13184	0.56051	0.21882
1.99	1.67352	0.12981	0.55803	0.21724
2.00	1.68750	0.12780	0.55556	0.21567
2.01	1.70164	0.12583	0.55309	0.21412
2.02	1.71596	0.12389	0.55064	0.21259
2.03	1.73046	0.12197	0.54819	0.21107
2.04	1.74513	0.12009	0.54576	0.20957
2.05	1.75998	0.11823	0.54333	0.20808
2.06	1.77501	0.11640	0.54092	0.20661
2.07	1.79022	0.11460	0.53851	0.20515
2.08	1.80561	0.11282	0.53611	0.20371
2.09	1.82118	0.11107	0.53373	0.20229
2.10	1.83694	0.10935	0.53135	0.20088
2.11	1.85288	0.10766	0.52898	0.19948
2.12	1.86901	0.10599	0.52663	0.19809
2.13	1.88533	0.10434	0.52428	0.19672
2.14	1.90184	0.10273	0.52194	0.19537
2.15	1.91854	0.10113	0.51962	0.19403
2.16	1.93543	0.09956	0.51730	0.19270
2.17	1.95252	0.09802	0.51499	0.19138
2.18	1.96980	0.09650	0.51269	0.19008
2.19	1.98729	0.09500	0.51041	0.18879
2.20	2.00497	0.09352	0.50813	0.18751
2.21	2.02285	0.09270	0.50586	0.18624
2.22	2.04094	0.09064	0.50361	0.18499
2.23	2.05923	0.08923	0.50136	0.18375
2.24	2.07773	0.08785	0.49912	0.18252

Mach	A/A^*	P/P_0	T/T_0	$(P/P_0) \times (A/A^*)$
2.25	2.09643	0.08648	0.49689	0.18130
2.26	2.11534	0.08514	0.49468	0.18010
2.27	2.13447	0.08382	0.49247	0.17890
2.28	2.15380	0.08252	0.49027	0.17772
2.29	2.17336	0.08123	0.48809	0.17655
2.30	2.19312	0.07997	0.48591	0.17539
2.31	2.21311	0.07873	0.48374	0.17424
2.32	2.23332	0.07751	0.48158	0.17310
2.33	2.25375	0.07631	0.47944	0.17198
2.34	2.27439	0.07512	0.47730	0.17086
2.35	2.29527	0.07396	0.47517	0.16975
2.36	2.31638	0.07281	0.47306	0.16866
2.37	2.33771	0.07168	0.47095	0.16757
2.38	2.35927	0.07057	0.46885	0.16649
2.39	2.38106	0.06948	0.46676	0.16543
2.40	2.40309	0.06840	0.46468	0.16437
2.41	2.42536	0.06734	0.46262	0.16332
2.42	2.44786	0.06630	0.46056	0.16229
2.43	2.47061	0.06527	0.45851	0.16126
2.44	2.49359	0.06426	0.45647	0.16024
2.45	2.51683	0.06327	0.45444	0.15923
2.46	2.54031	0.06229	0.45242	0.15823
2.47	2.56403	0.06133	0.45041	0.15724
2.48	2.58800	0.06038	0.44841	0.15626
2.49	2.61223	0.05945	0.44642	0.15529
2.50	2.63671	0.05853	0.44444	0.15432
2.51	2.66145	0.05762	0.44247	0.15336
2.52	2.68645	0.05674	0.44051	0.15242
2.53	2.71170	0.05586	0.43856	0.15148
2.54	2.73722	0.05500	0.43662	0.15055
2.55	2.76300	0.05415	0.43469	0.14963
2.56	2.78906	0.05332	0.43277	0.14871
2.57	2.81537	0.05250	0.43085	0.14780
2.58	2.84196	0.05169	0.42895	0.14691
2.59	2.86883	0.05090	0.42706	0.14602
2.60	2.89597	0.05012	0.42517	0.14513
2.61	2.92339	0.04935	0.42330	0.14426
2.62	2.95108	0.04859	0.42143	0.14339
2.63	2.97906	0.04784	0.41957	0.14253
2.64	3.00732	0.04711	0.41773	0.14168

Mach	A/A^*	P/P_0	T/T_0	$(P/P_0) \times (A/A^*)$
2.65	3.03587	0.04639	0.41589	0.14083
2.66	3.06471	0.04568	0.41406	0.13999
2.67	3.09384	0.04498	0.41224	0.13916
2.68	3.12327	0.04429	0.41043	0.13834
2.69	3.15298	0.04362	0.40863	0.13752
2.70	3.18300	0.04295	0.40684	0.13671
2.71	3.21332	0.04229	0.40505	0.13591
2.72	3.24394	0.04165	0.40328	0.13511
2.73	3.27487	0.04102	0.40151	0.13432
2.74	3.30610	0.04039	0.39976	0.13354
2.75	3.33765	0.03978	0.39801	0.13276
2.76	3.36950	0.03917	0.39627	0.13199
2.77	3.40168	0.03858	0.39454	0.13123
2.78	3.43417	0.03799	0.39282	0.13047
2.79	3.46698	0.03742	0.39111	0.12972
2.80	3.50012	0.03685	0.38941	0.12897
2.81	3.53357	0.03629	0.38771	0.12823
2.82	3.56736	0.03574	0.38603	0.12750
2.83	3.60148	0.03520	0.38435	0.12678
2.84	3.63592	0.03467	0.38268	0.12605
2.85	3.67071	0.03415	0.38103	0.12534
2.86	3.70584	0.03363	0.37937	0.12463
2.87	3.74130	0.03312	0.37773	0.12393
2.88	3.77711	0.03263	0.37610	0.12323
2.89	3.81326	0.03213	0.37447	0.12254
2.90	3.84976	0.03165	0.37286	0.12185
2.91	3.88661	0.03118	0.37125	0.12117
2.92	3.92382	0.03071	0.36965	0.12049
2.93	3.96139	0.03025	0.36806	0.11982
2.94	3.99931	0.02980	0.36647	0.11916
2.95	4.03759	0.02935	0.36490	0.11850
2.96	4.07625	0.02891	0.36333	0.11785
2.97	4.11527	0.02848	0.36177	0.11720
2.98	4.15465	0.02805	0.36022	0.11655
2.99	4.19442	0.02764	0.35868	0.11591
3.00	4.23456	0.02722	0.35714	0.11528
3.10	4.65730	0.02345	0.34223	0.10921
3.20	5.12095	0.02023	0.32808	0.10359
3.30	5.62864	0.01748	0.31466	0.09837
3.40	6.18368	0.01512	0.30193	0.09353

Mach	A/A^*	P/P_0	T/T_U	$(P/P_0) \times (A/A^*)$
3.50	6.78960	0.01311	0.28986	0.08902
3.60	7.45009	0.01138	0.27840	0.08482
3.70	8.16904	0.00990	0.26752	0.08090
3.80	8.95055	0.00863	0.25720	0.07723
3.90	9.79895	0.00753	0.24740	0.07381
4.00	10.71873	0.00659	0.23810	0.07059
4.10	11.71461	0.00577	0.22925	0.06758
4.20	12.79160	0.00506	0.22085	0.06475
4.30	13.95486	0.00445	0.21286	0.06209
4.40	15.20981	0.00392	0.20525	0.05959
4.50	16.56213	0.00346	0.19802	0.05723
4.60	18.01778	0.00305	0.19113	0.05500
4.70	19.58282	0.00270	0.18457	0.05290
4.80	21.26366	0.00239	0.17832	0.05091
4.90	23.06709	0.00213	0.17235	0.04903
5.00	25.00000	0.00189	0.16667	0.04725
6.00	53.17967	0.00063	0.12195	0.03368
7.00	104.14294	0.00024	0.09259	0.02516
8.00	190.10928	0.00010	0.07246	0.01947
9.00	327.18921	0.00005	0.05814	0.01550
10.00	535.93652	0.00002	0.04762	0.01263

MACH NUMBER FUNCTIONS FOR NORMAL SHOCK FLOW (PERFECT GAS, $K = 1.400$)

M_X	M_Y	P_Y/P_X	T_Y/T_X	P_{0Y}/P_{0X}	P_{0Y}/P_X
1.00	1.00000	1.00000	1.00000	1.00000	1.89293
1.01	0.99013	1.02345	1.00664	1.00000	1.91521
1.02	0.98052	1.04713	1.01325	0.99999	1.93790
1.03	0.97115	1.07105	1.01981	0.99997	1.96097
1.04	0.96203	1.09520	1.02634	0.99993	1.98442
1.05	0.95313	1.11958	1.03284	0.99986	2.00826
1.06	0.94445	1.14420	1.03931	0.99976	2.03245
1.07	0.93598	1.16905	1.04575	0.99961	2.05702
1.08	0.92771	1.19413	1.05217	0.99944	2.08194
1.09	0.91965	1.21945	1.05856	0.99921	2.10722
1.10	0.91177	1.24500	1.06494	0.99893	2.13284
1.11	0.90408	1.27078	1.07129	0.99860	2.15882
1.12	0.89656	1.29680	1.07763	0.99822	2.18513
1.13	0.88922	1.32304	1.08396	0.99777	2.21179
1.14	0.88204	1.34953	1.09027	0.99726	2.23877
1.15	0.87502	1.37625	1.09658	0.99670	2.26608
1.16	0.86816	1.40320	1.10287	0.99605	2.29373
1.17	0.86145	1.43038	1.10916	0.99535	2.32169
1.18	0.85488	1.45780	1.11544	0.99457	2.34997
1.19	0.84846	1.48545	1.12172	0.99372	2.37859
1.20	0.84217	1.51333	1.12799	0.99280	2.40750
1.21	0.83601	1.54144	1.13427	0.99181	2.43674
1.22	0.82999	1.56980	1.14054	0.99073	2.46628
1.23	0.82408	1.59838	1.14682	0.98959	2.49613
1.24	0.81830	1.62720	1.15309	0.98836	2.52629

M_X	M_Y	P_Y/P_X	T_Y/T_X	P_{0Y}/P_{0X}	P_{0Y}/P_X
1.25	0.81264	1.65625	1.15937	0.98706	2.55676
1.26	0.80709	1.68553	1.16566	0.98568	2.58753
1.27	0.80164	1.71505	1.17195	0.98422	2.61859
1.28	0.79631	1.74480	1.17825	0.98269	2.64997
1.29	0.79108	1.77478	1.18456	0.98107	2.68163
1.30	0.78596	1.80500	1.19087	0.97938	2.71359
1.31	0.78093	1.83545	1.19720	0.97760	2.74585
1.32	0.77600	1.86613	1.20353	0.97575	2.77841
1.33	0.77116	1.89705	1.20988	0.97383	2.81125
1.34	0.76641	1.92819	1.21624	0.97183	2.84438
1.35	0.76175	1.95958	1.22261	0.96974	2.87781
1.36	0.75718	1.99120	1.22900	0.96758	2.91152
1.37	0.75269	2.02305	1.23540	0.96535	2.94552
1.38	0.74829	2.05513	1.24181	0.96304	2.97980
1.39	0.74396	2.08745	1.24824	0.96066	3.01437
1.40	0.73971	2.12000	1.25469	0.95820	3.04924
1.41	0.73554	2.15278	1.26116	0.95567	3.08438
1.42	0.73144	2.18579	1.26764	0.95307	3.11980
1.43	0.72741	2.21904	1.27414	0.95040	3.15551
1.44	0.72345	2.25253	1.28066	0.94765	3.19149
1.45	0.71956	2.28625	1.28720	0.94484	3.22776
1.46	0.71574	2.32019	1.29376	0.94196	3.26430
1.47	0.71198	2.35438	1.30034	0.93902	3.30113
1.48	0.70829	2.38879	1.30695	0.93600	3.33824
1.49	0.70466	2.42345	1.31357	0.93293	3.37562
1.50	0.70109	2.45833	1.32021	0.92979	3.41328
1.51	0.69758	2.49344	1.32688	0.92659	3.45120
1.52	0.69413	2.52880	1.33357	0.92333	3.48942
1.53	0.69073	2.56438	1.34029	0.92001	3.52791
1.54	0.68739	2.60020	1.34702	0.91663	3.56666
1.55	0.68410	2.63624	1.35379	0.91319	3.60569
1.56	0.68087	2.67253	1.36057	0.90970	3.64500
1.57	0.67769	2.70905	1.36738	0.90615	3.68459
1.58	0.67455	2.74580	1.37422	0.90255	3.72444
1.59	0.67147	2.78278	1.38108	0.89891	3.76458
1.60	0.66844	2.81999	1.38797	0.89520	3.80497
1.61	0.66545	2.85744	1.39488	0.89146	3.84564
1.62	0.66251	2.89513	1.40182	0.88766	3.88659
1.63	0.65962	2.93304	1.40879	0.88381	3.92780
1.64	0.65677	2.97119	1.41578	0.87993	3.96927

M_X	M_Y	P_Y/P_X	T_Y/T_Y	P_{OY}/P_{OX}	P_{OY}/P_X
1.65	0.65396	3.00958	1.42280	0.87599	4.01103
1.66	0.65119	3.04820	1.42985	0.87202	4.05306
1.67	0.64847	3.08704	1.43693	0.86800	4.09534
1.68	0.64579	3.12613	1.44403	0.86395	4.13790
1.69	0.64315	3.16544	1.45117	0.85986	4.18074
1.70	0.64054	3.20500	1.45833	0.85573	4.22384
1.71	0.63798	3.24478	1.46552	0.85156	4.26720
1.72	0.63545	3.28479	1.47274	0.84736	4.31084
1.73	0.63296	3.32505	1.47999	0.84313	4.35474
1.74	0.63051	3.36553	1.48727	0.83886	4.39890
1.75	0.62809	3.40625	1.49458	0.83457	4.44335
1.76	0.62570	3.44719	1.50191	0.83025	4.48805
1.77	0.62335	3.48838	1.50928	0.82589	4.53301
1.78	0.62104	3.52980	1.51669	0.82152	4.57825
1.79	0.61875	3.57145	1.52412	0.81711	4.62375
1.80	0.61650	3.61333	1.53158	0.81269	4.66952
1.81	0.61428	3.65545	1.53907	0.80824	4.71554
1.82	0.61209	3.69779	1.54659	0.80377	4.76185
1.83	0.60993	3.74038	1.55414	0.79927	4.80841
1.84	0.60780	3.78319	1.56173	0.79477	4.85524
1.85	0.60570	3.82624	1.56935	0.79024	4.90236
1.86	0.60363	3.86953	1.57700	0.78569	4.94970
1.87	0.60158	3.91304	1.58468	0.78113	4.99733
1.88	0.59957	3.95679	1.59239	0.77655	5.04520
1.89	0.59758	4.00078	1.60014	0.77196	5.09337
1.90	0.59562	4.04499	1.60791	0.76736	5.14178
1.91	0.59368	4.08945	1.61572	0.76275	5.19046
1.92	0.59177	4.13412	1.62357	0.75812	5.23940
1.93	0.58988	4.17904	1.63144	0.75349	5.28861
1.94	0.58802	4.22419	1.63935	0.74885	5.33808
1.95	0.58618	4.26958	1.64729	0.74420	5.38783
1.96	0.58437	4.31519	1.65527	0.73954	5.43781
1.97	0.58258	4.36104	1.66327	0.73488	5.48807
1.98	0.58082	4.40713	1.67132	0.73022	5.53861
1.99	0.57907	4.45345	1.67939	0.72555	5.58939
2.00	0.57735	4.50000	1.68750	0.72088	5.64046
2.01	0.57565	4.54678	1.69564	0.71620	5.69176
2.02	0.57397	4.59379	1.70382	0.71153	5.74332
2.03	0.57232	4.64104	1.71203	0.70686	5.79518
2.04	0.57068	4.68853	1.72027	0.70218	5.84726

M_X	M_Y	P_Y/P_X	T_Y/T_X	P_{0Y}/P_{0X}	P_{0Y}/P_X
2.05	0.56906	4.73624	1.72855	0.69751	5.89964
2.06	0.56747	4.78419	1.73686	0.69284	5.95225
2.07	0.56589	4.83238	1.74520	0.68817	6.00514
2.08	0.56433	4.88079	1.75358	0.68351	6.05829
2.09	0.56280	4.92944	1.76200	0.67886	6.11171
2.10	0.56128	4.97833	1.77045	0.67420	6.16537
2.11	0.55978	5.02744	1.77893	0.66956	6.21931
2.12	0.55829	5.07679	1.78745	0.66492	6.27351
2.13	0.55683	5.12637	1.79600	0.66029	6.32796
2.14	0.55538	5.17619	1.80459	0.65567	6.38267
2.15	0.55395	5.22624	1.81322	0.65105	6.43766
2.16	0.55254	5.27653	1.82187	0.64645	6.49290
2.17	0.55115	5.32704	1.83057	0.64186	6.54841
2.18	0.54977	5.37779	1.83930	0.63727	6.60418
2.19	0.54840	5.42878	1.84803	0.63270	6.66021
2.20	0.54706	5.47999	1.85686	0.62814	6.71648
2.21	0.54572	5.53144	1.86569	0.62359	6.77302
2.22	0.54441	5.58313	1.87456	0.61905	6.82982
2.23	0.54311	5.63504	1.88347	0.61453	6.88690
2.24	0.54182	5.68719	1.89241	0.61002	6.94423
2.25	0.54055	5.73958	1.90138	0.60553	7.00183
2.26	0.53930	5.79219	1.91040	0.60105	7.05970
2.27	0.53805	5.84504	1.91944	0.59659	7.11779
2.28	0.53683	5.89813	1.92853	0.59214	7.17615
2.29	0.53561	5.95144	1.93764	0.58771	7.23479
2.30	0.53441	6.00499	1.94680	0.58330	7.29369
2.31	0.53322	6.05878	1.95599	0.57890	7.35283
2.32	0.53205	6.11279	1.96522	0.57452	7.41226
2.33	0.53089	6.16704	1.97448	0.57015	7.47193
2.34	0.52974	6.22152	1.98378	0.56581	7.53185
2.35	0.52861	6.27624	1.99311	0.56149	7.59206
2.36	0.52749	6.33119	2.00249	0.55718	7.65250
2.37	0.52638	6.38638	2.01189	0.55289	7.71322
2.38	0.52528	6.44179	2.02133	0.54862	7.77418
2.39	0.52419	6.49744	2.03081	0.54437	7.83541
2.40	0.52312	6.55332	2.04033	0.54015	7.89691
2.41	0.52206	6.60944	2.04988	0.53594	7.95866
2.42	0.52100	6.66579	2.05947	0.53175	8.02070
2.43	0.51996	6.72237	2.06910	0.52758	8.08296
2.44	0.51894	6.77919	2.07876	0.52344	8.14548

M_X	M_Y	P_Y/P_X	T_Y/T_X	P_{0Y}/P_{0X}	P_{0Y}/P_X
2.45	0.51792	6.83624	2.08846	0.51931	8.20827
2.46	0.51691	6.89352	2.09819	0.51521	8.27133
2.47	0.51592	6.95104	2.10796	0.51113	8.33464
2.48	0.51493	7.00879	2.11777	0.50707	8.39822
2.49	0.51395	7.06678	2.12762	0.50303	8.46206
2.50	0.51299	7.12500	2.13750	0.49902	8.52613
2.51	0.51203	7.18344	2.14742	0.49502	8.59051
2.52	0.51109	7.24213	2.15737	0.49105	8.65508
2.53	0.51015	7.30104	2.16736	0.48711	8.71996
2.54	0.50923	7.36019	2.17739	0.48318	8.78510
2.55	0.50831	7.41957	2.18746	0.47928	8.85049
2.56	0.50741	7.47919	2.19756	0.47540	8.91613
2.57	0.50651	7.53904	2.20770	0.47155	8.98203
2.58	0.50562	7.59913	2.21788	0.46772	9.04821
2.59	0.50474	7.65944	2.22809	0.46391	9.11463
2.60	0.50387	7.71999	2.23834	0.46013	9.18132
2.61	0.50301	7.78077	2.24863	0.45636	9.24825
2.62	0.50216	7.84179	2.25896	0.45263	9.31545
2.63	0.50131	7.90304	2.26932	0.44891	9.38293
2.64	0.50048	7.96452	2.27971	0.44522	9.45064
2.65	0.49965	8.02624	2.29015	0.44156	9.51862
2.66	0.49883	8.08819	2.30063	0.43792	9.58687
2.67	0.49802	8.15037	2.31113	0.43430	9.65536
2.68	0.49722	8.21279	2.32168	0.43071	9.72411
2.69	0.49642	8.27544	2.33227	0.42714	9.79312
2.70	0.49563	8.33833	2.34289	0.42359	9.86240
2.71	0.49485	8.40144	2.35355	0.42007	9.93194
2.72	0.49408	8.46479	2.36424	0.41657	10.00173
2.73	0.49332	8.52837	2.37498	0.41310	10.07178
2.74	0.49256	8.59219	2.38575	0.40965	10.14210
2.75	0.49181	8.65624	2.39656	0.40623	10.21267
2.76	0.49107	8.72052	2.40741	0.40283	10.28349
2.77	0.49033	8.78504	2.41829	0.39945	10.35458
2.78	0.48960	8.84979	2.42922	0.39610	10.42593
2.79	0.48888	8.91477	2.44018	0.39277	10.49754
2.80	0.48817	8.97999	2.45117	0.38946	10.56936
2.81	0.48746	9.04544	2.46220	0.38618	10.64150
2.82	0.48676	9.11112	2.47328	0.38293	10.71390
2.83	0.48606	9.17704	2.48439	0.37970	10.78655
2.84	0.48538	9.24319	2.49553	0.37649	10.85945

M_X	M_Y	P_Y/P_X	T_Y/T_X	P_{0Y}/P_{0X}	P_{0Y}/P_X
2.85	0.48469	9.30957	2.50672	0.37330	10.93258
2.86	0.48402	9.37619	2.51794	0.37014	11.00601
2.87	0.48335	9.44304	2.52920	0.36700	11.07969
2.88	0.48269	9.51012	2.54050	0.36389	11.15360
2.89	0.48203	9.57744	2.55183	0.36080	11.22779
2.90	0.48138	9.64499	2.56320	0.35773	11.30223
2.91	0.48073	9.71278	2.57462	0.35469	11.37699
2.92	0.48010	9.78079	2.58606	0.35167	11.45195
2.93	0.47946	9.84904	2.59755	0.34867	11.52717
2.94	0.47884	9.91752	2.60907	0.34570	11.60263
2.95	0.47821	9.98624	2.62064	0.34275	11.67838
2.96	0.47760	10.05519	2.63223	0.33982	11.75439
2.97	0.47699	10.12437	2.64387	0.33692	11.83065
2.98	0.47638	10.19379	2.65555	0.33404	11.90714
2.99	0.47578	10.26344	2.66726	0.33118	11.98394
3.00	0.47519	10.33333	2.67901	0.32834	12.06100
3.10	0.46953	11.04499	2.79860	0.30121	12.84551
3.20	0.46435	11.77999	2.92199	0.27623	13.65592
3.30	0.45959	12.53832	3.04919	0.25328	14.49213
3.40	0.45520	13.31999	3.18020	0.23223	15.35414
3.50	0.45115	14.12499	3.31505	0.21295	16.24200
3.60	0.44741	14.95332	3.45372	0.19531	17.15562
3.70	0.44395	15.80499	3.59624	0.17919	18.09505
3.80	0.44073	16.67996	3.74260	0.16447	19.06032
3.90	0.43774	17.57831	3.89281	0.15103	20.05128
4.00	0.43496	18.49998	4.04687	0.13876	21.06813
4.10	0.43236	19.44498	4.20479	0.12756	22.11058
4.20	0.42994	20.41330	4.36657	0.11733	23.17896
4.30	0.42767	21.40497	4.53221	0.10800	24.27312
4.40	0.42554	22.41997	4.70171	0.09948	25.39301
4.50	0.42355	23.45831	4.87508	0.09170	26.53865
4.60	0.42168	24.51997	5.05232	0.08459	27.71007
4.70	0.41992	25.60497	5.23343	0.07809	28.90730
4.80	0.41826	26.71330	5.41841	0.07214	30.13016
4.90	0.41670	27.84497	5.60727	0.06670	31.37898
5.00	0.41523	28.99997	5.80000	0.06172	32.65350
6.00	0.40416	41.83330	7.94058	0.02965	46.81523
7.00	0.39736	56.99995	10.46938	0.01535	63.55260
8.00	0.39289	74.49995	13.38671	0.00849	82.86559
9.00	0.38980	94.33327	16.69270	0.00496	104.75233
10.00	0.38758	116.49992	20.38747	0.00304	129.21661

PRANDTL–MEYER FUNCTION AND MACH ANGLE (PERFECT GAS, $K = 1.400$)

M	δ (deg.)	α (deg.)
1.00	0.000	90.000
1.01	0.045	81.931
1.02	0.126	78.635
1.03	0.229	76.138
1.04	0.351	74.058
1.05	0.487	72.247
1.06	0.637	70.630
1.07	0.797	69.160
1.08	0.968	67.808
1.09	1.148	66.553
1.10	1.336	65.380
1.11	1.532	64.277
1.12	1.735	63.234
1.13	1.944	62.246
1.14	2.160	61.306
1.15	2.381	60.408
1.16	2.607	59.550
1.17	2.838	58.727
1.18	3.074	57.936
1.19	3.314	57.176
1.20	3.558	56.443
1.21	3.806	55.735
1.22	4.057	55.052
1.23	4.312	54.391
1.24	4.569	53.751

M	δ (deg.)	α (deg.)
1.25	4.830	53.130
1.26	5.093	52.528
1.27	5.359	51.943
1.28	5.627	51.375
1.29	5.898	50.823
1.30	6.170	50.285
1.31	6.445	49.761
1.32	6.721	49.251
1.33	7.000	48.753
1.34	7.279	48.268
1.35	7.561	47.795
1.36	7.843	47.332
1.37	8.128	46.880
1.38	8.413	46.439
1.39	8.699	46.007
1.40	8.987	45.585
1.41	9.276	45.171
1.42	9.565	44.767
1.43	9.855	44.371
1.44	10.146	43.983
1.45	10.438	43.603
1.46	10.730	43.230
1.47	11.023	42.865
1.48	11.317	42.507
1.49	11.611	42.155
1.50	11.905	41.810
1.51	12.200	41.472
1.52	12.495	41.140
1.53	12.790	40.813
1.54	13.086	40.493
1.55	13.381	40.178
1.56	13.677	39.868
1.57	13.973	39.564
1.58	14.269	39.265
1.59	14.564	38.971
1.60	14.860	38.682
1.61	15.156	38.398
1.62	15.452	38.118
1.63	15.747	37.843
1.64	16.043	37.572

M	δ (deg.)	α (deg.)
1.65	16.338	37.305
1.66	16.633	37.043
1.67	16.928	36.784
1.68	17.222	36.530
1.69	17.516	36.279
1.70	17.810	36.032
1.71	18.103	35.789
1.72	18.396	35.549
1.73	18.689	35.312
1.74	18.981	35.080
1.75	19.273	34.850
1.76	19.565	34.624
1.77	19.855	34.400
1.78	20.146	34.180
1.79	20.436	33.963
1.80	20.725	33.749
1.81	21.014	33.538
1.82	21.302	33.329
1.83	21.590	33.124
1.84	21.877	32.921
1.85	22.163	32.720
1.86	22.449	32.523
1.87	22.734	32.328
1.88	23.019	32.135
1.89	23.303	31.945
1.90	23.586	31.757
1.91	23.869	31.571
1.92	24.151	31.388
1.93	24.432	31.207
1.94	24.712	31.028
1.95	24.992	30.852
1.96	25.271	30.677
1.97	25.549	30.505
1.98	25.827	30.335
1.99	26.104	30.166
2.00	26.380	30.000
2.01	26.655	29.836
2.02	26.930	29.673
2.03	27.203	29.512
2.04	27.476	29.353

M	δ (deg.)	α (deg.)
2.05	27.748	29.196
2.06	28.020	29.041
2.07	28.290	28.888
2.08	28.560	28.736
2.09	28.829	28.585
2.10	29.097	28.437
2.11	29.364	28.290
2.12	29.631	28.145
2.13	29.896	28.001
2.14	30.161	27.859
2.15	30.425	27.718
2.16	30.688	27.578
2.17	30.951	27.441
2.18	31.212	27.304
2.19	31.473	27.169
2.20	31.733	27.036
2.21	31.991	26.903
2.22	32.249	26.773
2.23	32.507	26.643
2.24	32.763	26.515
2.25	33.018	26.388
2.26	33.273	26.262
2.27	33.527	26.138
2.28	33.780	26.014
2.29	34.032	25.892
2.30	34.283	25.771
2.31	34.533	25.652
2.32	34.782	25.533
2.33	35.031	25.416
2.34	35.279	25.300
2.35	35.526	25.184
2.36	35.771	25.070
2.37	36.017	24.957
2.38	36.261	24.845
2.39	36.504	24.734
2.40	36.747	24.624
2.41	36.988	24.515
2.42	37.229	24.407
2.43	37.469	24.301
2.44	37.708	24.195

M	δ (deg.)	α (deg.)
2.45	37.946	24.089
2.46	38.183	23.985
2.47	38.420	23.882
2.48	38.655	23.780
2.49	38.890	23.679
2.50	39.124	23.578
2.51	39.357	23.479
2.52	39.589	23.380
2.53	39.820	23.282
2.54	40.050	23.185
2.55	40.280	23.089
2.56	40.508	22.993
2.57	40.736	22.899
2.58	40.963	22.805
2.59	41.189	22.712
2.60	41.415	22.620
2.61	41.639	22.528
2.62	41.863	22.438
2.63	42.085	22.348
2.64	42.307	22.259
2.65	42.528	22.170
2.66	42.749	22.082
2.67	42.968	21.995
2.68	43.187	21.909
2.69	43.404	21.823
2.70	43.621	21.738
2.71	43.838	21.654
2.72	44.053	21.571
2.73	44.267	21.488
2.74	44.481	21.405
2.75	44.694	21.324
2.76	44.906	21.243
2.77	45.117	21.162
2.78	45.327	21.083
2.79	45.537	21.003
2.80	45.746	20.925
2.81	45.954	20.847
2.82	46.161	20.770
2.83	46.367	20.693
2.84	46.573	20.617

M	δ (deg.)	α (deg.)
2.85	46.778	20.541
2.86	46.982	20.466
2.87	47.185	20.391
2.88	47.388	20.318
2.89	47.589	20.244
2.90	47.790	20.171
2.91	47.990	20.099
2.92	48.190	20.027
2.93	48.388	19.956
2.94	48.586	19.885
2.95	48.783	19.815
2.96	48.980	19.745
2.97	49.175	19.676
2.98	49.370	19.607
2.99	49.564	19.539
3.00	49.757	19.471
3.10	51.650	18.819
3.20	53.470	18.210
3.30	55.222	17.640
3.40	56.908	17.105
3.50	58.530	16.602
3.60	60.091	16.128
3.70	61.595	15.680
3.80	63.044	15.258
3.90	64.439	14.857
4.00	65.785	14.478
4.10	67.082	14.117
4.20	68.333	13.774
4.30	69.541	13.448
4.40	70.706	13.137
4.50	71.832	12.840
4.60	72.919	12.556
4.70	73.970	12.284
4.80	74.986	12.025
4.90	75.969	11.776
5.00	76.920	11.537
6.00	84.955	9.594
7.00	90.973	8.213
8.00	95.625	7.181
9.00	99.318	6.379
10.00	102.316	5.739

MACH NUMBER FUNCTIONS FOR FANNO FLOW (PERFECT GAS, $K = 1.400$)

Mach	T/T^*	P/P^*	P_0/P_0^*	F/F^*	FL_{MAX}/D
0.00	1.20000	INF	INF	INF	INF
0.01	1.19998	109.54343	57.87372	45.64949	7134.41016
0.02	1.19990	54.77008	28.94209	22.83363	1778.45093
0.03	1.19978	36.51155	19.30052	15.23233	787.08154
0.04	1.19962	27.38174	14.48148	11.43462	440.35229
0.05	1.19940	21.90341	11.59145	9.15837	280.02051
0.06	1.19914	18.25084	9.66591	7.64285	193.03133
0.07	1.19882	15.64155	8.29153	6.56203	140.65518
0.08	1.19847	13.68431	7.26162	5.75289	106.71832
0.09	1.19806	12.16177	6.46134	5.12487	83.49620
0.10	1.19760	10.94351	5.82184	4.62364	66.92165
0.11	1.19710	9.94657	5.29922	4.21461	54.68797
0.12	1.19655	9.11560	4.86431	3.87474	45.40802
0.13	1.19596	8.41230	4.49686	3.58805	38.20702
0.14	1.19531	7.80932	4.18240	3.34317	32.51132
0.15	1.19462	7.28659	3.91035	3.13172	27.93198
0.16	1.19389	6.82907	3.67273	2.94743	24.19785
0.17	1.19310	6.42525	3.46351	2.78551	21.11519
0.18	1.19227	6.06619	3.27793	2.64223	18.54266
0.19	1.19140	5.74480	3.11225	2.51464	16.37515
0.20	1.19048	5.45545	2.96352	2.40040	14.53328
0.21	1.18951	5.19355	2.82929	2.29758	12.95603
0.22	1.18849	4.95537	2.70761	2.20464	11.59607
0.23	1.18744	4.73781	2.59681	2.12029	10.41611
0.24	1.18633	4.53829	2.49556	2.04344	9.38649

Mach	T/T^*	P/P^*	P_0/P_0^*	F/F^*	FL_{MAX}/D
0.25	1.18518	4.35465	2.40271	1.97320	8.48341
0.26	1.18399	4.18505	2.31729	1.90880	7.68758
0.27	1.18276	4.02795	2.23847	1.84960	6.98318
0.28	1.18147	3.88199	2.16555	1.79503	6.35722
0.29	1.18015	3.74602	2.09793	1.74462	5.79892
0.30	1.17878	3.61906	2.03507	1.69794	5.29926
0.31	1.17737	3.50022	1.97651	1.65464	4.85067
0.32	1.17592	3.38874	1.92185	1.61440	4.44675
0.33	1.17442	3.28396	1.87074	1.57693	4.08206
0.34	1.17288	3.18528	1.82288	1.54200	3.75196
0.35	1.17130	3.09219	1.77797	1.50938	3.45246
0.36	1.16968	3.00422	1.73578	1.47888	3.18012
0.37	1.16802	2.92094	1.69608	1.45032	2.93198
0.38	1.16632	2.84200	1.65869	1.42356	2.70545
0.39	1.16457	2.76706	1.62343	1.39845	2.49828
0.40	1.16279	2.69582	1.59014	1.37487	2.30849
0.41	1.16097	2.62801	1.55867	1.35270	2.13437
0.42	1.15911	2.56338	1.52890	1.33185	1.97437
0.43	1.15721	2.50171	1.50072	1.31221	1.82715
0.44	1.15527	2.44280	1.47400	1.29371	1.69153
0.45	1.15329	2.38648	1.44867	1.27627	1.56643
0.46	1.15128	2.33256	1.42463	1.25982	1.45091
0.47	1.14923	2.28089	1.40180	1.24428	1.34414
0.48	1.14714	2.23135	1.38010	1.22962	1.24534
0.49	1.14502	2.18378	1.35947	1.21577	1.15385
0.50	1.14286	2.13809	1.33984	1.20267	1.06906
0.51	1.14066	2.09415	1.32117	1.19030	0.99041
0.52	1.13843	2.05187	1.30339	1.17860	0.91742
0.53	1.13617	2.01116	1.28645	1.16753	0.84963
0.54	1.13387	1.97192	1.27032	1.15705	0.78663
0.55	1.13154	1.93407	1.25494	1.14715	0.72805
0.56	1.12918	1.89755	1.24029	1.13777	0.67357
0.57	1.12678	1.86228	1.22633	1.12890	0.62287
0.58	1.12435	1.82820	1.21301	1.12050	0.57568
0.59	1.12189	1.79524	1.20031	1.11256	0.53174
0.60	1.11940	1.76336	1.18820	1.10504	0.49082
0.61	1.11688	1.73250	1.17665	1.09793	0.45271
0.62	1.11433	1.70261	1.16565	1.09120	0.41720
0.63	1.11175	1.67364	1.15515	1.08484	0.38412
0.64	1.10914	1.64556	1.14515	1.07883	0.35330

Mach	T/T^*	P/P^*	P_0/P_0^*	F/F^*	FL_{MAX}/D
0.65	1.10650	1.61831	1.13562	1.07314	0.32459
0.66	1.10383	1.59187	1.12653	1.06777	0.29785
0.67	1.10114	1.56620	1.11789	1.06270	0.27295
0.68	1.09842	1.54126	1.10965	1.05792	0.24978
0.69	1.09567	1.51702	1.10182	1.05340	0.22820
0.70	1.09290	1.49345	1.09437	1.04915	0.20814
0.71	1.09010	1.47053	1.08729	1.04514	0.18948
0.72	1.08727	1.44823	1.08057	1.04137	0.17215
0.73	1.08442	1.42651	1.07419	1.03783	0.15605
0.74	1.08155	1.40537	1.06814	1.03449	0.14112
0.75	1.07865	1.38478	1.06242	1.03137	0.12728
0.76	1.07573	1.36470	1.05700	1.02844	0.11447
0.77	1.07279	1.34514	1.05188	1.02570	0.10262
0.78	1.06982	1.32605	1.04705	1.02314	0.09167
0.79	1.06684	1.30744	1.04251	1.02075	0.08158
0.80	1.06383	1.28928	1.03823	1.01853	0.07229
0.81	1.06080	1.27155	1.03422	1.01646	0.06376
0.82	1.05775	1.25423	1.03046	1.01455	0.05593
0.83	1.05469	1.23732	1.02696	1.01278	0.04878
0.84	1.05160	1.22080	1.02369	1.01115	0.04226
0.85	1.04849	1.20466	1.02067	1.00965	0.03633
0.86	1.04537	1.18888	1.01787	1.00829	0.03096
0.87	1.04223	1.17344	1.01530	1.00704	0.02613
0.88	1.03907	1.15835	1.01294	1.00591	0.02179
0.89	1.03589	1.14358	1.01080	1.00489	0.01793
0.90	1.03270	1.12913	1.00886	1.00399	0.01451
0.91	1.02950	1.11499	1.00713	1.00318	0.01151
0.92	1.02627	1.10114	1.00560	1.00248	0.00891
0.93	1.02304	1.08758	1.00426	1.00187	0.00669
0.94	1.01978	1.07430	1.00311	1.00136	0.00482
0.95	1.01652	1.06129	1.00214	1.00093	0.00328
0.96	1.01324	1.04854	1.00136	1.00059	0.00206
0.97	1.00995	1.03604	1.00076	1.00033	0.00114
0.98	1.00664	1.02379	1.00034	1.00014	0.00049
0.99	1.00333	1.01178	1.00008	1.00004	0.00012
1.00	1.00000	1.00000	1.00000	1.00000	0.00000
1.01	0.99666	0.98845	1.00008	1.00003	0.00012
1.02	0.99331	0.97711	1.00033	1.00013	0.00046
1.03	0.98995	0.96598	1.00074	1.00030	0.00101
1.04	0.98658	0.95507	1.00130	1.00053	0.00177

Mach	T/T^*	P/P^*	P_0/P_0^*	F/F^*	FL_{MAX}/D
1.05	0.98320	0.94435	1.00203	1.00081	0.00271
1.06	0.97982	0.93383	1.00290	1.00116	0.00384
1.07	0.97642	0.92350	1.00393	1.00155	0.00513
1.08	0.97302	0.91335	1.00511	1.00200	0.00658
1.09	0.96960	0.90338	1.00644	1.00250	0.00819
1.10	0.96618	0.89359	1.00792	1.00305	0.00993
1.11	0.96276	0.88397	1.00954	1.00365	0.01182
1.12	0.95933	0.87451	1.01131	1.00429	0.01382
1.13	0.95589	0.86522	1.01322	1.00497	0.01595
1.14	0.95244	0.85608	1.01526	1.00569	0.01819
1.15	0.94899	0.84710	1.01745	1.00646	0.02053
1.16	0.94554	0.83827	1.01978	1.00726	0.02298
1.17	0.94208	0.82958	1.02224	1.00810	0.02551
1.18	0.93862	0.82104	1.02484	1.00897	0.02814
1.19	0.93515	0.81263	1.02757	1.00988	0.03085
1.20	0.93168	0.80436	1.03044	1.01081	0.03364
1.21	0.92820	0.79623	1.03344	1.01178	0.03650
1.22	0.92473	0.78822	1.03657	1.01278	0.03943
1.23	0.92125	0.78034	1.03983	1.01381	0.04242
1.24	0.91777	0.77258	1.04323	1.01486	0.04547
1.25	0.91429	0.76495	1.04675	1.01594	0.04858
1.26	0.91080	0.75743	1.05040	1.01705	0.05174
1.27	0.90732	0.75003	1.05419	1.01818	0.05495
1.28	0.90383	0.74274	1.05810	1.01933	0.05820
1.29	0.90035	0.73556	1.06214	1.02050	0.06150
1.30	0.89686	0.72848	1.06630	1.02170	0.06483
1.31	0.89338	0.72152	1.07059	1.02291	0.06820
1.32	0.88989	0.71465	1.07502	1.02414	0.07161
1.33	0.88641	0.70789	1.07956	1.02539	0.07504
1.34	0.88293	0.70123	1.08423	1.02666	0.07850
1.35	0.87944	0.69466	1.08904	1.02795	0.08199
1.36	0.87596	0.68818	1.09396	1.02925	0.08550
1.37	0.87249	0.68180	1.09901	1.03056	0.08903
1.38	0.86901	0.67551	1.10419	1.03189	0.09259
1.39	0.86554	0.66931	1.10949	1.03323	0.09615
1.40	0.86207	0.66320	1.11492	1.03459	0.09974
1.41	0.85860	0.65717	1.12048	1.03596	0.10333
1.42	0.85514	0.65122	1.12616	1.03733	0.10694
1.43	0.85168	0.64536	1.13197	1.03872	0.11056
1.44	0.84823	0.63958	1.13790	1.04012	0.11419

Mach	T/T^*	P/P^*	P_0/P_0^*	F/F^*	FL_{MAX}/D
1.45	0.84477	0.63387	1.14396	1.04153	0.11782
1.46	0.84133	0.62825	1.15014	1.04295	0.12146
1.47	0.83788	0.62269	1.15646	1.04438	0.12511
1.48	0.83445	0.61722	1.16290	1.04581	0.12875
1.49	0.83101	0.61181	1.16947	1.04725	0.13240
1.50	0.82759	0.60648	1.17616	1.04870	0.13605
1.51	0.82417	0.60122	1.18299	1.05016	0.13970
1.52	0.82075	0.59602	1.18994	1.05162	0.14335
1.53	0.81734	0.59089	1.19702	1.05309	0.14699
1.54	0.81393	0.58583	1.20423	1.05456	0.15063
1.55	0.81054	0.58084	1.21157	1.05604	0.15427
1.56	0.80715	0.57591	1.21904	1.05752	0.15790
1.57	0.80376	0.57104	1.22664	1.05900	0.16152
1.58	0.80038	0.56623	1.23437	1.06049	0.16514
1.59	0.79701	0.56148	1.24224	1.06198	0.16875
1.60	0.79365	0.55679	1.25023	1.06348	0.17236
1.61	0.79030	0.55217	1.25836	1.06497	0.17595
1.62	0.78695	0.54759	1.26662	1.06647	0.17954
1.63	0.78361	0.54308	1.27502	1.06797	0.18311
1.64	0.78028	0.53862	1.28355	1.06948	0.18667
1.65	0.77695	0.53421	1.29221	1.07098	0.19023
1.66	0.77364	0.52986	1.30102	1.07249	0.19377
1.67	0.77033	0.52556	1.30996	1.07399	0.19729
1.68	0.76703	0.52131	1.31903	1.07550	0.20081
1.69	0.76374	0.51711	1.32825	1.07700	0.20431
1.70	0.76046	0.51297	1.33760	1.07851	0.20780
1.71	0.75718	0.50887	1.34710	1.08002	0.21128
1.72	0.75392	0.50482	1.35673	1.08152	0.21474
1.73	0.75067	0.50082	1.36651	1.08302	0.21819
1.74	0.74742	0.49686	1.37643	1.08453	0.22162
1.75	0.74419	0.49295	1.38649	1.08603	0.22504
1.76	0.74096	0.48909	1.39669	1.08753	0.22844
1.77	0.73774	0.48527	1.40705	1.08903	0.23182
1.78	0.73454	0.48149	1.41754	1.09053	0.23519
1.79	0.73134	0.47776	1.42819	1.09202	0.23855
1.80	0.72816	0.47407	1.43898	1.09351	0.24189
1.81	0.72498	0.47042	1.44992	1.09500	0.24521
1.82	0.72181	0.46681	1.46101	1.09649	0.24851
1.83	0.71866	0.46324	1.47225	1.09798	0.25180
1.84	0.71551	0.45972	1.48364	1.09946	0.25507

Mach	T/T^*	P/P^*	P_0/P_0^*	F/F^*	FL_{MAX}/D
1.85	0.71238	0.45623	1.49519	1.10094	0.25832
1.86	0.70925	0.45278	1.50689	1.10241	0.26156
1.87	0.70614	0.44937	1.51875	1.10389	0.26478
1.88	0.70304	0.44600	1.53076	1.10536	0.26798
1.89	0.69995	0.44266	1.54292	1.10682	0.27116
1.90	0.69686	0.43936	1.55525	1.10828	0.27433
1.91	0.69379	0.43610	1.56774	1.10974	0.27748
1.92	0.69074	0.43287	1.58039	1.11120	0.28061
1.93	0.68769	0.42967	1.59320	1.11265	0.28372
1.94	0.68465	0.42651	1.60617	1.11409	0.28681
1.95	0.68163	0.42339	1.61931	1.11554	0.28989
1.96	0.67861	0.42030	1.63261	1.11697	0.29295
1.97	0.67561	0.41724	1.64608	1.11841	0.29599
1.98	0.67262	0.41421	1.65971	1.11984	0.29901
1.99	0.66964	0.41121	1.67352	1.12126	0.30201
2.00	0.66667	0.40825	1.68750	1.12268	0.30500
2.01	0.66371	0.40532	1.70164	1.12410	0.30796
2.02	0.66076	0.40241	1.71597	1.12551	0.31091
2.03	0.65783	0.39954	1.73046	1.12691	0.31384
2.04	0.65491	0.39670	1.74514	1.12831	0.31676
2.05	0.65200	0.39388	1.75998	1.12971	0.31965
2.06	0.64910	0.39110	1.77501	1.13110	0.32253
2.07	0.64621	0.38834	1.79022	1.13249	0.32538
2.08	0.64334	0.38562	1.80561	1.13387	0.32822
2.09	0.64047	0.38292	1.82118	1.13524	0.33105
2.10	0.63762	0.38024	1.83694	1.13661	0.33385
2.11	0.63478	0.37760	1.85288	1.13797	0.33664
2.12	0.63195	0.37498	1.86901	1.13933	0.33940
2.13	0.62914	0.37239	1.88533	1.14069	0.34215
2.14	0.62633	0.36982	1.90184	1.14203	0.34489
2.15	0.62354	0.36728	1.91854	1.14338	0.34760
2.16	0.62076	0.36476	1.93544	1.14471	0.35030
2.17	0.61799	0.36227	1.95252	1.14605	0.35298
2.18	0.61523	0.35980	1.96980	1.14737	0.35564
2.19	0.61249	0.35736	1.98729	1.14869	0.35828
2.20	0.60976	0.35494	2.00497	1.15001	0.36091
2.21	0.60704	0.35255	2.02285	1.15131	0.36352
2.22	0.60433	0.35017	2.04094	1.15262	0.36611
2.23	0.60163	0.34782	2.05923	1.15392	0.36868
2.24	0.59895	0.34550	2.07773	1.15521	0.37124

Mach	T/T^*	P/P^*	P_0/P_0^*	F/F^*	FL_{MAX}/D
2.25	0.59627	0.34319	2.09643	1.15649	0.37378
2.26	0.59361	0.34091	2.11534	1.15777	0.37631
2.27	0.59096	0.33865	2.13447	1.15905	0.37881
2.28	0.58833	0.33641	2.15381	1.16032	0.38130
2.29	0.58570	0.33420	2.17336	1.16158	0.38377
2.30	0.58309	0.33200	2.19313	1.16284	0.38623
2.31	0.58049	0.32983	2.21311	1.16409	0.38867
2.32	0.57790	0.32767	2.23332	1.16533	0.39109
2.33	0.57532	0.32554	2.25375	1.16657	0.39350
2.34	0.57276	0.32342	2.27439	1.16780	0.39589
2.35	0.57021	0.32133	2.29527	1.16903	0.39826
2.36	0.56767	0.31925	2.31638	1.17025	0.40062
2.37	0.56514	0.31720	2.33771	1.17147	0.40296
2.38	0.56262	0.31516	2.35927	1.17268	0.40529
2.39	0.56011	0.31314	2.38107	1.17388	0.40759
2.40	0.55762	0.31114	2.40309	1.17508	0.40989
2.41	0.55514	0.30916	2.42536	1.17627	0.41217
2.42	0.55267	0.30720	2.44786	1.17746	0.41443
2.43	0.55021	0.30525	2.47061	1.17864	0.41668
2.44	0.54777	0.30332	2.49360	1.17981	0.41891
2.45	0.54533	0.30141	2.51683	1.18098	0.42112
2.46	0.54291	0.29952	2.54031	1.18214	0.42332
2.47	0.54050	0.29765	2.56403	1.18330	0.42551
2.48	0.53810	0.29579	2.58800	1.18445	0.42768
2.49	0.53571	0.29394	2.61223	1.18559	0.42984
2.50	0.53333	0.29212	2.63671	1.18673	0.43198
2.51	0.53097	0.29031	2.66145	1.18786	0.43410
2.52	0.52862	0.28852	2.68645	1.18899	0.43621
2.53	0.52627	0.28674	2.71171	1.19011	0.43831
2.54	0.52394	0.28498	2.73723	1.19123	0.44039
2.55	0.52163	0.28323	2.76301	1.19234	0.44246
2.56	0.51932	0.28150	2.78906	1.19344	0.44451
2.57	0.51702	0.27978	2.81538	1.19454	0.44655
2.58	0.51474	0.27808	2.84197	1.19563	0.44858
2.59	0.51247	0.27640	2.86883	1.19672	0.45059
2.60	0.51020	0.27473	2.89597	1.19780	0.45259
2.61	0.50795	0.27307	2.92339	1.19888	0.45457
2.62	0.50572	0.27143	2.95108	1.19995	0.45654
2.63	0.50349	0.26980	2.97906	1.20101	0.45850
2.64	0.50127	0.26818	3.00732	1.20207	0.46044

Mach	T/T^*	P/P^*	P_0/P_0^*	F/F^*	FL_{MAX}/D
2.65	0.49906	0.26658	3.03588	1.20312	0.46237
2.66	0.49687	0.26500	3.06472	1.20417	0.46429
2.67	0.49469	0.26342	3.09384	1.20521	0.46619
2.68	0.49251	0.26186	3.12327	1.20625	0.46808
2.69	0.49035	0.26032	3.15299	1.20728	0.46995
2.70	0.48820	0.25878	3.18301	1.20830	0.47182
2.71	0.48606	0.25726	3.21332	1.20932	0.47367
2.72	0.48393	0.25576	3.24395	1.21033	0.47551
2.73	0.48182	0.25426	3.27487	1.21134	0.47733
2.74	0.47971	0.25278	3.30611	1.21235	0.47915
2.75	0.47761	0.25131	3.33765	1.21334	0.48095
2.76	0.47553	0.24985	3.36951	1.21433	0.48273
2.77	0.47345	0.24840	3.40168	1.21532	0.48451
2.78	0.47139	0.24697	3.43417	1.21630	0.48627
2.79	0.46933	0.24555	3.46698	1.21728	0.48802
2.80	0.46729	0.24414	3.50012	1.21825	0.48976
2.81	0.46526	0.24274	3.53358	1.21921	0.49149
2.82	0.46323	0.24135	3.56737	1.22017	0.49321
2.83	0.46122	0.23998	3.60148	1.22113	0.49491
2.84	0.45922	0.23861	3.63592	1.22208	0.49660
2.85	0.45723	0.23726	3.67072	1.22302	0.49828
2.86	0.45525	0.23592	3.70584	1.22396	0.49995
2.87	0.45328	0.23459	3.74130	1.22489	0.50161
2.88	0.45132	0.23326	3.77711	1.22582	0.50326
2.89	0.44937	0.23195	3.81326	1.22674	0.50489
2.90	0.44743	0.23066	3.84976	1.22766	0.50651
2.91	0.44550	0.22937	3.88662	1.22857	0.50813
2.92	0.44358	0.22809	3.92382	1.22948	0.50973
2.93	0.44167	0.22682	3.96139	1.23039	0.51132
2.94	0.43977	0.22556	3.99932	1.23128	0.51290
2.95	0.43788	0.22431	4.03760	1.23218	0.51447
2.96	0.43600	0.22307	4.07625	1.23306	0.51603
2.97	0.43413	0.22185	4.11527	1.23395	0.51758
2.98	0.43226	0.22063	4.15466	1.23483	0.51911
2.99	0.43041	0.21942	4.19442	1.23570	0.52064
3.00	0.42857	0.21822	4.23457	1.23657	0.52216
3.10	0.41068	0.20672	4.65730	1.24499	0.53678
3.20	0.39370	0.19608	5.12096	1.25295	0.55044
3.30	0.37760	0.18621	5.62864	1.26048	0.56323
3.40	0.36232	0.17704	6.18370	1.26759	0.57521

Mach	T/T^*	P/P^*	P_0/P_0^*	F/F^*	FL_{MAX}/D
3.50	0.34783	0.16851	6.78962	1.27432	0.58643
3.60	0.33408	0.16055	7.45011	1.28068	0.59695
3.70	0.32103	0.15313	8.16906	1.28670	0.60683
3.80	0.30864	0.14620	8.95058	1.29240	0.61612
3.90	0.29688	0.13971	9.79897	1.29779	0.62485
4.00	0.28571	0.13363	10.71875	1.30290	0.63306
4.10	0.27510	0.12793	11.71466	1.30774	0.64080
4.20	0.26502	0.12257	12.79165	1.31233	0.64810
4.30	0.25543	0.11753	13.95491	1.31668	0.65499
4.40	0.24631	0.11279	15.20987	1.32081	0.66149
4.50	0.23762	0.10833	16.56218	1.32474	0.66763
4.60	0.22936	0.10411	18.01781	1.32846	0.67345
4.70	0.22148	0.10013	19.58282	1.33201	0.67895
4.80	0.21398	0.09637	21.26370	1.33538	0.68417
4.90	0.20683	0.09281	23.06714	1.33858	0.68911
5.00	0.20000	0.08944	25.00003	1.34164	0.69380
6.00	0.14634	0.06376	53.17987	1.36548	0.72988
7.00	0.11111	0.04762	104.14316	1.38095	0.75280
8.00	0.08696	0.03686	190.11009	1.39148	0.76819
9.00	0.06977	0.02935	327.19043	1.39894	0.77899
10.00	0.05714	0.02390	535.93750	1.40439	0.78683

MACH NUMBER FUNCTIONS FOR RAYLEIGH FLOW (PERFECT GAS, $K = 1.400$)

Mach	T_0/T_0^*	T/T^*	P/P^*	P_0/P_0^*
0.00	0	0	2.40000	1.2679
0.01	0.00048	0.00058	2.39966	1.26778
0.02	0.00192	0.00230	2.39866	1.26752
0.03	0.00431	0.00517	2.39698	1.26707
0.04	0.00765	0.00917	2.39464	1.26646
0.05	0.01192	0.01430	2.39163	1.26566
0.06	0.01712	0.02053	2.38797	1.26470
0.07	0.02322	0.02784	2.38365	1.26356
0.08	0.03022	0.03621	2.37869	1.26225
0.09	0.03807	0.04562	2.37309	1.26078
0.10	0.04678	0.05602	2.36686	1.25914
0.11	0.05630	0.06739	2.36002	1.25735
0.12	0.06661	0.07970	2.35257	1.25539
0.13	0.07767	0.09290	2.34453	1.25328
0.14	0.08947	0.10695	2.33590	1.25103
0.15	0.10196	0.12181	2.32671	1.24862
0.16	0.11511	0.13743	2.31696	1.24608
0.17	0.12888	0.15377	2.30667	1.24340
0.18	0.14324	0.17078	2.29586	1.24059
0.19	0.15814	0.18841	2.28454	1.23765
0.20	0.17355	0.20661	2.27273	1.23459
0.21	0.18943	0.22533	2.26044	1.23142
0.22	0.20574	0.24452	2.24770	1.22813
0.23	0.22244	0.26413	2.23451	1.22474
0.24	0.23948	0.28411	2.22091	1.22125

Mach	T_0/T_0^*	T/T^*	P/P^*	P_0/P_0^*
0.25	0.25684	0.30440	2.20690	1.21767
0.26	0.27446	0.32496	2.19250	1.21400
0.27	0.29231	0.34573	2.17774	1.21025
0.28	0.31035	0.36667	2.16263	1.20641
0.29	0.32855	0.38774	2.14719	1.20251
0.30	0.34686	0.40887	2.13144	1.19855
0.31	0.36525	0.43004	2.11539	1.19452
0.32	0.38369	0.45119	2.09908	1.19044
0.33	0.40214	0.47228	2.08250	1.18632
0.34	0.42056	0.49327	2.06569	1.18215
0.35	0.43894	0.51413	2.04866	1.17795
0.36	0.45723	0.53482	2.03142	1.17371
0.37	0.47541	0.55529	2.01400	1.16945
0.38	0.49346	0.57553	1.99641	1.16517
0.39	0.51134	0.59549	1.97866	1.16088
0.40	0.52903	0.61515	1.96078	1.15658
0.41	0.54651	0.63448	1.94279	1.15227
0.42	0.56376	0.65346	1.92468	1.14796
0.43	0.58076	0.67205	1.90649	1.14365
0.44	0.59748	0.69026	1.88822	1.13936
0.45	0.61393	0.70804	1.86989	1.13508
0.46	0.63007	0.72538	1.85151	1.13082
0.47	0.64589	0.74228	1.83310	1.12659
0.48	0.66139	0.75871	1.81466	1.12238
0.49	0.67655	0.77466	1.79622	1.11819
0.50	0.69136	0.79012	1.77778	1.11405
0.51	0.70581	0.80509	1.75935	1.10994
0.52	0.71990	0.81955	1.74095	1.10588
0.53	0.73361	0.83351	1.72258	1.10186
0.54	0.74695	0.84695	1.70426	1.09789
0.55	0.75991	0.85987	1.68599	1.09397
0.56	0.77249	0.87227	1.66778	1.09011
0.57	0.78467	0.88416	1.64964	1.08630
0.58	0.79648	0.89552	1.63159	1.08255
0.59	0.80789	0.90637	1.61362	1.07887
0.60	0.81892	0.91670	1.59574	1.07525
0.61	0.82957	0.92653	1.57797	1.07170
0.62	0.83982	0.93584	1.56031	1.06822
0.63	0.84970	0.94466	1.54275	1.06481
0.64	0.85920	0.95298	1.52532	1.06147

Mach	T_0/T_0^*	T/T^*	P/P^*	P_0/P_0^*
0.65	0.86833	0.96081	1.50801	1.05821
0.66	0.87708	0.96815	1.49083	1.05503
0.67	0.88547	0.97503	1.47379	1.05193
0.68	0.89350	0.98144	1.45688	1.04890
0.69	0.90118	0.98739	1.44011	1.04596
0.70	0.90850	0.99290	1.42349	1.04310
0.71	0.91548	0.99796	1.40701	1.04033
0.72	0.92212	1.00260	1.39069	1.03764
0.73	0.92843	1.00681	1.37452	1.03504
0.74	0.93442	1.01062	1.35851	1.03252
0.75	0.94009	1.01403	1.34266	1.03010
0.76	0.94546	1.01706	1.32696	1.02777
0.77	0.95051	1.01970	1.31143	1.02552
0.78	0.95528	1.02198	1.29606	1.02337
0.79	0.95975	1.02390	1.28086	1.02131
0.80	0.96395	1.02548	1.26582	1.01934
0.81	0.96787	1.02672	1.25095	1.01746
0.82	0.97152	1.02763	1.23625	1.01569
0.83	0.97492	1.02824	1.22171	1.01400
0.84	0.97807	1.02853	1.20734	1.01240
0.85	0.98097	1.02854	1.19314	1.01091
0.86	0.98363	1.02826	1.17911	1.00951
0.87	0.98607	1.02771	1.16524	1.00820
0.88	0.98828	1.02689	1.15154	1.00699
0.89	0.99028	1.02583	1.13801	1.00587
0.90	0.99207	1.02452	1.12465	1.00486
0.91	0.99366	1.02297	1.11145	1.00393
0.92	0.99506	1.02120	1.09842	1.00311
0.93	0.99627	1.01922	1.08555	1.00238
0.94	0.99729	1.01702	1.07285	1.00175
0.95	0.99814	1.01463	1.06030	1.00121
0.96	0.99883	1.01205	1.04792	1.00078
0.97	0.99935	1.00929	1.03571	1.00043
0.98	0.99971	1.00636	1.02365	1.00019
0.99	0.99993	1.00326	1.01174	1.00005
1.00	1.00000	1.00000	1.00000	1.00000
1.01	0.99993	0.99659	0.98841	1.00004
1.02	0.99973	0.99304	0.97698	1.00019
1.03	0.99940	0.98936	0.96569	1.00043
1.04	0.99895	0.98554	0.95456	1.00078

Mach	T_0/T_0^*	T/T^*	P/P^*	P_0/P_0^*
1.05	0.99837	0.98161	0.94358	1.00121
1.06	0.99769	0.97756	0.93275	1.00175
1.07	0.99690	0.97339	0.92206	1.00238
1.08	0.99601	0.96913	0.91152	1.00311
1.09	0.99501	0.96477	0.90113	1.00393
1.10	0.99392	0.96031	0.89087	1.00486
1.11	0.99274	0.95577	0.88075	1.00587
1.12	0.99148	0.95115	0.87078	1.00699
1.13	0.99013	0.94646	0.86094	1.00820
1.14	0.98871	0.94169	0.85123	1.00952
1.15	0.98721	0.93685	0.84166	1.01092
1.16	0.98564	0.93196	0.83222	1.01243
1.17	0.98400	0.92701	0.82292	1.01403
1.18	0.98230	0.92200	0.81374	1.01572
1.19	0.98054	0.91695	0.80468	1.01752
1.20	0.97872	0.91185	0.79576	1.01941
1.21	0.97684	0.90671	0.78695	1.02140
1.22	0.97492	0.90153	0.77827	1.02349
1.23	0.97294	0.89632	0.76971	1.02567
1.24	0.97092	0.89108	0.76127	1.02794
1.25	0.96886	0.88581	0.75294	1.03032
1.26	0.96675	0.88052	0.74473	1.03280
1.27	0.96461	0.87521	0.73664	1.03537
1.28	0.96243	0.86988	0.72865	1.03803
1.29	0.96022	0.86453	0.72078	1.04080
1.30	0.95798	0.85917	0.71301	1.04366
1.31	0.95571	0.85381	0.70536	1.04661
1.32	0.95341	0.84843	0.69780	1.04967
1.33	0.95108	0.84305	0.69036	1.05283
1.34	0.94873	0.83766	0.68301	1.05608
1.35	0.94636	0.83227	0.67577	1.05943
1.36	0.94398	0.82689	0.66863	1.06287
1.37	0.94157	0.82151	0.66158	1.06642
1.38	0.93914	0.81613	0.65464	1.07006
1.39	0.93671	0.81076	0.64778	1.07381
1.40	0.93425	0.80539	0.64103	1.07765
1.41	0.93179	0.80004	0.63436	1.08159
1.42	0.92931	0.79469	0.62779	1.08563
1.43	0.92683	0.78936	0.62130	1.08977
1.44	0.92434	0.78405	0.61491	1.09400

Mach	T_0/T_0^*	T/T^*	P/P^*	P_0/P_0^*
1.45	0.92184	0.77874	0.60860	1.09834
1.46	0.91933	0.77346	0.60237	1.10278
1.47	0.91682	0.76819	0.59624	1.10732
1.48	0.91431	0.76294	0.59018	1.11196
1.49	0.91179	0.75771	0.58421	1.11670
1.50	0.90927	0.75250	0.57831	1.12154
1.51	0.90676	0.74732	0.57250	1.12648
1.52	0.90424	0.74215	0.56677	1.13153
1.53	0.90172	0.73701	0.56111	1.13668
1.54	0.89920	0.73189	0.55552	1.14193
1.55	0.89669	0.72680	0.55002	1.14728
1.56	0.89418	0.72174	0.54458	1.15274
1.57	0.89167	0.71669	0.53922	1.15830
1.58	0.88917	0.71168	0.53393	1.16396
1.59	0.88668	0.70669	0.52871	1.16973
1.60	0.88419	0.70174	0.52356	1.17561
1.61	0.88170	0.69680	0.51848	1.18159
1.62	0.87922	0.69190	0.51346	1.18767
1.63	0.87675	0.68703	0.50851	1.19387
1.64	0.87429	0.68219	0.50363	1.20016
1.65	0.87184	0.67738	0.49881	1.20657
1.66	0.86939	0.67259	0.49405	1.21309
1.67	0.86696	0.66784	0.48935	1.21971
1.68	0.86453	0.66312	0.48472	1.22644
1.69	0.86211	0.65843	0.48014	1.23328
1.70	0.85971	0.65377	0.47562	1.24023
1.71	0.85731	0.64914	0.47117	1.24729
1.72	0.85493	0.64455	0.46677	1.25447
1.73	0.85256	0.63999	0.46242	1.26175
1.74	0.85019	0.63545	0.45813	1.26915
1.75	0.84784	0.63095	0.45390	1.27666
1.76	0.84551	0.62649	0.44972	1.28428
1.77	0.84318	0.62205	0.44560	1.29202
1.78	0.84087	0.61765	0.44152	1.29987
1.79	0.83857	0.61328	0.43750	1.30784
1.80	0.83628	0.60894	0.43353	1.31592
1.81	0.83400	0.60464	0.42960	1.32412
1.82	0.83174	0.60036	0.42573	1.33244
1.83	0.82949	0.59612	0.42191	1.34088
1.84	0.82726	0.59191	0.41813	1.34943

Mach	T_0/T_0^*	T/T^*	P/P^*	P_0/P_0^*
1.85	0.82504	0.58774	0.41440	1.35810
1.86	0.82283	0.58359	0.41072	1.36690
1.87	0.82063	0.57948	0.40708	1.37582
1.88	0.81845	0.57540	0.40349	1.38485
1.89	0.81629	0.57136	0.39994	1.39401
1.90	0.81414	0.56734	0.39643	1.40330
1.91	0.81200	0.56336	0.39297	1.41270
1.92	0.80987	0.55941	0.38955	1.42223
1.93	0.80776	0.55549	0.38617	1.43189
1.94	0.80567	0.55160	0.38283	1.44168
1.95	0.80358	0.54774	0.37954	1.45159
1.96	0.80152	0.54392	0.37628	1.46163
1.97	0.79946	0.54012	0.37306	1.47180
1.98	0.79742	0.53636	0.36988	1.48210
1.99	0.79540	0.53263	0.36674	1.49253
2.00	0.79339	0.52893	0.36364	1.50309
2.01	0.79139	0.52525	0.36057	1.51379
2.02	0.78941	0.52161	0.35754	1.52461
2.03	0.78744	0.51800	0.35454	1.53558
2.04	0.78549	0.51442	0.35158	1.54667
2.05	0.78355	0.51087	0.34866	1.55791
2.06	0.78162	0.50735	0.34577	1.56928
2.07	0.77971	0.50386	0.34291	1.58079
2.08	0.77781	0.50040	0.34009	1.59244
2.09	0.77593	0.49696	0.33730	1.60422
2.10	0.77406	0.49356	0.33454	1.61616
2.11	0.77221	0.49018	0.33182	1.62823
2.12	0.77037	0.48683	0.32912	1.64044
2.13	0.76854	0.48352	0.32646	1.65280
2.14	0.76673	0.48023	0.32382	1.66531
2.15	0.76493	0.47696	0.32122	1.67796
2.16	0.76314	0.47373	0.31865	1.69076
2.17	0.76137	0.47052	0.31610	1.70370
2.18	0.75961	0.46734	0.31359	1.71679
2.19	0.75787	0.46418	0.31110	1.73004
2.20	0.75613	0.46106	0.30864	1.74344
2.21	0.75442	0.45796	0.30621	1.75699
2.22	0.75271	0.45488	0.30381	1.77069
2.23	0.75102	0.45184	0.30143	1.78456
2.24	0.74934	0.44881	0.29908	1.79857

Mach	T_0/T_0^*	T/T^*	P/P^*	P_0/P_0^*
2.25	0.74768	0.44582	0.29675	1.81275
2.26	0.74602	0.44285	0.29446	1.82708
2.27	0.74438	0.43990	0.29218	1.84157
2.28	0.74276	0.43698	0.28993	1.85622
2.29	0.74114	0.43409	0.28771	1.87104
2.30	0.73954	0.43122	0.28551	1.88602
2.31	0.73795	0.42838	0.28334	1.90116
2.32	0.73638	0.42555	0.28118	1.91647
2.33	0.73481	0.42276	0.27905	1.93194
2.34	0.73326	0.41998	0.27695	1.94758
2.35	0.73172	0.41723	0.27487	1.96340
2.36	0.73020	0.41451	0.27281	1.97938
2.37	0.72868	0.41181	0.27077	1.99554
2.38	0.72718	0.40913	0.26875	2.01186
2.39	0.72569	0.40647	0.26676	2.02837
2.40	0.72421	0.40384	0.26478	2.04505
2.41	0.72274	0.40122	0.26283	2.06191
2.42	0.72129	0.39864	0.26090	2.07894
2.43	0.71985	0.39607	0.25899	2.09616
2.44	0.71841	0.39352	0.25710	2.11355
2.45	0.71699	0.39100	0.25522	2.13114
2.46	0.71558	0.38850	0.25337	2.14890
2.47	0.71419	0.38602	0.25154	2.16685
2.48	0.71280	0.38356	0.24973	2.18498
2.49	0.71142	0.38112	0.24793	2.20331
2.50	0.71006	0.37870	0.24615	2.22183
2.51	0.70870	0.37630	0.24440	2.24053
2.52	0.70736	0.37392	0.24266	2.25943
2.53	0.70603	0.37157	0.24093	2.27853
2.54	0.70471	0.36923	0.23923	2.29782
2.55	0.70340	0.36691	0.23754	2.31730
2.56	0.70210	0.36461	0.23587	2.33698
2.57	0.70080	0.36233	0.23422	2.35687
2.58	0.69952	0.36007	0.23258	2.37695
2.59	0.69825	0.35783	0.23096	2.39724
2.60	0.69699	0.35561	0.22936	2.41773
2.61	0.69574	0.35341	0.22777	2.43844
2.62	0.69451	0.35122	0.22620	2.45934
2.63	0.69327	0.34906	0.22464	2.48046
2.64	0.69205	0.34691	0.22310	2.50179

Mach	T_0/T_0^*	T/T^*	P/P^*	P_0/P_0^*
2.65	0.69084	0.34478	0.22158	2.52333
2.66	0.68964	0.34266	0.22007	2.54509
2.67	0.68845	0.34057	0.21857	2.56706
2.68	0.68727	0.33849	0.21709	2.58925
2.69	0.68610	0.33643	0.21562	2.61166
2.70	0.68493	0.33439	0.21417	2.63428
2.71	0.68378	0.33236	0.21273	2.65713
2.72	0.68264	0.33035	0.21131	2.68020
2.73	0.68150	0.32836	0.20990	2.70351
2.74	0.68037	0.32638	0.20850	2.72703
2.75	0.67925	0.32442	0.20712	2.75079
2.76	0.67815	0.32248	0.20575	2.77477
2.77	0.67705	0.32055	0.20439	2.79900
2.78	0.67595	0.31864	0.20305	2.82345
2.79	0.67487	0.31674	0.20172	2.84815
2.80	0.67380	0.31486	0.20040	2.87307
2.81	0.67273	0.31299	0.19910	2.89824
2.82	0.67167	0.31114	0.19780	2.92365
2.83	0.67062	0.30931	0.19652	2.94931
2.84	0.66958	0.30749	0.19525	2.97521
2.85	0.66855	0.30568	0.19399	3.00135
2.86	0.66752	0.30389	0.19275	3.02775
2.87	0.66651	0.30211	0.19151	3.05439
2.88	0.66550	0.30035	0.19029	3.08129
2.89	0.66449	0.29860	0.18908	3.10844
2.90	0.66350	0.29687	0.18788	3.13585
2.91	0.66252	0.29515	0.18669	3.16352
2.92	0.66154	0.29344	0.18552	3.19144
2.93	0.66057	0.29175	0.18435	3.21963
2.94	0.65960	0.29007	0.18319	3.24808
2.95	0.65865	0.28841	0.18205	3.27680
2.96	0.65770	0.28675	0.18091	3.30578
2.97	0.65676	0.28512	0.17979	3.33504
2.98	0.65582	0.28349	0.17867	3.36456
2.99	0.65490	0.28188	0.17757	3.39437
3.00	0.65398	0.28028	0.17647	3.42445
3.10	0.64516	0.26495	0.16604	3.74084
3.20	0.63699	0.25078	0.15649	4.08711
3.30	0.62941	0.23766	0.14773	4.46549
3.40	0.62236	0.22549	0.13966	4.87830

Mach	T_0/T_0^*	T/T^*	P/P^*	P_0/P_0^*
3.50	0.61580	0.21419	0.13223	5.32803
3.60	0.60970	0.20369	0.12537	5.81729
3.70	0.60401	0.19390	0.11901	6.34884
3.80	0.59870	0.18478	0.11312	6.92556
3.90	0.59373	0.17627	0.10765	7.55050
4.00	0.58909	0.16831	0.10256	8.22684
4.10	0.58473	0.16086	0.09782	8.95794
4.20	0.58065	0.15388	0.09340	9.74728
4.30	0.57682	0.14734	0.08927	10.59853
4.40	0.57321	0.14119	0.08540	11.51553
4.50	0.56982	0.13540	0.08177	12.50226
4.60	0.56663	0.12996	0.07837	13.56288
4.70	0.56362	0.12483	0.07517	14.70174
4.80	0.56078	0.12000	0.07217	15.92335
4.90	0.55809	0.11543	0.06934	17.23244
5.00	0.55555	0.11111	0.06667	18.63390
6.00	0.53633	0.07849	0.04669	38.94594
7.00	0.52438	0.05826	0.03448	75.41383
8.00	0.51647	0.04491	0.02649	136.62375
9.00	0.51098	0.03565	0.02098	233.88435
10.00	0.50702	0.02897	0.01702	381.61450

INTRODUCTION TO INDEX NOTATION

For our purposes, index notation is a convenient shorthand for writing expressions in compact form. There is no complex theory involved, and a review of this appendix, together with a little practice, will provide a proficiency that is adequate for following the developments in this text. For a more complete grounding in the subject of index notation, the reader may wish to consult the literature.*

The need for compact notation arises from the consideration of tensor quantities. A scalar requires no notational embellishment. Thus symbols such as a, q, β, T, 7, etc., are taken to represent scalars; a single quantity is sufficient. Vectors, on the other hand, require two quantities to complete their definition, and some sort of special notation, such as $\mathbf{q}$, is required. Magnitudes and directions of vectors may be explicitly provided by their three components:

$$\mathbf{q} = u\hat{\imath} + v\hat{\jmath} + w\hat{k}$$

If the three Cartesian coordinate directions are represented by the subscript i, where i has the "range" (1, 2, 3), then this expression may also be written

$$\mathbf{q} = u_1\hat{e}_1 + u_2\hat{e}_2 + u_3\hat{e}_3$$

where the $\hat{e}_i$ are the unit vectors in each of the three coordinate directions.

Extensions to more complex quantities, requiring further information for their description, can lead to notations that are cumbersome. Adding a further degree of specification to the vector, in addition to magnitude and direction, leads to a

*See, for instance, H. Jeffreys, *Cartesian Tensors,* Cambridge University Press, Cambridge, 1965.

second-order tensor. Second-order Cartesian tensors have two indices, each with three possible designations, so that there are $(3)^2$ possible entities; second-order tensors are composed of nine quantities. For example, the stress tensor τ_{ji} has a physical significance that is defined by oriented surfaces:

$$\tau_{ji} = \text{a stress acting on a } j \text{ surface in the } i \text{ direction}$$

A j surface is one whose outward-drawn normal points in a direction parallel to the j axis of the coordinate system in use.

Other examples of second-order tensors are the moment of inertia tensor in dynamics, the strain tensor in strength of materials, and the rate-of-strain tensor in fluid mechanics. Tensors of higher order are found in other physical situations, and tensors of order n have $(3)^n$ components. Extensions of the tensor concept can lead to a need to keep track of a large number of quantities. In addition, we are usually concerned with the rates of variations of these quantities in the three coordinate directions. The need for a convenient bookkeeping system is apparent.

SUMMATION CONVENTION

Expressions can often be written in a more compact form using the summation convention. The method requires the adaption of only one principle: *Every time an index is repeated in any term in an expression, that term is to be summed over every value (i.e., 1, 2, 3) of that index.* Thus,

$$x_i x_i = \sum_{i=1}^{3} (x_i x_i) = x_1 x_1 + x_2 x_2 + x_3 x_3$$

or, similarly,

$$a_j b_j = \sum_{m=1}^{3} (a_m b_m) = a_1 b_1 + a_2 b_2 + a_3 b_3$$

Note that since the repeated index takes on specific integer values in the summation conventions, it is not a variable in the usual sense and it does not matter what symbol is used to designate it. Such indices lose their identities upon taking on the values of 1, 2, and then 3, in accordance with the summation conventions, and they are called "dummy" indices.

As an example of the utility of the summation convention, consider the dot product of two vectors, $\mathbf{a}$ and $\mathbf{b}$:

$$\mathbf{a} \cdot \mathbf{b} = a_1 b_1 + a_2 b_2 + a_3 b_3 = \sum_{i=1}^{3} a_i b_i$$

where the a_i and b_i are the components (three each) of the two vectors. In index notation this is simply

$$\mathbf{a} \cdot \mathbf{b} = a_i b_i$$

As a further example, consider the continuity equation for an incompressible flow:

$$\nabla \cdot \mathbf{q} = \frac{\partial u}{\partial x} + \frac{\partial v}{\partial y} + \frac{\partial w}{\partial z} = \frac{\partial u_1}{\partial x_1} + \frac{\partial u_2}{\partial x_2} + \frac{\partial u_3}{\partial x_3} = 0$$

In index notation this would be written

$$\frac{\partial u_i}{\partial x_i} = 0$$

The abbreviating effect of the summation convention is apparent in these examples.

KRONECKER DELTA

A very useful second-order tensor, called the Kronecker delta, is defined as follows:

$$\delta_{ji} = 1 \qquad \text{if } j = i$$

and

$$\delta_{ji} = 0 \qquad \text{if } j \neq i$$

For example, consider the following:

$$\delta_{ji} x_i = \delta_{j1} x_1 + \delta_{j2} x_2 + \delta_{j3} x_3$$

which is three equations, one for each value of j:

$$\delta_{1i} x_i = \delta_{11} x_1 + \delta_{12} x_2 + \delta_{13} x_3 = x_1$$

$$\delta_{2i} x_i = \delta_{21} x_1 + \delta_{22} x_2 + \delta_{23} x_3 = x_2$$

$$\delta_{3i} x_i = \delta_{31} x_1 + \delta_{32} x_2 + \delta_{33} x_3 = x_3$$

These may be recompacted to give

$$\delta_{ji} x_i = x_j$$

In other words, the expression has a nonzero value only when $j = i$, so the effect of contracting with (multiplying by) δ_{ji} is to replace i with j. Note that the expression $\delta_{ji} x_i$ has only three components. It is a first-order system since it has only one "free" (not dummy) index—in this case, j.

A FEW MORE RELEVANT EXAMPLES

The general mass continuity equation, including unsteady and compressibility effects, is written in index notation as

$$\frac{\partial \rho}{\partial t} + \frac{\partial}{\partial x_j} (\rho u_j) = 0$$

This may be expanded in the usual way to give

$$\frac{\partial \rho}{\partial t} + \rho \frac{\partial u_j}{\partial x_j} + u_j \frac{\partial \rho}{\partial x_j} = 0$$

For steady flow the first term in this expression is zero. If the flow is incompressible, both the first and last terms are zero.

The balance of forces on a fluid element of volume δV gives for the x direction ($i = 1$):

$$\frac{\Sigma F_1}{\delta V} = \frac{\partial}{\partial x_1} (\tau_{11}) + \frac{\partial}{\partial x_2} (\tau_{21}) + \frac{\partial}{\partial x_3} (\tau_{31}) = \frac{\partial}{\partial x_j} (\tau_{j1})$$

which is easily extended to all dimensions:

$$\frac{\Sigma F_i}{\delta V} = \frac{\partial}{\partial x_j} (\tau_{ji})$$

In the Navier-Stokes equations the viscous terms appear as follows:

$$\frac{\partial}{\partial x_j} \left[\mu \left(\frac{\partial u_j}{\partial x_i} + \frac{\partial u_i}{\partial x_j} \right) \right] - \frac{2}{3} \frac{\partial}{\partial x_j} \left[\mu \left(\frac{\partial u_k}{\partial x_k} \right) \delta_{ji} \right]$$

These viscous forces per unit volume are due to the angular and linear deformation, respectively, of a viscous fluid element. The appearance of the Kronecker delta in the last term causes the $\partial/\partial x_j$ term to become $\partial/\partial x_i$. If the viscosity may be taken as constant, the first term expands to

$$\mu \frac{\partial^2 u_j}{\partial x_j \partial x_i} + \mu \frac{\partial^2 u_i}{\partial x_j \partial x_j}$$

which may be written as

$$\mu \frac{\partial}{\partial x_i} \left(\frac{\partial u_j}{\partial x_j} \right) + \mu \nabla^2 u_i$$

Now there is no difference between $\partial u_j/\partial x_j$ and $\partial u_k/\partial x_k$ in index notation, and the full expression may therefore be written as

$$\mu \nabla^2 u_i + \frac{1}{3} \mu \frac{\partial}{\partial x_i} \left(\frac{\partial u_m}{\partial x_m} \right)$$

In incompressible flow, of course, this last term is zero.

As a final demonstration of the notation, it will be noted that the derivative moving with the flow, Du_i/Dt, is given by

$$\frac{Du_i}{Dt} = \frac{\partial u_i}{\partial t} + u_j \frac{\partial u_i}{\partial x_j}$$

EPILOGUE

COMPUTER APPLICATIONS AND COMPUTATIONAL FLUID DYNAMICS

The points of view developed in this text, together with their mathematical representations, serve to illustrate the fundamental principles underlying the three categories of fluid mechanics: ideal flow, compressible flow, and viscous flow. This partitioning of the subject of fluid mechanics has aided in placing emphasis on the dominant feature of each of the categories: irrotational motion, variations in fluid density, and the actions of shear stresses, respectively. On the other hand, such a treatment is inherently less than general in that the methods of analysis often involve an assumption that a given dominant feature is unaffected by the presence of the others.

Generality has been further limited within each of the categories. Thus the discussion of ideal fluid flows has been limited to incompressible two-dimensional cases, the treatment of compressible flows has been restricted to one-dimensional approximations, and examples of viscous flows have been simplified in similar ways. In most cases the effects of unsteadiness have also been left to further studies. In spite of these restrictions, there are many instances in which they do not prevent the solution of practical engineering problems, at least in approximate ways, and several of these have been illustrated in this text.

Most of the principles presented here have been well known for many decades—even centuries. But from an engineering point of view the restrictions discussed above have often been of a practical necessity because of limited powers of solution. It has long been possible to state fluid mechanical problems in more or less general ways—with governing partial differential equations and boundary conditions. Only within the last few decades, however, has it become practical for engineers to pursue problem solutions to a degree commensurate with the most general problem statements.

What has occurred is, of course, the advent of the high-speed digital computer. This has marked the beginning of a revolution in the solution of scientific and engineering problems, and the area of fluid mechanics has been one of the chief beneficiaries of this revolution. We have remarked about this "new" capability in computational power in several portions of this text, and it is certainly a legitimate area of coverage within the scope of intermediate fluid mechanics. But because the subject of computer methods in fluid mechanics has, in itself, evolved into a distinct and important subarea in fluid mechanics, anything approaching complete coverage is deserving of an entire text.

In fact the last several decades have given birth to a steadily increasing flow of contributions to the international literature on computational methods and their applications to the solution of problems in fluid mechanics. The discipline has matured to the extent that entire journals, texts, and reference volumes have been dedicated to the subject (see Bibliography). In this text, therefore, only this brief Epilogue has been devoted to the subject, lest the reader complete his or her study of intermediate fluid mechanics without an appreciation of the challenges and opportunities that lie ahead in this important area of engineering analysis.

In what follows we provide a general view of the features of computer methods in engineering analysis. In addition, a brief overview is given of the discretized forms of the pertinent governing differential equations, together with some of the problems that can develop in seeking their solutions via computer methods.

COMPUTERS AS ENGINEERING TOOLS

The so-called scientific method has traditionally been thought of as a judicial blend of theoretical and experimental endeavors. Theories formulated from available physical principles have been cast into mathematical forms in such a way that they lead to predictions of the behavior of physical systems. These predictions, which reflect the assumptions of their theoretical bases, are then subject to verification through experimentation. In this way the predictive theoretical models are gradually expanded and improved until further refinement is constrained by the limits of knowledge. These limits may relate to the availability of known physical principles, the adequacy of mathematical solution methods, or the extent to which quantities can be measured with adequate precision.

Computers do not generate physical principles, but they do contribute significantly to understanding the physical phenomena from which such principles may often be deduced. In addition, computers have found widespread use in extending the frontiers that have limited both mathematical solution methods and the measurement and interpretation of physical events. Thus the advent of the computer age has brought with it a significant modification to the traditional view of the scientific method. Today the scientific method might be thought of as a trilogy in which the computer is brought to bear as an indispensable adjunct to both aspects of the judicial blend of theoretical and experimental endeavors.

In the study of fluid mechanics the most prominent role of the computer has been played in the area that has come to be known as computational fluid dynamics (CFD). This application involves direct solution of the most general forms of the relevant governing equations, and is discussed in a subsequent section. Before this, however, some mention will be made of the use of the computer in its primitive role as a number cruncher.

Computer Applications as Aids to Analysis

In this text we have developed several well-formulated problems and, in selected cases, we have presented their solutions. Even with solutions in hand, in virtually every case the computer could have been brought to bear in performing repeated calculations. Tables and/or graphs could have been generated that reflect the sensitivity of the results to changes in parameter values. More sophisticated methods could have been used to seek maximal, minimal, or "optimal" conditions according to the nature of the problem. Thus the computer provides an important if somewhat pedantic capability of simply repeating a multitude of tasks that might otherwise have been found to be beyond the limits of human patience.

We have seen examples of computer applications that are considerably more sophisticated. In Chapter 5, for instance, the principles of ideal flow, which led to the linear Laplace equation, were used to form integral equations. The conversion of these integrals into summations led to formulations of linear algebraic equations, the solutions to which, again, would try human patience. With the computer at hand, however, limits on patience are effectively replaced with limits on computational precision. (As a practical matter, these latter limits usually amount to constraints on the allowable expenses arising from the use of the computer.)

Another important way that the computer is frequently used is to combine simplified theoretical models to form more complex ones. By way of illustration, recall that in Chapter 7 we formulated the equations governing the change in Mach number due to area changes in an isentropic compressible flow. In Chapter 9 the isentropic constraint was removed, but constant-area conditions were invoked. With the computer it is a relatively simple matter to remove both these constraints and calculate the effects of area change in flows with changing entropy. Such a procedure is outlined below.

From Eqs. (7.5) and (7.8), a differential expression for the change in Mach number with area in an isentropic flow may be developed for a given Mach number and area. Call this $(\partial M/\partial A)_s$. Equation (9.13) for Fanno flow gives a differential relationship for Mach number change along a constant-area duct with friction, $(\partial M/\partial x)_A$. The two effects might be combined for computational purposes as follows:

$$\Delta M = \left(\frac{\partial M}{\partial A}\right)_s \Delta A + \left(\frac{\partial M}{\partial x}\right)_A \Delta x = (\Delta M)_1 + (\Delta M)_2 = fct(M, x)$$

with the duct area variation $A(x)$ specified. Using the cited theoretical results, a computational algorithm could be constructed that calculates the change in Mach number $(\Delta M)_1$ for a given area change, followed by a change in Mach number $(\Delta M)_2$ due to friction acting for a short distance at a given area. The cumulative effect of area and friction would be calculated by proceeding down the duct in small steps, Δx, interspersed with small area changes, ΔA. Concomittant changes in physical properties could likewise be computed from the equations developed in Chapters 7 and 9. Since the area change and friction effects are presumed to take place in the absence of each other, the validity of the calculation would improve with smaller step sizes. (A rapidly varying area or a large friction factor would dictate smaller step sizes.) This step-size limitation is a feature common to all discretized calculations such as this.

Of course, the calculation would be subject to the constraints discussed in Chapters 7–9, such as the boundary influences of pressure, the possibility of shocks, and the dependence of the friction factor on Reynolds number. (With the computer, this variation of the friction factor might also be taken into account.) The point here is that the computer often has a direct role to play in removing limitations otherwise found necessary for computational reasons.

As a final example, one might cite the solution of the Blasius equation, Eq. (12.17), for the laminar boundary layer. This was accomplished (circa 1908) by Blasius using a power series valid near the wall, an asymptotic expansion satisfying the free stream boundary condition, and a matching criterion joining the two within the boundary layer. Today, with a few lines of computer code, the equation may be solved in seconds (see Chapter 12). This example further affords an opportunity to again remark that the influence of the computer is not always a positive one. The efforts of Blasius constituted a remarkable mathematical feat in addition to lending insights into the messages contained in the equation and its boundary conditions. In fact, if Blasius had had access to a computer, the reduced form of the boundary layer equation that he solved might never have been discovered.

Computer Applications as Aids to Experimentation

The most direct benefit that the computer offers to experimentation lies in its role as an adjunct to the scientific method. Effective experimentation depends to a great extent on preliminary analysis and the establishment of an analytical model of the process under evaluation. To the extent that the analytical model is accurate and comprehensive, experimentation is more likely to yield useful results.

Analysis using the computer allows better definition of the parameters dominating a physical process, and this understanding has direct and beneficial effects on the design of experiments. Quantities to be measured are better defined, as well as their expected ranges of values and the necessary precision of measurement. In addition, selections of experimental facilities and instrumentation systems are more likely to be optimal (neither too much nor too little) for the experimental task.

Experimental facilities themselves may be designed using computer methods so that they provide the necessary environment for the process to be tested. For instance, the design of wind tunnels that are compensated for the effects of wall boundary layers is now a routine matter when using the computer. In some specialized applications tunnel contours may also be adjusted to compensate for the blockage effects of equipment and models mounted within the tunnel.

A further important benefit is that the generality of an analytical model, which is usually enhanced through computer analysis, tends to reduce the amount and expense of experimental testing. With an analytical model capable of predicting the behavior of the process under investigation over a wide range of conditions, the required number of test points may be reduced—often by a significant amount.

Computers have also come to play a prominent role in the control of experiments, the acquisition of data, and the processing of experimental information. This area is again one that constitutes an entire field of engineering endeavor: computer-based testing methodology. A common example is the on-line processing of sensor signals to provide the investigator with quantities that relate directly to system performance. In the measurements of the forces and moments on a body suspended in a fluid flow, for instance, accounting for six degrees of freedom typically requires a complex combination of several strain-gage signals. This is often accomplished by computer so that forces in the drag, lift, and sway directions, together with their associated moments (roll, yaw, and pitch), are provided directly to the test engineer. If necessary, the computation can be accomplished at a speed that allows interactive adjustment of experimental conditions in response to the computer-processed signals.

A final example of modern experimental technology based on the computer is what is sometimes called system identification (SI). The basis of this method is the fact that while system models often describe the interrelation of various causes and effects with relative precision (e.g., $F = ma$), there are often significant uncertainties concerning those quantities that modulate or control these interrelationships. Examples are resistance, capacitance, and inductance as they may be defined in electrical, mechanical, thermal, or fluid systems.

In the SI procedure the analytical model is constructed as in the usual case, but the terms with high uncertainty are considered to be variables of the model design. The goal of the experimental procedure is to conduct tests in such a way that the measured response of the system to known stimuli leads to insights regarding the system parameters. As an example, consider the thermal response of a cool object suddenly immersed in a high-temperature, high-speed flow.

A theoretical model of the body may be constructed that predicts the time-dependent temperature at a particular point on the body. Such a model would involve the thermal resistances acting at the body boundaries, and in the problem at hand these may be known only in a very approximate way. The thermal conduction and storage characteristics of the body, on the other hand, might be relatively well defined. In the SI method the body and its boundaries would be thought of as a "black box" containing the various thermal properties mentioned above together with the model linking them together. In other words, an input-output model is envisioned, with the thermal-fluid events forming a "transfer function" giving the body temperature (output) as a function of the kinetic and thermal energy of the flow (input).

With the measured thermal response and the predictive analytical model, the SI method uses the computer to deduce what the parameter values (i.e., the boundary thermal resistance) would have had to be in order to match the model prediction to the observed response. Clearly, this approach is potentially very powerful from an engineering analysis point of view. In effect, it allows a direct extension of experimental measurements to *deduce* quantities that cannot themselves be measured or theoretically specified. The SI method is widely used in the analysis of dynamic mechanical and electrical systems, especially in the field of automatic controls. Applications to fluid-thermal systems, such as that described here, are relatively new.

DISCRETIZED FORMS OF EQUATIONS OF MOTION

The most common computational methods for solution of complex flow problems begin by approximating the equations of motion in forms using what are called finite differences. In order to illustrate this procedure here, we address our attention to the equations governing ideal fluid flow, which, as we have seen in Part I, lead to the Laplace equation. Thus, in the analysis of ideal flows, finite difference methods rely on the approximation of the partial derivatives in the Laplace equation by difference terms. The resulting equations are then applied to a computational mesh of nodal points, superimposed on the flow field, in order to calculate ϕ or ψ while satisfying the specified boundary conditions.

Theoretical Development

Approximation of derivatives by finite differencing can be explained by reference to Fig. 1. From the calculus, the definition of the derivative of a function $F(x)$ is given by

$$F'(x) = \lim_{\Delta x \to 0} \frac{F(x + \Delta x) - F(x)}{\Delta x}$$

According to the finite difference notation of Fig. 1, the derivative of the stream function ψ approximated at point x_i is

$$\psi'(x_i) \equiv \left(\frac{d\psi}{dx}\right)_{x_i} \approx \frac{\psi_{i+1} - \psi_i}{\Delta x}$$

To approximate the value of $\psi''(x_i)$, consider the Taylor series expansion of ψ_{i+1}

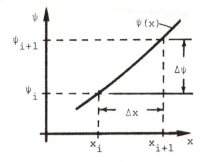

Figure 1 Approximation of the derivative of the stream function.

about the point x_i:

$$\psi_{i+1} = \psi_i + \psi_i' \frac{\Delta x}{1!} + \psi_i'' \frac{(\Delta x)^2}{2!} + \psi_i''' \frac{(\Delta x)^3}{3!} + O[(\Delta x)^4] \tag{1}$$

Repeating the expansion for ψ_{i-1}:

$$\psi_{i-1} = \psi_i - \psi_i' \frac{\Delta x}{1!} + \psi_i'' \frac{(\Delta x)^2}{2!} - \psi_i''' \frac{(\Delta x)^3}{3!} + O[(\Delta x)^4] \tag{2}$$

All terms that are of order $[(\Delta x)^4]$ and higher are collectively contained in the last terms (called the remainders) of these equations. With Δx made sufficiently small, these terms of higher order may be neglected. Equations (1) and (2) may be added together to obtain a simplified expression for the second derivative, ψ_i'':

$$\psi_i'' = \frac{\psi_{i+1} - 2\psi_i + \psi_{i-1}}{(\Delta x)^2} + O[(\Delta x)^4] \tag{3}$$

Again, in practice, a suitably accurate approximation is obtained by neglecting the remainder term for sufficiently small values of Δx.

The Laplace equation may now be solved by using Eq. (3) in both coordinate directions. A two-dimensional Cartesian grid, such as that illustrated in Fig. 2,

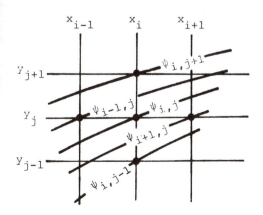

Figure 2 Grid arrangement for evaluating the stream function at points in a flow field.

may be used for point-by-point calculations for ψ. (Of course, as we have seen in Part One, the same procedures apply for the solution of the Laplace equation for the velocity potential.) The application of Eq. (3) at the point x_i, y_j gives the following finite difference form of the Laplace equation for the stream function:

$$\frac{\psi_{i+1,j} - 2\psi_{i,j} + \psi_{i-1,j}}{(\Delta x)^2} + \frac{\psi_{i,j+1} - 2\psi_{i,j} + \psi_{i,j-1}}{(\Delta y)^2} = 0$$

Note that each of the terms are expressed at a fixed value of j and i, respectively. This is a reflection of the meaning of a partial derivative in finite difference terms. If the grid spacings are made equal, $\Delta x = \Delta y$, and the Laplace equation becomes

$$\psi_{i,j} = \frac{1}{4}(\psi_{i+1,j} + \psi_{i-1,j} + \psi_{i,j+1} + \psi_{i,j-1}) \tag{4}$$

From Eq. (4) it will be noted that the value of ψ at any interior grid point is merely the arithmetic average of ψ at its four nearest nodal neighbors (plus higher-order, but negligible, terms).

Boundary Conditions

The appropriate boundary condition for Eq. (4) is that the flow at the boundaries must be parallel to those boundaries (assuming, as mentioned previously, that the boundaries are impermeable). If a boundary direction is designated by the coordinate direction s, then the velocity component normal to that boundary, which must be zero, is given by $\partial\psi/\partial s$. The mathematical expression of the boundary condition is, therefore,

$$\frac{\partial\psi}{\partial s} = 0 \qquad \text{(at a solid boundary)} \tag{5}$$

In other words, and as we have found in previous work with the stream function, its value must be constant along a boundary, and the boundary is itself a streamline. For the situation illustrated in Fig. 3, for instance, the finite difference interpretation of Eq. (5) is that

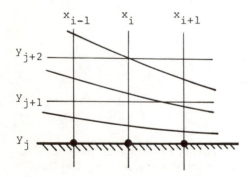

Figure 3 Regular nodal points lying on the solid boundary.

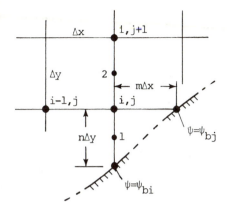

Figure 4 Irregular nodal points near a boundary.

$$\cdots = \psi_{i-1,j} = \psi_{i,j} = \psi_{i+1,j} = \cdots = \text{constant}$$

When the shape of a computational grid precludes the location of intersections (nodes) at points where values are specified (such as along bounding walls), satisfaction of the boundary conditions becomes somewhat more complicated. The problem now becomes one of estimating the value of ψ at so-called irregular points that lie in the vicinity of such boundaries. (More sophisticated grid generation schemes employ body-fitted grid systems that permit the location of nodes on irregular boundaries.) Such a point is shown in Fig. 4 at the node designated by x_i, y_j. The distance from this point to the next lower node is truncated due to the presence of a boundary. Expressing the extent of this truncation by the fraction n, the derivative of ψ at point 1, midway between x_i, y_j and the boundary point, may be approximated as follows:

$$\left(\frac{\partial \psi}{\partial y}\right)_1 \approx \frac{\psi_{i,j} - \psi_{bi}}{n\,\Delta y}$$

where ψ_{bi} refers to the value of the stream function on the boundary at given value of x. Likewise, for point 2, midway between the point x_i, y_j and the point next above,

$$\left(\frac{\partial \psi}{\partial y}\right)_2 \approx \frac{\psi_{i,j+1} - \psi_{i,j}}{\Delta y}$$

The second partial derivative of ψ at the point in question may now be approximated as

$$\left(\frac{\partial^2 \psi}{\partial y^2}\right)_{i,j} \approx \frac{(\partial \psi/\partial y)_2 - (\partial \psi/\partial y)_1}{(1 + n)\,\Delta y/2}$$

and, upon substitution of the relationships above,

$$\left(\frac{\partial^2 \psi}{\partial y^2}\right)_{i,j} \approx \frac{2}{(\Delta y)^2(1 + n)}\left[\psi_{i,j+1} - \psi_{i,j} + \frac{1}{n}(\psi_{bi} - \psi_{i,j})\right]$$

A similar procedure with respect to changes in the x direction from the point x_i, y_j will give

$$\left(\frac{\partial^2 \psi}{\partial x^2}\right)_{i,j} \approx \frac{2}{(\Delta x)^2(1+m)}\left[\psi_{j,i-1} - \psi_{i,j} + \frac{1}{m}(\psi_{bj} - \psi_{i,j})\right]$$

where the fraction m expresses the distance of a boundary, in the x direction, from the point in question. Note that if the boundary is a continuous impermeable wall, as implied in Fig. 4, then it itself is a streamline and $\psi_{bi} = \psi_{bj}$.

The sum of the two second partial derivatives of ψ must add up to zero, in accordance with the Laplace equation, so that $(\nabla^2\psi)_{i,j} = 0$. When this is done, retaining the simplification that $\Delta x = \Delta y$, the result may be rearranged to give the stream function value at the point x_i, y_j, as follows:

$$\psi_{i,j} = \frac{mn}{m+n}\left[\frac{\psi_{bi}}{n(1+n)} + \frac{\psi_{bi}}{m(1+m)} + \frac{\psi_{i,j+1}}{1+n} + \frac{\psi_{i-1,j}}{1+m}\right] \qquad (6)$$

Note that the stream function is again expressed in terms of the values at neighboring points, but in this case the contributions of values on the boundary are weighted by the spacing fractions, m and n. [If both m and n are unity, then Eq. (4) is obtained.] A useful approximation to Eq. (6) is obtained by noting that if $(1 + m)/(1 + n) \approx 1$, acceptable results may be obtained from

$$\psi_{i,j} = \frac{mn}{m+2mn+n}\left(\frac{\psi_{bj}}{n} + \frac{\psi_{bi}}{m} + \psi_{i,j+1} + \psi_{i-1,j}\right) \qquad (6a)$$

Problem Formulation and Solution Methods

In this section the finite difference equations are used to formulate a simple example problem. Two common methods of solution of the resulting equation set will also be discussed briefly. Consider flow in a duct that has a ramp convergence, as shown in Fig. 5. The flow rate through the duct, per unit depth, is given as 300 m²/s, and it is desired to estimate the stream function values at five interior nodal points, as shown. (This is a very course grid, and is used only for explanatory purposes.) The difference in ψ between the two solid boundaries must be equal to the flow, $Q' = 300$ m²/s; for convenience, the stream function value on the lower wall is arbitrarily assigned a value of zero. On the opposite wall, therefore, $\psi = 300$ m²/s.

In order to proceed, it is necessary to specify values of ψ at the duct inlet and outlet. The assumption of uniform flow in these regions allows the nodal values of ψ to be obtained from simple linear interpolation between the two wall values. (In reality, the flow following the steep ramp and at the duct exit would probably be quite nonuniform, but ideal flow modeling cannot account for this.) Thus:

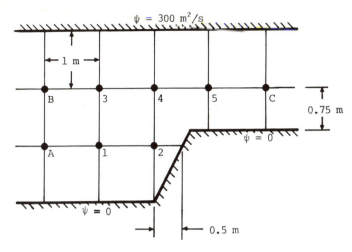

Figure 5 Grid layout and boundaries for the example problem.

$$\psi_A = 100 \text{ m}^2/\text{s} \qquad \psi_B = 200 \text{ m}^2/\text{s}$$

$$\psi_C = 300\left(\frac{0.75}{1.75}\right) \approx 129 \text{ m}^2/\text{s}$$

At regular interior nodal points, ψ is expressed by means of Eq. (4). At the irregular points (2 and 5 in Fig. 5), Eq. (6a) will be applied. [The reader may wish to compare the results with the more exact values obtained from Eq. (6).] For each of the five nodal points we have

$$\psi_1 = \frac{\psi_2 + 0 + 100 + \psi_3}{4}$$

$$\psi_2 = \frac{0/0.5 + 0/1 + \psi_1 + \psi_4}{5}$$

$$\psi_3 = \frac{\psi_4 + \psi_1 + 200 + 300}{4}$$

$$\psi_4 = \frac{\psi_5 + \psi_2 + \psi_3 + 300}{4}$$

$$\psi_5 = \frac{0/0.75 + 129/1 + \psi_4 + 300}{4.33} \qquad (7)$$

In implicit form these five simultaneous linear equations have the following arrangement:

$$
\begin{aligned}
4\psi_1 \quad -\psi_2 \quad -\psi_3 \qquad\qquad\qquad &= 100 \\
-\psi_1 \quad 5\psi_2 \qquad\qquad -\psi_4 \qquad &= 0 \\
-\psi_1 \qquad\qquad 4\psi_3 \quad -\psi_4 \qquad &= 500 \\
-\psi_2 \quad -\psi_3 \quad 4\psi_4 \quad -\psi_5 &= 300 \\
-\psi_4 \quad 4.33\psi_5 &= 429
\end{aligned} \tag{8}
$$

All that remains is to solve this set of equations for the five values of ψ at the interior nodal points. Among the several methods that are available for doing this, we demonstrate both matrix inversion and relaxation techniques.

Matrix inversion. Linear equation sets, such as those developed here, naturally lend themselves to solution by matrix methods. In matrix notation, Eq. (8) may be written $\mathbf{A}\psi = \mathbf{C}$ where the matrices $\mathbf{A}$, ψ, and $\mathbf{C}$ are illustrated in the following expanded form:

$$
\begin{bmatrix}
4 & -1 & -1 & 0 & 0 \\
-1 & 5 & 0 & -1 & 0 \\
-1 & 0 & 4 & -1 & 0 \\
0 & -1 & -1 & 4 & -1 \\
0 & 0 & 0 & -1 & 4.33
\end{bmatrix}
\begin{bmatrix}
\psi_1 \\ \psi_2 \\ \psi_3 \\ \psi_4 \\ \psi_5
\end{bmatrix}
=
\begin{bmatrix}
100 \\ 0 \\ 500 \\ 300 \\ 429
\end{bmatrix} \tag{9}
$$

From Eq. (9) the solution vector, ψ, may be written as follows:

$$
\psi = \mathbf{A}^{-1}(\mathbf{A}\psi) = \mathbf{A}^{-1}\mathbf{C}
$$

Thus the problem reduces to that of finding the inverse of the $\mathbf{A}$ matrix. Gaussian elimination is an efficient technique for such purposes, and the following approximate values are obtained for the stream function at the nodal points:

$$
\begin{bmatrix}
\psi_1 \\ \psi_2 \\ \psi_3 \\ \psi_4 \\ \psi_5
\end{bmatrix}
=
\begin{bmatrix}
85 \\ 51 \\ 186 \\ 169 \\ 138
\end{bmatrix}
$$

Iteration. For the inversion of the matrix given in the example problem, the program using Gaussian elimination requires enough memory to store only $5 \times 5 = 25$ real numbers. When more complicated flow patterns, containing thousands of nodal points, are analyzed, it may become necessary to store millions of real numbers in order to accommodate matrix inversion. In such cases, if computer memory is at a premium, iterative methods are often preferred over matrix inversion schemes.

To formulate an iterative procedure, the nodal equations are first written in their primitive form, as in Eq. (7). Again, note that these equations express each stream function value in terms of the values at adjacent nodal points. The iteration process is begun by guessing values of ψ at each nodal point. (The number of iterations required to obtain a good solution will depend directly on the accuracy of these initial values). Next, new values of ψ are calculated from a set of so-called relaxation equations, which, for the problem at hand, appear as follows (the superscript N refers to new values of ψ that are calculated on the basis of the old values, superscript O):

$$\psi_1^{(N)} = \psi_1^{(O)} + \frac{w}{4}\left[\psi_2^{(O)} + 100 + \psi_3^{(O)} - 4\psi_1^{(O)}\right]$$

$$\psi_2^{(N)} = \psi_2^{(O)} + \frac{w}{5}\left[\psi_1^{(O)} + \psi_4^{(O)} - 5\psi_2^{(O)}\right]$$

$$\psi_3^{(N)} = \psi_3^{(O)} + \frac{w}{4}\left[\psi_4^{(O)} + \psi_1^{(O)} + 500 - 4\psi_3^{(O)}\right]$$

$$\psi_4^{(N)} = \psi_4^{(O)} + \frac{w}{4}\left[\psi_5^{(O)} + \psi_2^{(O)} + \psi_3^{(O)} + 300 - 4\psi_4^{(O)}\right]$$

$$\psi_5^{(N)} = \psi_5^{(O)} + \frac{w}{4.33}\left[\psi_4^{(O)} + 429 - 4.33\psi_5^{(O)}\right]$$

Note that if the old values are the true values of ψ, then the correcting quantities in brackets become zero and $\psi^{(N)} = \psi^{(O)}$. The iteration sequence is repeated until the new and old values of ψ agree within a specified tolerance band.

In the form given here, the quantity w, called a relaxation parameter, is included as a device for reducing the number of iterations required for convergence to within the tolerance band. For values of w greater than unity, overrelaxation occurs and solutions are extrapolated from earlier iterations. Generally, a choice of w between 1.2 and 1.4 leads to rapid and stable convergence. For values of w less than unity, underrelaxation is obtained and solutions are interpolated from earlier solutions.

Convergence of iterative processes is not guaranteed for all situations and depends on, among other things, the nature of the matrix of coefficients. It is also to be noted that the convergence criterion, or tolerance band, refers only to the difference between two successive iterations and does not indicate precision with respect to the exact solution of the flow field. The finite difference method will produce solution values for the stream function that are close approximations to the true values if the number of nodal points is sufficiently large. Interested readers may find a great deal of additional information on these issues in the literature. For applications of numerical methods to engineering problems, Crandall (1956) will be useful. General treatments of numerical methods such as those noted here will be found, for instance, in Southworth and Deleeuw (1965) and Pipes and Hovanessian (1969).

The Laplace partial differential equation expresses the motion of one of the most simple types of fluid flow—incompressible, steady, irrotational—and this simplicity is reflected in the discrete forms and solution procedures that have been described above. In such ideal flow problems the solution procedures are largely dependent on the complexity of the boundary conditions, and the sophistication of applicable computer methods generally reflects their ability to handle diverse and complex boundary configurations. As might be expected, there are many packaged computer codes that handle problems in ideal fluid flow in very general ways, including three-dimensional grid-generation schemes and provisions for steady and unsteady conditions. In addition, the solution methods usually involve standard matrix manipulation schemes that are well understood and widely available in computer software products. Compressibility effects may also be treated, although these can introduce significant difficulties due to shock waves and singular effects near Mach numbers of unity.

As we have seen, introduction of the dissipative action of viscosity leads to the Navier-Stokes equations, which, together with their approximate boundary layer forms, are nonlinear and much more difficult to solve. In addition, the satisfaction of the no-slip boundary conditions in viscous flow requires that the flow velocity at the boundary must be the same as the boundary velocity. (In ideal flow this condition constrains only the component of velocity normal to the boundary.) In the next section we briefly describe some of the features of viscous flows that are of particular importance when solutions are sought by computer means.

GENERAL FORMS AND COMPUTATIONAL FLUID DYNAMICS

Previous sections have introduced the concept of discretization of partial differential equations and, for the purposes of illustration, the Laplace equation has been solved using numerical methods. As we have seen in Part Three, the most general form of the equations of motion is embodied in the Navier-Stokes equations. If necessary, these equations may be written to reflect the effects of turbulence if a suitable turbulence model is available (a big "if"). Unsteady effects may also be incorporated, and in addition, if the energy equation is brought into play, it is possible to account for temperature variations in the flow. Compressible flows require the addition of equation-of-state information.

In this section we complete our Epilogue by providing some insights into the sort of complexities to be expected in seeking computer-based solutions for flows of a general nature. As a beginning, consider the form of the Navier-Stokes equations valid for steady, incompressible, laminar flow. In terms of the stream function, we found in Chapter 10 that

$$-\frac{\partial \psi}{\partial y}\frac{\partial}{\partial x}(\nabla^2\psi) + \frac{\partial \psi}{\partial x}\frac{\partial}{\partial y}(\nabla^2\psi) = \upsilon\,\nabla^4\psi \qquad (10)$$

Note that $\nabla^2\psi = 0$ is a particular solution to this expression, which removes the dependence on viscosity and returns us to the irrotational case discussed previously. But the Laplace equation is linear and of second order in the stream function, whereas Eq. (10) is nonlinear and of fourth order.

In concept, the partial derivatives in Eq. (10) can be expressed in discrete forms by employing schemes similar to those described above. If this were done, the difference relationships would contain higher-order terms because of the necessity to evaluate the fourth-order partial derivative. Whereas evaluation of the Laplacian at a point in a computational grid requires knowledge of values at each of the four grid points in the immediate vicinity, the "Laplacian of the Laplacian" would involve values at the points next further displaced from the point in question. Because the errors inherent in the finite difference approximations for values at a point are proportional to powers of the distances from that point, smaller grids (and more computations) are required to control accuracy.

Further severe difficulties arise from the fact that the left-hand side of Eq. (10) is not zero in a flow with rotation. Because of this, it is not possible to solve the difference equations explicitly for each value of ψ in terms of values at the surrounding grid points. That is, equations of the form of Eq. (4) do not evolve naturally, and numerical methods requiring iteration must be used of necessity. The applicable iteration schemes involve more simultaneous algebraic equations to be solved, and the likelihood of rapid convergence to a solution is thereby reduced.

The reader will also observe that the left-hand side of Eq. (10) contains the nonlinear terms. Whereas the absence of these terms is the paramount benefit of an irrotationial motion, their presence, which is carried over into the discretized form of Eq. (10), serves to exacerbate an already complicated situation as far as numerical solution schemes are concerned.

From the above discussion it is apparent that the availability of the computer does not remove the difficulties inherent in the complex motions that arise because of the presence of viscous action. In any case, when other effects are to be taken into account (unsteadiness, compressibility, turbulence, heat transfer, variable fluid properties, etc.), Eq. (10) is inadequate to describe the motion. These factors are among the reasons that modern computational methods seldom deal with fourth-order expressions such as Eq. (10). Instead, it is common practice to address the problem of solution of the basic governing equations (sometimes called the primitive forms) using numerical approaches that rely heavily on physical interpretation of the terms that appear in these relationships.

General Forms

The development of the field of computational fluid dynamics (CFD) has often focused on the formulation of the finite difference forms of the governing equations in ways that lead to computational efficiency. A spin-off of this work has been to underline the physical interpretation of the various terms in the governing equations and, in doing so, enahance the basic understanding of fluid flow phenomena. (Many of these phenomena, it turns out, are reflected in the behavior of numerical solution methods.)

Recall that the equation for mass continuity may be written

$$\frac{\partial \rho}{\partial t} + \frac{\partial}{\partial x_j}(\rho u_j) = 0 \tag{11}$$

When this form is incorporated into the Cartesian tensor form of the Navier-Stokes equation for laminar flow with constant viscosity, Eq. (10.18), the result is as follows:

$$\frac{\partial}{dt}(\rho u_i) + \frac{\partial}{\partial x_j}(\rho u_j u_i) = -\gamma \frac{\partial h}{\partial x_i} - \frac{\partial p}{\partial x_i} + \frac{\partial}{\partial x_j}\left(\mu \frac{\partial u_i}{\partial x_i}\right) - \frac{2}{3}\frac{\partial}{\partial x_i}\left(\mu \frac{\partial u_m}{\partial x_m}\right) \tag{12}$$

On the right-hand of this expression we see that forces per unit volume are given. Recognizing that velocity is momentum per unit mass, the expressions on the left-hand side are seen to be the local and convective rates of change in momentum on a per-unit-volume basis. Thus we have written an expression that says, in effect, that the rate of change of the momentum contained within a fluid volume, plus the net rate of efflux of fluid momentum across that volume, equals the sum of the various forces acting on the fluid while it traverses the volume—all on a per-unit-volume basis. Because forces and velocities are vectors, Eq. (12) represents the ith component of the forces, and the changes in momentum with which they are associated.

Consider now the following expression:

$$\frac{\partial}{\partial t}(\rho\phi) + \frac{\partial}{\partial x_j}(\rho u_j \phi) = \frac{\partial}{\partial x_j}\left(\Gamma\frac{\partial\phi}{\partial x_j}\right) + S \tag{13}$$

If the quantity ϕ is considered to be u_i, the momentum per unit mass in the ith component direction, then the terms on the left-hand side are seen to be the local and convective change in the momentum per unit volume in that direction, respectively. On the right-hand side, the first term represents the *diffusion* of momentum, through the action of viscosity, which, as we have seen, leads to the presence of shear forces acting on the surface of a fluid volume. The final term, S is called a *source* term, which in this case contains a number of effects that may be identified by comparison with Eq. (12): the body force (weight, in this case), the pressure force arising from the inviscid part of the normal stress, and the force due to viscous effects over and above the nominal diffusion term—all on a per-unit-volume basis. Actually, all the terms on the right-hand side may be thought of as sources (or sinks) of momentum. The nominal diffusion term is given explicitly because of its common occurrence in conservation expressions. In practice, the pressure term is also excluded from the residual source term because pressure is needed as a dependent variable to be found by means other than those used to approximate the source terms.

The general nature of the form expressed in Eq. (13) may be further demonstrated by considering the quantity of interest to be mass, so that the general dependent variable ϕ is unity (mass per unit mass) and the diffusion coefficient Γ is zero. With this substitution, Eq. (13) becomes Eq. (11), expressing the conservation of mass on a per-unit-volume basis. (The absence of mass sources is assumed in both expressions.) Setting ϕ equal to the specific enthalpy and Γ equal to the thermal conductivity, Eq. (13) yields an expression for conservation of energy with various source terms (including, if appropriate, heat transfer, shaft work, kinetic energy, and the thermal effect of viscous dissipation within the volume). Applications of Eq. (13) to other conserved quantities, such as chemical species and turbulence variables, may be found in Patankar (1980). Other general and compact vector forms of the equations governing conserved quantities may be found, for instance, in Anderson et al. (1984, chap. 5).

The problem of solving an entire flow field is that of integrating equations such as Eqs. (10), (11), and (12), subject to specified boundary conditions. When this is not possible by analytical means, numerical methods offer powerful alternatives. In the case of the Laplace equation, the finite difference form, Eq. (4), leads to relatively simple and linear closed-form numerical schemes, which, when applied to a computational grid, lead to integrated solutions for the entire flow field.

The field of CFD has arisen to address the numerical solution of more general cases, and many of the methods used in CFD rely on the physical insights expressed in formulations such as Eq. (13). These formulations are mathematical representations of the general and intuitive requirement that within a volume in a field of flow (a "control" volume), the rate of change of any conserved quantity must be equal to the rate at which that quantity is carried into the volume with the flow, plus the rate at which it is created at or within the volume boundaries. In CFD, this principle is satisfied at each point in a computational grid (usually in an approximate way), by using finite difference expressions to estimate the local, convective, diffusive, and source effects. Naturally, these expressions involve the values of quantities at neighboring points and, in the case of unsteady flows, at previous steps in time. In effect, the equations of motion are "integrated" around each tiny volume enclosed within a computational grid, paying due attention to grid points near boundaries where special techniques are needed to account for the applicable boundary conditions. The cumulative effect of each of these discrete integrations is the solution of the entire flow field.

Although the physical principles governing fluid motion are expressed by Eq. (13) in a fairly straightforward way, the generation of efficient numerical algorithms for solution at each grid point is by no means a simple matter, and it is this effort that has caused CFD to develop into an entire field of engineering endeavor. Because digital computers ultimately deal with linear elements (bytes), it is always desirable, and often most efficient, to express the various finite difference expressions in approximate linear forms. In CFD, therefore, the treatment of the nonlinearities inherent in the convective terms, and other nonlinearities that arise in particular problems (such as viscous energy dissipation quantities, and turbulent momentum transport terms), gives rise to a considerable amount of creative effort.

A further important feature of CFD is that the governing equations differ in their fundamental mathematical form, depending on the nature of the flow described, and this is reflected in the configuration of the applicable finite difference algorithms. Thus we have seen that evaluation of the Laplace equation at a point suggests an equal influence of stream function values at each of the four surrounding points. In such cases, sometimes called "jury" problems, the influence of values at the boundaries will propagate to all points within the computational grid. This behavior is characteristic of partial differential equations of the elliptic type, such as the Laplace equation.

The term "parabolic" is often applied to problems that are dominated by the effects of diffusion. In the boundary layer, for instance, the diffusion of momentum across the thin shear layer is by far the predominant feature of the flow, and equations of the boundary layer type are sometimes said to be parabolic, semiparabolic, or parabolized Navier-Stokes equations. Numerical solution of the boundary layer equation may be accomplished by "marching" in the streamwise direction, with the values along successive cross-stream lines obtained from those at downstream stations. At the completion of the march, there is no need to conduct further sweeps across the computational field, so that efficiencies in computational time and storage are realized.

In our discussion of supersonic flows we have seen that there are regions in which changes cannot propagate upstream. The equations governing such behavior, in which solutions at a point are influenced by values in a limited domain of dependence, are said to be hyperbolic in character. Another feature of supersonic flows is the possible existence of shocks. Such abrupt occurrences in a flow are an anathema to finite difference schemes, since most of these depend on well-behaved derivatives within a computational region. For this reason, special "shock-capturing" methods are employed in computations where shocks may be expected.

The elliptic, parabolic, or hyperbolic nature of governing partial differential equations has a direct impact on the types of numerical schemes that may be effectively brought to bear. In many cases the equations governing a fluid flow problem may be a mixture of several types, with attendant complications in the numerical procedures. In other cases the nature of the flow itself (compressible, unsteady, etc.) may cause the same basic equation to display several different behaviors. The steady-flow Euler equation, for instance, may be classified as elliptic, parabolic, or hyperbolic depending on whether the flow domain is subsonic, sonic, or supersonic. (Strictly speaking, the Euler equation for unsteady flow is hyperbolic in all cases.) In many cases, the flow region itself consists of subregions where different forms of the governing equations are applicable. A common example is the region containing the interaction of a shock wave with a turbulent shear layer such as at a wall. With mixed regions it is often necessary to apply grid refinement schemes in order to enhance computational efficiency.

Needless to say, there are many areas of research in the field of CFD in which new approaches are steadily forthcoming. For the reasons outlined above, all computational methods are restricted in their application to certain types of problems, and there is no single CFD package that will encompass all flow possibilities, nor is there likely to be one in the near future. Though massive computer codes have been constructed in efforts to provide general-use capabilities, many of these are extremely inefficient and require enormous computing power.

SUMMARY

In this Epilogue an attempt has been made to inform the reader of the scope and nature of the methods that may be brought to bear in the use of the computer to solve problems in fluid mechanics. Although, at the bottom line, the computer only gives the capability to perform voluminous calculations at incredible speeds, such applications have come to play a prominent role in the scientific method. The traditional blend of theory and experiment has now been expanded to include computational studies.

Although many of the physical principles governing fluid motions have long been known, the solutions of many engineering problems have perforce been based on simplified mathematical models. Such solutions are of the type illustrated in the main body of this text, and in many cases the computer may be used to combine these efforts to produce more general results. Other computer applications are of the number-crunching type, in which known relationships are used to produce tabular, graphical, or other forms of information that describe the nature of solutions for wide ranges of the parameters applicable to a given problem. We have also briefly discussed the ways that computers may be used to enhance experimental work, both in the design, control, and optimization of experiments, and in the processing of data.

A separate and distinct role of the computer in the study of fluid mechanics is found in the field of computational fluid dynamics. Here the basic governing equations are themselves cast into forms suitable for numerical evaluation. In doing so, many of the most general relationships, including the effects of compressibility, unsteadiness, turbulence, and multidimensionality, have been modeled and solved by CFD experts. Only a brief introduction to this work has been provided here, with some discussion of discretized forms and their treatment by numerical means. It has also been noted that the vast range of types of fluid mechanical problems is reflected in the spectrum of CFD approaches to their solutions. For further study in this area, the reader is again referred to the references and bibliography.

REFERENCES

Anderson, D. A., J. C. Tannehill, and R. H. Pletcher: *Computational Fluid Mechanics and Heat Transfer,* Hemisphere, Washington, D.C., 1984.

Crandall, S. D.: *Engineering Analysis,* McGraw-Hill, New York, 1956.

Patankar, S. V.: *Numerical Heat Transfer and Fluid Flow,* Hemisphere, Washington, D.C., 1980.

Pipes, L. A., and S. A. Hovanessian: *Matrix-Computer Methods in Engineering,* John Wiley & Sons, New York, 1969.

Southworth, R. W., and S. L. Deleeuw: *Digital Computation and Numerical Methods,* McGraw-Hill, New York, 1965.

BIBLIOGRAPHY

Ames, W. F.: *Numerical Methods for Partial Differential Equations,* 2nd ed., Academic Press, New York, 1977.

Carnahan, B., H. A. Luther, and J. O. Wilkes: *Applied Numerical Methods,* John Wiley & Sons, New York, 1969.

Cebici, T. and P. Bradshaw: *Momentum Transfer in Boundary Layers,* Hemisphere, Washington, D.C., 1977.

Chow, C. Y.: *An Introduction to Computational Fluid Mechanics,* John Wiley & Sons, New York, 1979.

Chung, T. J.: *Finite Element Analysis in Fluid Dynamics,* McGraw-Hill, New York, 1978.

Forsythe, G. E., and W. Wasow: *Finite Difference Methods for Partial Differential Equations,* John Wiley & Sons, New York, 1960.

Gosman, A. D., W. M. Pun, A. K. Runchal, D. B. Spalding, and M. Wolfshtein: *Heat and Mass Transfer in Recirculating Flows,* Academic Press, New York, 1969.

Launder, B. E., and D. B. Spalding: *Mathematical Models of Turbulence,* Academic Press, New York, 1972.

Peyret, R., and H. Viviand: *Computation of Viscous Compressible Flows Based on the Navier-Stokes Equations,* AGARD-AG-212, NASA, Langley Field, Va., 1975.

Richtmyer, R. D.: *Difference Methods for Initial-Value Problems,* Interscience, New York, 1957.

Richtmyer, R. D., and K. W. Morton: *Difference Methods for Initial Value Problems,* 2nd ed., Interscience, New York, 1967.

Roache, P. J., *Computational Fluid Dynamics,* Hermosa, Albuquerque, N.M., 1976.